TIME

Time

A Vocabulary of the Present

Edited by Joel Burges and Amy J. Elias

NEW YORK UNIVERSITY PRESS
New York

NEW YORK UNIVERSITY PRESS
New York
www.nyupress.org

© 2016 by New York University
All rights reserved

References to Internet websites (URLs) were accurate at the time of writing. Neither the author nor New York University Press is responsible for URLs that may have expired or changed since the manuscript was prepared.
Library of Congress Cataloging-in-Publication Data

Names: Burges, Joel, editor. | Elias, Amy J., 1961- editor.
Title: Time : a vocabulary of the present / edited by Joel Burges and Amy J. Elias.
Description: New York : New York University Press, [2016] | Includes bibliographical references and index.
Identifiers: LCCN 2016010261 | ISBN 978-1-4798-2170-9 (cl : alk. paper) | ISBN 978-1-4798-7484-2 (pb : alk. paper)
Subjects: LCSH: Time. | Horology. | Time—Psychological aspects. | Time—Philosophy.
Classification: LCC QB209 .T484 2016 | DDC 529—dc23
LC record available at https://lccn.loc.gov/2016010261

New York University Press books are printed on acid-free paper, and their binding materials are chosen for strength and durability. We strive to use environmentally responsible suppliers and materials to the greatest extent possible in publishing our books.

Manufactured in the United States of America

10 9 8 7 6 5 4 3 2 1

Also available as an ebook

CONTENTS

Acknowledgments vii

Introduction: Time Studies Today. 1
   Joel Burges and Amy J. Elias

PART I. TIME AS HISTORY: PERIODIZING TIME

1. Past / Future 35
   Amy J. Elias

2. Extinction / Adaptation. 51
   Ursula K. Heise

3. Modern / Altermodern 66
   David James

4. Obsolescence / Innovation 82
   Joel Burges

5. Anticipation / Unexpected 97
   Mark Currie

PART II. TIME AS CALCULATION: MEASURING TIME

6. Clock / Lived. 113
   Jimena Canales

7. Synchronic / Anachronic 129
   Elizabeth Freeman

8. Human / Planetary 144
   Heather Houser

9. Serial / Simultaneous 161
   Jared Gardner

10. Emergency / Everyday 177
   Ben Anderson

11. Labor / Leisure...................... 192
    Aubrey Anable
12. Real / Quality........................ 209
    Mark McGurl

PART III. TIME AS CULTURE: MEDIATING TIME

13. Aesthetic / Prosthetic............... 225
    Jesse Matz
14. Analepsis / Prolepsis................. 240
    James Phelan
15. Embodied / Disembodied............... 255
    Michelle Stephens and Sandra Stephens
16. Theological / Worldly................ 281
    Stanley Hauerwas
17. Authentic / Artificial............... 294
    Anthony Reed
18. Batch / Interactive.................. 309
    Nick Montfort
19. Transmission / Influence............. 323
    Rachel Haidu
20. Silence / Beat....................... 337
    Paul D. Miller, aka DJ Spooky, That Subliminal Kid

Time Studies: A Bibliographical Reading List........... 345
About the Contributors 355
Index 361

ACKNOWLEDGMENTS

As co-editors, we have different sources to thank, but together we thank our contributors for their exciting contributions and patient involvement in this project. We would also thank Eric Zinner, senior editor, and Alicia Nadkarni, assistant editor, at New York University Press for their faith in this project and the four anonymous readers who evaluated it early on. We would also like to thank William Kentridge and Anne McIlleron for their permission to reproduce a still from Kentridge's *The Refusal of Time* on the book cover. And finally, we both want to thank the copy editors at NYU Press for their excellent work, and Andrew Todd for his help with the bibliography.

Joel Burges: I have spent a lot of time thinking about temporality over the years. The time spent, however, would have been impossible without the support of numerous institutions and people. This collection began as an idea at a bar in Germany while I was a Mellon postdoctoral fellow in the humanities at the Massachusetts Institute of Technology, so both the Mellon Foundation and MIT, especially the literature faculty there, deserve special thanks. My current home, the Department of English at the University of Rochester (ably headed by two wonderful chairs, John Michael and Rosemary Kegl), has been an amazing place to be a junior faculty member. Without the intellectual and financial support of my department and university, this book could not have come into being. I also wish to thank my students in the courses The Poetics of Television and Clocks and Computers: Visualizing Cultural Time, all of whom have added to how I understand temporality enormously. ASAP: The Association for the Study of the Arts of the Present hosted two panels related to this book, giving it stimulating audiences ahead of publication. And the Susan and Donald Newhouse Center for the Humanities provided a great place to complete the collection (thanks in particular to Carol Dougherty and Jane Jackson). I can't praise my co-editor, Amy Elias, enough. Working with her taught me how to be a better member of the profession, and how to do things I didn't yet know how to do—and in less time than I thought I could! Two singularly remarkable individuals came into my life while I was working on this project: Rachel Haidu and Nathan Johnson. Both have caringly clarified why I care so much about time, especially alternative temporalities. And my family—Nancy Burges, Shirley Fray, and Ann Burges, not to mention Dirk Dixon and my dog, Anya—have shown enormous patience, care, and enthusiasm as I made my way toward an intellectual life over the years. My mother deserves to be singled out: her support often gives me the time I need to live that life. I dedicate the work I have done here to her, with love.

Amy J. Elias: I would like to thank and acknowledge the support of the University of Tennessee's Department of English and the Hodges Better English Fund for research time and monetary support that led to this collection's completion, particularly a Hodges Research Grant in summer 2012; the University of Tennessee Humanities Center, headed at the time by Thomas J. Heffernan, for a research fellowship in 2013 that granted time for collection editing and production; Stanton Garner, head of the UT English Department during the time of this project and a staunch supporter of the contemporary arts as well as a friend; and the College of Arts and Sciences at the University of Tennessee for a 2012 SARIF travel grant enabling work toward this project, and the college's dean, Theresa Lee, for unflagging support of humanities initiatives. ASAP: The Association for the Study of the Arts of the Present provided opportunities at its conferences and symposia to explore and shape ideas expressed in these pages and to gather together some of its splendid contributors, as well as my co-editor (yes, the bar in Germany). I am grateful especially for the generosity, collegiality, and brilliance of fellow scholars and arts practitioners throughout the world, too many to name here, who have discussed and debated the contemporary arts with me and helped to generate exciting new conversations about them. I thank the members of the UT Critical Theory Reading Group and the UT Committee on Social Theory for collegial interchange and valuable intellectual conversations; the members of the Contemporary Arts and Society Research Seminar at UT for providing rare and thoughtful community across arts fields; and my girls at UT (you know who you are), who are a deeply appreciated source of female solidarity and friendship. Jonathan Barnes is thanked always for everything without reserve or limit. And Joel Burges is sent sincere thanks and gratitude for sharing his intellectual gifts and friendship in this endeavor; it's a pleasure to work with such brilliant scholars who will shape the field in the years to come.

Introduction

Time Studies Today

JOEL BURGES AND AMY J. ELIAS

In short, *geopolitics* has its ideological foundations in *chronopolitics*.
—Johannes Fabian, *Time and the Other*

Time does not pass, it accumulates, and as it accumulates it deposits an ever greater freight of material within the cargo holds of a present that is, in this sense, eternally after the Enlightenment present.
—Ian Baucom, *Specters of the Atlantic*

Time eludes us. Since Aristotle and Augustine posed their paradoxical questions about time to the Western world in the *Physics* and the *Confessions* respectively, we have been trying to determine what it is that we talk about when we talk about time. The terms "past," "present," and "future" seem too static, too thin to express our full experience of temporality. They capture neither our sense of the ephemerality of the instant nor our anxieties about the long unfurlings of time that exceed human lifespans and comprehension: geological time, evolutionary time, the time of climate change, or the time of the universe.[1] On the one hand we know that this trio of terms describing something as long and deep as time is necessarily reductive. On the other hand, we are driven again and again to track those lengths and depths as best as we can. We are compelled to track the terms because "past," present," and "future" are examples of *keywords* (originally "a word that acts as the key to a cipher or code") defined differently at different moments in history. If the terms do nothing else, then, as ciphers they unlock the historicity of time: tracking them shows us that our conceptions of time, and our enactments of it, are rooted in specific social contexts and grow from historical transformations, some of which are finished, many of which are not.[2]

If concepts of time are historically situated, mutating as societies and their science, economics, values, language, and institutions change, then it is logical to ask what are our own twenty-first-century concepts of time. When we try to track how time is conceptualized in our own moment, will we see a pattern to our societal priorities—to our present anxieties, past miscalculations, and future hopes? This collection addresses this question by presenting essays about time that construct what we are calling a "vocabulary of the present." They are a selection of keyword essays that talk about time as it has been experienced in developed nations after World War II, and they talk about time today in ways that, in part, reveal us to ourselves.

This is the critical condition and historical motivation behind what we are calling "time studies." In naming this field of discourse, we do not assume that the always complex and largely hidden meanings of current sociocultural time can be made transparent or be fully understood by those living at the same moment. The present is usually illegible to itself, and its assumptions can be glimpsed only by hermeneutics other than its own.[3] However, time studies compels attention because its findings are part of what anthropologist David Scott has recently described as "a new time-consciousness . . . emerging everywhere in contemporary theory" now, if "in a still inchoate way."[4] In the humanities after the turn of the millennium, for example, new critical attention to temporality is seen in the claims of affect studies, studies of the "everyday," posthumanism, ecocriticism, and the expanding territory of media studies. Thus what follows aims to guide readers historically and critically toward understanding (especially in the study of the arts, technology, and culture) how the postwar period—our present—is animated by certain kinds of time consciousness.

However, while we take the decades since World War II as our focus, we are not interested in engaging debates about how to define "the present" or "the contemporary" as a period marker. Certainly, we understand 1945 to be a date of significant social, economic, and political change—marked not only by the dropping of the atom bomb (heralding the advancement of technology on a scale never seen before in human history) but also the dismantling of many colonial empires, the rise of finance capitalism, the acceleration of the effects of the Anthropocene, and the beginnings of the Cold War that would eventually reshape the political landscape of the world. Instead, we are focused on the *semantics* of the present since 1945, particularly in relation to time.

That semantics—or vocabulary, as our subtitle has it—becomes visible to readers in this volume through a series of essays that investigate paired keywords addressing the specificity of time today, in the interest of glimpsing the unique textures that postwar temporality has assumed. We therefore begin the volume with a pair of terms that have come to dominate our sense of the present: multiplicity and simultaneity. Over the course of modernity, and with continued momentum in our time, the present has emerged as an experience of simultaneity in which temporalities multiply because they are synchronized as simultaneous on economic, cultural, technological, ecological, and planetary registers. Thus while simultaneity is often understood as a reduction of that multiplicity, creating a singular time beholden to capital, the present is actually animated by a tension between the simultaneous and the multiple, variously contracting and protracting a sense of contemporaneity in which times conjoin. The essays in this book are devoted to grappling with that conjuncture.

Multiplicity / Simultaneity

Critical discussions of the post-1945 period are often concerned with a debate that can easily become predictable. The debate focuses on what comes after postmodernism, when "the contemporary period" begins, and how the arts uniquely embody those questions.[5] Our interest is otherwise. Given the etymology of the term, we assume that "the contemporary" is a historically deictic term, indicative of a sense of presentness that has been felt by cultures of the historical past as well as those of the current moment. However, we take that term to uniquely signify *today* an emerging structure of temporal *multiplicity* embedded in a history that has constructed the present as an experience of *simultaneity*.

Roger Luckhurst and Peter Marks suggest that "the difference at the heart of the 'now' can be seen as a *constitutive* and *productive* heterogeneity, a circulation of multiple times within the single instant. We might take this to be what the 'contemporary,' *con-temporarius*, literally suggests: 'joined times' or 'times together.'"[6] Time is multiply measured both within a society or culture and between cultures: "What if we were to get into a companionable habit of looking at our watches to ask, not what time it is, or where we are in time, but whose time it is?"[7] This dovetails with Terry Smith's claim that periodization of the contemporary is actually impossible, since "contemporaneity consists precisely in the acceleration, ubiquity, and constancy of radical disjunctures of perception, of mismatching ways of seeing and valuing the same world, in the actual coincidence of asynchronous temporalities, in the jostling contingency of various cultural and social multiplicities all thrown together in ways that highlight the fast-growing inequalities within and between them."[8] The contemporary is not "our time" because of its heterogeneity and opacity, and it is not "a time" or period because it is defined by antinomies. The contemporary present is a conjuncture of times that take time.

The following essays each address a pair of interrelated terms—past/future, real/quality, labor/leisure, synchronic/anachronic, transmission/influence—to describe, individually and collectively, that conjuncture as multiple and simultaneous. The "present" performs the semantic and conceptual work of coordinating coetaneous temporalities as a sensory horizon, an affective technology, a psychic matter, a market duration, a political economy, a global *mentalité*, and a natural history. Understood this way, the present may be grasped as textured and stretched, latent and current—a mediation of presence and distance in time. The present is both in and out of time and takes time; it is made up of multiple temporalities. Philosophers of the late nineteenth and twentieth centuries such as Martin Heidegger and Henri Bergson recognized this mulitplicity as they sought to grapple with conceptions of time as an infinite series of instants, on the one hand, and as a directional flow, on the other.

No recent artwork addresses this notion of the multiplicity of present time—the present as times that take time together—more than Christian Marclay's 2010 installation *The Clock*, one of the most critically acclaimed works of art since the turn of the millennium. A blockbuster event drawing lines of spectators who wait hours to enter the exhibit, *The Clock* runs twenty-four straight hours and splices together scenes from movies and television programs that either incidentally show clocks and watches noting the time of day, or present characters reflecting on time in various forms. As it runs, the film becomes a kind of real-time clock, its scenes deliberately synchronized to correspond to times of the twenty-four-hour day within the time zone in which it is being projected. The extreme length of the film and its showing as a continuous run in the space of the museum not only thematizes the relentless movement of time (visitors are aware that they are unlikely to sit through the entire film as it keeps running throughout the day and night), but also provides an at first funny and then fascinating experience for spectators (they collectively participate in a kind of chronophilia).[9] It treats time as a formal device and aesthetic means, as a social experience gesturing toward a whole way of temporal life, and as an artwork that playfully, hypnotically invites us to reflect on the times—day and night, death and life—of our lives.[10] But while this artwork both is centered on time and happens in time, it is made of clips that are moments taken out of time, out of the context of their own televisual and filmic origins, out of the history surrounding their own production, and out of the temporal sequence of the stories to which they originally belonged. *High Noon* (1952) predictably and pleasurably appears when Marclay's impractical clock strikes 12:00 p.m., while a later midday moment (12:05 p.m., to be exact) from *American Gigolo* (1980), happily fueled by cocaine and couture, is juxtaposed with the same 12:05 p.m. (or is it?) in *The Exorcist* (1973), when Max von Sydow, a world away and a film apart from a young Richard Gere, sees a clock come to a demonic stop, foreshadowing his own death at the hands of a possessed little girl later in this classic of horror.[11] Thus on the one hand, *The Clock* makes us hyperaware of time as historical and narrative multiplicity, with distinct times of day embodying divergent qualities, encouraging differing actions, and engendering differential rhythms. On the other, it flattens different times as all equally synchronizable according to the chronometer, homogenizing different *kinds* of time into one video presentation. *The Clock* thus implies a contradiction: the "time that takes time" is portrayed as multiple, drawn from different moments in the history of film and television, and simultaneous, with those moments experienced as a singular event that we attend at a museum within the present created by the installation.

These are the very traits of a postmodernism that *The Clock* seems, as an artwork, both to challenge and exemplify. *The Clock*, for example, harkens back to the paradigm-shifting work of Marxist theorist Fredric Jameson, who claimed throughout his work on postmodernism from the 1980s onward that in the culture of "late capitalism," space has supplanted time. What he meant—citing examples in specific postwar art forms such as video, literature, and architecture—was that in today's technologically enhanced and globalized world, we see the triumphant late stage of capitalism, which restructures all levels of society (not just economics) according to market values, eradicating or repressing other values generated by other systems. The values of capital include speed enabled by ever-improving technology, globalized contact, uniformity (or at least extreme translatability) of world cultures, and the redefinition of all aspects of life and human perception to forms that can be exploited by a world market that infinitely expands and thrives on diversification—a very specific instance in which simultaneity and multiplicity meet. For Jameson, capitalism has little use for history defined as implicit or explicit critique of market values or class privilege, or for art that makes historically constructed power relations come into focus. It is dedicated to promoting art only as an investment commodity, a "commercial" for other products, or an affective *frisson* rather than a vehicle for thought (particularly political thought) or utopian desire. Thus historical images might appear in music videos, but they are used in the interest of "pastiche," a flattened, ahistorical montage that exists not to enable audiences to think critically and historically about their present but actually to shut down such thought, implying the idea that history is *nothing more* than a series of pictures that can be spliced together in any order in order to create an emotional effect for a purchasing public.[12]

Jameson's assertions spoke to an emerging conversation among mid-twentieth-century theorists about the nature of time in postmodernism, the postwar sociocultural and economic dominant based on the increasing power (and, by century's end, global triumph) of finance capitalism. In these theories, postmodernism is characterized by the dominance of space over time: temporality is suppressed in favor of implicitly or explicitly spatialized models. All places become the same in a tourist economy (all equally marketable images of the past, though they may depict different historical scenes), and Umberto Eco and Jean Baudrillard theorized how history itself becomes a commodity in the shopping mall hall of mirrors of consumer society. Charles Jencks, working in architecture, noted how new postmodern buildings raided the past for styles and spliced them together in a new urban aesthetic, much in the manner of Jameson's postmodern video, while Edward Soja wrote about how the postmodern city became a jumble of historical styles that created a new kind of postmodern space, a "thirdspace" that lay somewhere between real and imaginary space, reality and desire.[13] In these and many other critical statements, space and spatialization became the dominant conceptual frames for postmodernism.

Yet the 1980s and the 1990s also saw groundbreaking scholarship that was sharply focused on postmodern time and did not reduce history to a spatial model. In *The Condition of Postmodernity* (1989), for example, David Harvey argued that a central characteristic of the postmodern was in fact the annihilation of space by time. In the same vein, in *Time and Commodity Culture* (1997), John Frow rejected the spatialization of time and asserted that heterogeneous strands of time comprise historical movement. Ursula Heise argued in *Chronoschisms: Time, Narrative, and Postmodernism* (1997) that late-twentieth-century technological transformations foreshortened and fractured time, constructing a "culture of time" in a new way; Andreas Huyssen's *Twilight Memories: Marking Time in a Culture of Amnesia* (1995) argued that postwar society was obsessed with memory and history in reaction to the timelessness of electronic communications.[14] In these and other studies, postmillennial time studies as we defined it above was beginning to take hold in the work of critics who sought to break free of postmodernism's putative spatial dominant without naïvely bracketing the questions about time and space that theorists of the postmodern taught us to ask.

The Clock embodies and embraces many of these historical tendencies and theoretical queries. The installation, after all, is easily described as a twenty-four-hour music video that flattens rather than intensifies historicity by splicing together multiple moments from the history of film as if they were simultaneous, part of a singular now, in an artistic event akin to the opening weekend of a blockbuster movie. This is to say that the simultaneity of *The Clock* threatens to become a singularity. *Singularity* is a term denoting a kind of postmodern and post-postmodern space, the reduction by the capitalist system of all things to one model or one kind of thing—a commodity space or conceptual container defined by capital.[15] *Simultaneity* is the temporal register of the singularity. Within capitalism, if all things become, really, one thing (the commodity), then all times become, really, only one time: the present, the time of consumption that is also the time of the affective singularity. The singularity, however, should be distinguished from the event—the timespace happening (appearing in theory from Foucault to Lyotard to Badiou) that cannot be contained within existing social registers and structures. Jameson writes, for instance, that "the [art] installation and its kindred productions are made, not for posterity, nor even for the permanent collection, but rather for the now and for a temporality that may be rather different from the old modernist kind. This is indeed why it has become appropriate to speak of it [the artwork] not as a work or a style, nor even as the expression of something deeper, but rather as a strategy (or a recipe)—a strategy for producing an event, a recipe for events."[16] We therefore see a potential contradiction about time in the present that *The Clock* makes clear. On the one hand, contemporary time is a multiplicity that, in the Marclay artwork, pluralizes each minute of the day with many moments. On the other hand, contemporary time is a singular simultaneity concocted from a recipe that calls for aligning clips with the time of day and shown to a paying audience. Yet these poles are not as oppositional as they at first appear, and *The Clock* ultimately seems to exemplify how contemporary installation art sits on a shimmering line between singularity and event, working a Jamesonian territory of precarious and implicated critique.

To see how, consider a compressed history of the present as both simultaneous and multiple. The trajectory by which capitalist nations of the developed world arrived at a state of temporal simultaneity is a long one. The present—and presentism, it seems—has a past.[17] In the modern history of time, especially in the developed world, the experience of the present is one of increasing simultaneity as a result of technological, economic, and cultural developments that expand and extend our sense of the now, even though doing so has progressively attenuated past and future. According to some cultural historians, the experience of simultaneity can be traced to three events: the creation of the mechanical clock, the advent of capitalist modernity, and the development of the modern nation-state. The chronometer and capital appeared around the same historical moment. Mechanical clocks performed an important function. They abstracted time from cyclical, natural, and mythic time such as seasonal cycles, circadian rhythms, and liturgical calendars and replaced them with "a continuous succession of constant temporal units," such as the twenty-four-hour day and the sixty-minute hour. Time became tied to the demands of the workplace (the need for a regulated workday, scheduling the transfer of goods, and international trade)—that is, tied to a market society dominated by the commodity, the very embodiment of the present in its extinguishing of the past of production that went into its making.[18] The chronometer facilitated the arbitrary and abstract time of production: the workday and its redefinition of natural time into calculable time.

Eventually, as many studies have by now shown, the clock gave rise to that mode of time that Walter Benjamin famously called "homogenous, empty time."[19] According to Benedict Anderson, the newspaper, especially by the eighteenth century, likewise abstracted time but added calendrical coincidence around a given date, thus synchronizing disparate events for the reader as simultaneously occurring (much as *The Clock* uses a particular minute in the day to coordinate historically distinct moments of film experience). In the same century, the novel began to make of mass culture a similar horizon of temporal experience because it elaborated "meanwhile time,"[20] the time of reading that allowed different plot threads, sequentially narrated, to be perceived by readers as simultaneous in time. Yet if the novel and newspaper intensified the present in the eighteenth century, then the content of both made of that present a conjunctural time to the degree that the former depicted plots that involved backstories, flashbacks and flashforwards, and parallel storylines, and the latter reported events that began at different moments in the past and finished at distinct moments in the future. Indeed, one might say that in this respect newspaper and novel gave back to the commodity the past it extinguished, since both were themselves commodities of a particular sort: the regular novelty. As such, they marked change according to the cadences of consumption that were emerging as European societies, especially England, commercialized and modernized in the eighteenth and nineteenth centuries.

Anderson's work suggests that the present was increasingly a mediated experience of simultaneity, even if we would add that such simultaneity was not necessarily an experience of temporal homogeneity: the present was shot through with past and future in novel and newspaper alike. Moreover, both were part of a growing presentism that seized upon itself as radically and reflexively contemporary—a perspective, Peter Fritzsche has argued, that can be traced to the revolutionary events of the late eighteenth and early nineteenth centuries. These events forged "a newly felt sense of contemporaneity which pulled Europeans together into a circle of shared understanding and busy literary activity, one which newspapers attempted to bind in a more permanent way."[21] A "sense of contemporaneity" was the bid to make of the present a virtue, as modernity enacted a radical break with tradition. Pursuing the present as a virtue was in many respects a necessary temporal compensation in the face of those events by which, in Ian Baucom's words, "modernity [was coming] into a historicist awareness of itself" through the French Revolution and the Napoleonic wars as well as colonialism's activity (especially the global slave trade) worldwide.[22]

The nineteenth century saw other world-historical pursuits of the present (especially of a presentism in scientific research and information practices such as geology) and technological development of infrastructure such as railroads that demanded synchronized timetables so as to run on singularly chronometric time.[23] By the twentieth century, such presentism merged with a historically modern sense of contemporaneity that further elaborated the present as an experience of simultaneity. Stephen Kern has argued, for example, that steam power and wireless telegraphy were technologies that not only played an innovative role in travel and communication by the early 1900s, but also transmitted the present instantaneously *as such*. In this respect, they were descendants of the newspaper, offering their users "[t]he ability to experience many distant events at the same time," an ability that telephony also provided.[24] Likewise, cinema, a major technological development of this period, continued the "meanwhile time" of the novel through techniques such as parallel editing, which sequenced shots so that spectators felt two events to be happening at the same time. But the presentism of parallel editing was inflected by the past, if a past overwhelmed by an investment in the present. As Mary Ann Doane has shown, in exposing "people to aspects and events of the world that had previously been distant and inaccessible," early film produced "the sense of a present moment laden with historicity at the same time that [it encouraged] a belief in our access to pure presence, instantaneity."[25]

Though this breathless overview seems to give a developmental view of history, it would be a mistake to present a "history of presentism" as teleological. Understanding the present as a temporal conjuncture often linked to an experience of simultaneity might be thought rather as an *intensity* in our postmillennial moment. As an experience of simultaneity, the present of modernity has been an effect of increasing numbers of events and actors undergoing intensifying technological and economic synchronization across space, with local times becoming one time through the advent of developments such as the clock, the locomotive, and the telephone. This reached a nexus in the era of high modernism, which supposedly put into place a movement toward the postmodern experience of spatialized time. For postmodern theorists that experience is a symptom of "the end of temporality" and a "crisis in historicity."[26] In this model, there are few protensions or retentions, and time is nothing more than global capitalism extending its systemic tentacles around the world, seemingly replacing time with space. Yet if for Jürgen Habermas the "project of modernity is not yet finished," one might say the same for time's evolution in our present. Our experience of time is not yet complete, and it is not directed toward any one end.[27] Even as it is marked by an experience of simultaneity that veers toward singularity, multiplicity persists. To use Baucom's terms in the second epigraph to this introduction, time accumulates in many forms that crystallize both simultaneity and multiplicity.

The forms of high modernism often suggest this tendency as much as they do the movement toward spatialized time. Thus while cinema discovered how to create "meanwhile time" through parallel editing, most innovatively in the films of D. W. Griffith, modernist authors experimented with free indirect discourse (allowing a simultaneous subject/object positionality that accesses the private time of individual subjects, as in Virginia Woolf's *Mrs. Dalloway*) and multiple focalizations of events (allowing simultaneous experience of different minds and thus many psychic tempos). Philosophers such as William James, Edmund Husserl, Martin Heidegger, Henri Bergson, and Maurice Merleau-Ponty worked to reconcile the idea of the present with models of immanence and singularity (time as a multiplicity, a series of singular instants) or with durational models of time (a kind of simultaneity). Such perspectives persist in postmodern philosophies, such as Jean-Luc Nancy's concept of an "inoperative community" of singularities and the metaphysics of Gilles Deleuze and Félix Guattari.[28] These all point towards an alternative genealogy of the present in which multiplicity is just as significant as simultaneity, complicating any simple telos for our history of presentism.

However, the question remains: how does *simultaneity* intersect with *multiplicity* in a way that evades *singularity*? As noted above, seen through the lens of Marxist-inflected theories, the temporal terms are two aspects of capital itself: capitalism reduces all time to simultaneity, a presentism that demands, paradoxically, multiplicity defined as infinite creation of new markets and diversification. From this angle, multiplicity is a kind of dodge: what seems multiple and diverse is actually always part of the system, always feeding the logic of capital.[29] However, as a temporal term, "multiplicity" can mean something different: the recognition of multiple times operating subversively within or outside the seemingly airtight simultaneity and singularity of the world system. For example, some theorists note that there are still cultural and even economic elements as yet uncaptured by capitalism, and in fact there has been significant criticism of theories of postmodernism for their Anglo-European and American bias. Some claim that postmodernism is not, or not yet, a seamlessly triumphant global system, while other theorists note that we can discuss this system without importing putatively outmoded Marxist theoretical concepts.[30] Is it possible that some cultures or bodies themselves resist incorporation into capitalist time? Is it possible that capitalist time itself produces or calls forth obstinacies in the form of times that chafe against its relentlessly singular structuring of our days? Is it possible that temporality itself might be something that always eludes complete cooptation by capital, something on a different categorical or ontological level leading to multiple fractures and sites of resistance within (or outside of) that very system?

Such questions are powerful motivating inquiries in theory today. For instance, Peter Osborne writes of temporal multiplicity that "the distinctive conceptual grammar of con-temporaneity" is that of "a coming together not simply 'in' time, but *of* times: the present is increasingly characterized by a coming together of *different but equally 'present'* temporalities or 'times,' a temporal unity in disjunction, or a *disjunctive unity of present times*."[31] Using a number of metaphors to convey the "polytemporality of the present," the fluid and unpredictable nature of time in social contexts, Steven Connor asserts a compatible perspective:

> In contemporality, the thread of one duration is pulled constantly through the loop formed by another, one temporality is strained through another's mesh; but the resulting knot can itself be retied, and the filtered system also simultaneously refilters the system through which it is percolating. The scoring of time constituted by one temporality is played out on temporal instruments for which it may never have been intended, but which give it its music precisely in the way they change its metre and phrasings, and remix its elements.[32]

We are back to *The Clock* with this passage, for Marclay's installation does rely on sound and music to "remix its elements." Remixed into a temporal montage that constantly announces what time it is in the most impractical of ways, each minute is never one time, but multiple times shaped by multiple references and overlapping historicities. As much as it recalls the ahistorical pastiche decried as a temporal terminus by postmodern theory, *The Clock* also implies Heidegger's "Being-in-the-world": the private individual often mistakes the present for a moment that rapidly comes to an end, especially in postmodernity, but the public character of the world means that individuals will always come into contact with other times characterized by significance.[33] In contrast to singularity, such "significance" is what spectators discover when they sit with friends and among strangers viewing *The Clock*, watching time multiply simultaneously from moment to moment, minute to minute, and hour to hour. That temporal multiplication happens not only as a formal feature of the "remix" on screen, but also as a dimension of being in time with others as a temporal public constituted in the present and presence of *The Clock*. Thus as much as *The Clock* exists within a postmodernist temporality of static simultaneity associated with capital, the antinomic tension of the installation is how it also exists within a counter-temporality of multiplicity. Rather than contributing to the worlding of capital, this multiplicity may gesture toward a new, potentially deterritorializing, planetarity.[34]

The Emergence of Time Studies

How then is the multiplicity of temporality and counter-temporality theoretically defined today? The question is too large to address here in detail, so we have provided a bibliography at the end of this book to guide readers to the many sources they might consult to pursue further answers—which, of course, the keyword essays in this volume also aim to provide. One might nonetheless identify a few threads—or an interdisciplinary series of "turns"—in the humanities in relation to the postmillennial emergence of time studies: the global turn, the durational turn, the affective turn. But these are just a few of the ways to conceptualize how time studies is emerging and diversifying today. Readers should pursue others as well.

The "global turn" toward time is actually discussed above: the theorization of a time unique to capital on a global scale, and the possibilities of its refusal, fracturing, or deterritorialization. Yet the global turn takes different forms. It includes ecocriticism, new materialist theories, and posthumanisms that redefine time on a global or even cosmic scale. This global turn rests in part on redefining "globalization" within the frame of the Anthropocene or as part of "planetarity" (a social model that understands time as structured by the needs of living systems, ecologically entwined). The global turn also includes, however, theories that mesh these concerns with the functioning of capital and the world system. For example, modernity overseen and defined by capital depends upon speed of communication, speed of consumption, and speed of obsolescence and innovation.[35] The phenomenologist and urbanist Paul Virilio characterizes this technological culture of speed as "dromology" (taken from the Greek *dromos*, meaning race or racecourse); this is part of what he calls the "war model" of the modern city and of human society.[36] For Virilio, the logic of acceleration is the basis of modernity, and technology is a handmaiden to increasing societal control.[37] Moreover, we now may have "hit a wall of acceleration": with global digital data transmission and instantaneous communication, we may be now approaching a critical point at which further acceleration (and hence development) is no longer possible. Virilio wants to examine the consequences of this temporal state, and claims that speed as a way of calculating time now unifies perception as well as social, political, and military development.[38]

Virilio's unmasking of a culture of speed is one way to interrogate existing structure. But if speed characterizes modernity circumscribed by capital, then "slowness" also becomes a counter-temporality as well as a means of resistance. One thinks of "slow food movements," or the "Slow Movement" itself that "aims to address the issue of 'time poverty' through making connections."[39] And in point of fact, slowing down and falling asleep are being reclaimed by critics focused on the time of the present. That reclamation is unfolding as an explicit counter to the routinely pivotal role that acceleration has played in accounts of the historical movement toward time-space compression, with slowness and sleep both suggesting tactics to expand the present once again. In addition, the pleasures of speed and the malignancies of velocity, Lutz Koepnick argues, have blurred how much the present is an experience of "different stories, durations, movements, speeds that energize the present, including this present's notions of discontinuity and rupture, of promise and loss."[40] Slowness, argues Koepnick, "asks viewers, listeners, and readers to hesitate, not in order to step out of history," but rather to engage slowness as a contemporary tactic of historical reflexivity by which we can register and reflect on "the co-existence of multiple strands of time in our expanded present."[41] Less hopeful than Koepnick about the present as a waking horizon of experience, Jonathan Crary nonetheless lodges a related protest against a "24/7" temporality through which networked technologies insist that we stay perennially awake to always accelerating and expanding work time. Falling asleep for Crary is akin to the historical reflexivity Koepnick attributes to the "medium" of "slowness," because the time of sleep constitutes "a periodic release from individuation. . . . In the depersonalization of slumber, the sleeper inhabits a world in common, a shared enactment of withdrawal from the calamitous nullity and waste of 24/7 praxis."[42] Yet Sarah Sharma has argued that simply opposing slowness and sleep to the velocity of modernity may only solidify—rather than disrupt—the violent hierarchies structuring such modernity. Based on her ethnographic research with people who work in occupations defined by temporal parameters (e.g., taxi drivers, frequent-flyer business travelers, and devotees of the slow food movements), she defines "power-chronography" as a way in which capital uses temporality in the interest of biopolitical control.[43]

In a very different way, a new critical return to durational time is also positioned as a resistance to the unnatural time dictated by the chronometer and central to Western modernity, but also underlying the division of mind/body, subject/object, and thinking/feeling central to much Western philosophy. Now claiming widespread influence in fields as diverse as systems theory, posthumanism, and virtual aesthetics, the temporality theories of Henri Bergson, William James's radical empiricism, the materialist philosophy of Baruch Spinoza, the radical metaphysics of Gilles Deleuze and Félix Guattari, and the process philosophy of Alfred North Whitehead reconceive durational time as materialist temporality. In these theories, which side with realist philosophies, time is understood as haptic and coterminous with an undifferentiated flow of energy/matter that makes up a vitalist universe. Registered proprioceptively, in the body's own participation in the movement of time-matter, reality is a force—"event-ness" based in the relation of matter to all other matter, the nature of matter as energy flux and flow. Opposing the rationality and dichotomous thinking undergirding Anglo-European philosophy that parcels reality into discrete objects and forms, such philosophy emphasizes relationality and co-presence, the creative "becoming" of life that can be constrained neither by social forces nor biopolitically driven theories of cognition and perception. All objects are events, based in "virtual movement." Brian Massumi's example is instructive: a chair is not just an object but a marker of temporality ("the feeling in this chair of past and future chairs 'like' it"); a connection to life forces ("the feeling in this chair that life goes on"); a presentation of its own relation "to the flow not of action but of life itself, its dynamic unfolding"; and a marker of relation and activity ("It's how life feels when you see it can seat you").[44] Such philosophy asserts an alternative to the instrumentalization and objectification of reality, and its counter-hegemonic potential is seen, for example, in the anarchist vitalism of Deluze and Guattari's philosophy. What it offers to time studies is a definition of the present that is based in philosophical, material, and political assumptions about a reality and collectivity hidden from, and often antithetical to, the cognitive rationality organizing the dichotomous categories of modernity.

The "affective turn" in theory today often draws from the haptic and somatic grounding of such anarchic vitalism to assert a temporal register based in the body—the nonrational, affective somatics of "thinking-feeling" that stand in counter-relation to cognitive rationality and "conscious" thought. Affects link emotion and body, but in theoretical work by Lauren Berlant, Sianne Ngai, Sara Ahmed, Brian Massumi, Eugenie Brinkema, Richard Grusin, and others, they also register social and formal logics. Notes Ngai, "I'm interested in . . . 'minor' or non-cathartic feelings that index situations of suspended agency. . . . I'm interested in the surprising power these weak affects and aesthetic categories seem to have, in why they've become so paradoxically central to late capitalist culture."[45] Affect theory is the effort to understand the present as it plays out in somatic contexts, which become politicized as perceptive platforms that either can or cannot be coopted by capital. It is the sustained attempt to attend to "the contemporary moment from within that moment," as Lauren Berlant writes, calling attention to the present as a "mediated affect" or "a thing that is sensed and under constant revision, a temporal genre whose conventions emerge from the personal and public filtering of the situations and events that are happening in the extended now whose very parameters (when did 'the present' begin?) are also always there for debate."[46]

In these theories, affect becomes a medium not unlike slowness or sleep—a condition of relation and an activity that mediates the present reflexively in its ongoingness, allowing contemporaneity to take hold of us and produce potentially political effects. For example, joining many other affect theorists who reject Fredric Jameson's claim that postmodernism signals a "waning of affect," Kathleen Stewart contends that "the emergent present" is "weighted and reeling"—that "a present that began some time ago" is more than the effect of a structural totality (late capitalism), a conjuncture of events (globalization), or a recognizable and legible period (neoliberalism). Using the language of metaphysics, Stewart asserts that the present is "a scene of immanent force . . . composed out of heterogeneous and noncoherent singularities." Or, more concretely, the present is "[s]omething that surges into view like a snapped live wire sparking on a cold suburban street."[47] In Berlant's work, the present "surges into view" affectively: "the present is what makes itself present to us before it becomes anything else, such as an orchestrated collective event or an epoch on which we can look back." For Berlant, this means that the present "is also a thing that is sensed and under constant revision."[48]

As the above shows, the present is being rethought and politicized in new theories that ask how a reconception of temporality today can have social, ethical, and psychic effects. Moreover, as the example of *The Clock* has demonstrated, art often plays a central role in all of these new analyses: it is in literature, painting, photography, installation art, music, digital media, and film that new ways of seeing and experiencing time are captured and new ways of conceiving time in our moment are glimpsed. Certainly, such theories illustrate just some of the ways that "a new time-consciousness is emerging everywhere in contemporary theory" now, if "in a still inchoate way."

Time: A Vocabulary of the Present

The essays that follow examine how numerous postwar "contemporalities" come together in and as the present. In them, that present is definitively galvanized by multiple temporalities that collide dynamically and dialogically, effecting an experience of simultaneity that often—but not always—exceeds singularity of the kind that was the focus of postmodern theorists. This project is facilitated, we hope, by the keyword essay format, in which each author tackles two temporal terms. To explore the disjuncture of times that make up the historical present—the contemporary as an immediacy, a moment, an era, or a longer *durée*—each contributor to this book has juxtaposed two temporal terms important to postwar arts and society. These essays are grouped into three sections: "Time as History," "Time as Measurement," and "Time as Culture."

Time as History: Periodizing Time

Common sense tells us that time and history are natural siblings. And indeed, the "conception of history as an immanent and continuous process in chronological, or secular, time" is central to many if not most Anglo-European academic approaches to history.[49] Lorenz and Bevernage note, for example, that "Marc Bloch famously called history 'the science of men in time.' Similarly, Jacques Le Goff labels time the 'fundamental material' of historians," and conversely, Norbert Elias famously declared, "Wherever one operates with 'time,' people within their 'environment,' within social and physical processes, are always involved."[50] Moreover, write Lorenz and Bevernage, several historians—in particular Lucian Hölscher, François Hartog, and Peter Fritzsche—have started historicizing time conceptions previously taken for granted.[51]

But *is* history, "the science of men in time," the same as the science of time by men? Sexist language aside, the question is an important one: to what extent are historians actually conversant in the language, perspectives, and claims of the philosophy of time when they posit history *as time*? "All those people who think that historians are in some sense experts in the study of temporality," Hayden White has remarked, "should realize that farmers probably know more about time than historians do."[52] It might behoove us to understand history and time studies as fundamentally related fields of discourse but operating at different levels of conceptual construction. Jörn Rüsen writes, "History is a specific way in which humans deal with the experience of temporal change. The way they realize it essentially *depends upon pre-given or underlying ideas and concepts of time*."[53] This implies that history is not philosophy of time per se but rather is built *upon* time, upon already ensconced cultural understandings of what time is.[54] Predicated on time schemas, history is at least two removes from "actual" time, whatever that may be.[55] History is not time but a cultural mediation of time, which itself has already been mediated through our cognitive schemas, cultural vocabularies, moral systems, ontologies, and epistemologies (and perhaps even our mathematics, as the quantum revolution has shown).

In the essays that follow, time as history is examined in this light: as a social organization of time aligned with historical narrativization, accounting, or logic. Moreover, each of the essays attempts to illustrate how the post-1945 period in developed nations has created cultural conditions that shape the presentation of time-as-history in new, unique, or unexamined ways. This return to history crosses disciplinary boundaries and is different in tone from earlier postmodernist metahistoricity. Amelia Groom notes that in art and art history,

> early-21st-century art has seen a rising concern with *re-present-ing* the past. Many artists are embracing obsolete technologies, abandoned places and outmoded materials; resuscitating unfinished ideas; revisiting docu-

ments and testimonies; and restaging downtrodden possibilities. Rather than a winking postmodern pastiche of appropriated styles, or an earnest nostalgic immersion in a fixed, absent past, these new engagements with the remnant of previous times mark a thickening of the present to acknowledge its multiple, interwoven temporalities.[56]

One might recall Kara Walker's black cut-paper silhouettes, or Joe Sacco's graphic novel *The Great War*—both of which avoid pastiche as well as nostalgia in their depictions of the past.[57]

Thus the essays in "Time as History: Periodizing Time" assume that we need to rethink temporal terminology associated with the logic of historical understanding in light of sociocultural change. In her essay, "Past / Future," Amy J. Elias posits that we now live in a time of "techno-duration," in which duration has assumed a metaleptic character *inside* both planetary and historical time. She argues that in the twenty-first century, "past" and "future" have become redefined as "retrofuture" and "slipstream," terms that signal neither progress nor decline but rather a doubled movement of simultaneous futurity and historicity that provokes an image of "moving stasis" compatible with techno-durational "presentism."

What time *is* as history may, however, be discernable through the lenses of cultural analysis and art, both of which work to defamiliarize the unexamined assumptions of a society. In "Extinction / Adaptation," Ursula K. Heise illustrates how post-1960s film and literature expose these terms as processes in the biological world but also as narrative metaphors for different kinds of cultural engagement with ongoing processes of modernization and globalization. Ranging through paleobiology, environmentalism, ethnolinguistics, evolutionary theory, genetics, and biotechnology, she invites us to rethink the temporal directionality of evolution through both adaptations and extinctions.

A very different kind of historicism characterizes time in the arts—namely, periodization. In "Modern / Altermodern," David James asks whether modernism can still provide an appropriate interpretive lexicon for apprehending the arts of our time and how we account for the curious persistence and institutional acceptance of periodization itself. Periodization, he claims, offers a particular kind of "solace" to the contemporary imagination, inhibiting the interrogation of period concepts and paradigms.

Yet while the modern and the "post"-modern seem to indicate a logic of continuance between the past and the present, the "perpetual present" of contemporary global society actually depends centrally upon the values of innovation and obsolescence. At their most general, Joel Burges argues in "Obsolescence / Innovation," these terms refer to twinned processes of technological change that for us engender a temporal relation between past and future: innovation and obsolescence spark the emergence of a horizon of temporal sensation shaped by the commodified coming and going of technologies and products. This is a sensory horizon in which historical time unfolds by way of our ongoing sensitivities to the "product life cycle." What is produced for us as a result is a rhythmic experience of historical time.

Once the critical potentiality of innovation is exhausted today, however, what is the fate of emergence? In "Anticipation / Unexpected," Mark Currie turns to philosophy of time, specifically phenomenology, to examine the complex relation between "emergent terms" as modes of temporality. He argues that the category of the unexpected event is more prominent in a world that is methodical, accurate, and expert in its predictions, and for this reason it makes sense to explain emergent ideas about a world less predictable than it used to be as the product, and not the failure, of anticipation.

Time as Calculation: Measuring Time

If someone were to ask, "What do you think about when I say the word 'time'?" many of us might automatically think of time's numerical forms. We might think first of minutes and hours—clock time. We might then think of time calculators and time limits—stopwatches, media run times, class or work schedules. If we were attuned to our bodies for some reason at the moment we were asked, we might also think of aging, or a circadian rhythm that we conceptualize in chronometric terms as our "biological clock." Associating time with calculation in this way seems completely natural. It is not natural, of course. Time as calculation has its own history, like every other form of time. If the Egyptians were the first to create a twenty-four-hour day, it was the U.S. government that issued the Calder Act of 1918, which implemented standard time and daylight savings time throughout U.S. territory.

Moreover, time calculation has a long narrative of development and change. While horological devices of one kind or another (including sundials, water clocks, and hourglasses) are among the oldest of human technologies, early mechanical clocks were seen in Europe in the early Middle Ages and eventually became essential to Europe's great age of colonial expansion.[58] As noted above, chronometric temporality was bound not only to the invention and worldwide adoption of the mechanical clock, but also to new demands for schedules and time calculation by a growing intercontinental trade empire.

This is the temporal calculation discussed by Jimena Canales in her essay "Clock / Lived," which specifically examines the development of chronometric time in "the West." Starting with the fact that the development of the atomic clock after WWII gave such confidence to the scientific community that it changed the length of what constituted a second as an interval of time, Canales examines the twentieth-century divide between scientism's faith in chronometrics and competing versions of time based in lived and biological rhythms, such as theories of duration by Bergson. Both versions of time were mediated, she claims, by phenomenological notions of time by Heidegger and others and, later, by modifications in time calculation provided by filmic time, literary time, and historical time.

In "Synchronic / Anachronic," Elizabeth Freeman tackles two terms associated with durational time and argues that both synchronic and anachronic time calculation offer ways to conceptualize freedom as corporeal modes of time, sites through which bodies are dominated and resist domination. She argues, however, that both are also bound up in specific iterations of the social: under contemporary conditions of "chronobiopolitics" and instantaneity, anachronistic irruptions of other forms of time or historical moments can become sites of critique or alternative imagining.

In terms of scale, the keywords examined in "Human / Planetary" look to a very different mode of time calculation. In her essay, Heather Houser uses this scale to contrast human time to planetary time, noting that through the concept of the Anthropocene, the two times interpenetrate in complex ways. She introduces the notion of geological timekeeping in climate change calculators and argues that human time and technological time are interwoven with the inhuman time of instrumentation, computation, and mathematicization, a third kind of time that mediates the binary of human and planetary time.

Thus the role of visualization seems to play a key role also in how we understand time today—from our various timepieces that show us second to second where we are in time, to our online calculating technologies, to our art forms, which visualize time in new ways. In his essay, "Serial / Simultaneous," Jared Gardner discusses how Einstein's special theory of relativity highlighted the paradoxical relationship between seriality and simultaneity, two seemingly contradictory models of time. More than a century of experimentation since Einstein formulated his theories culminates in a growing number of opportunities in the popular narrative arts of the present, such as comics, to explore and experience time as both serial and simultaneous.

In "Emergency / Everyday," Ben Anderson draws our attention away from the scientific and the popular to how we calibrate time socially and politically: how do we measure the difference between the everyday, slow-moving time of the quotidian and the rushed, instantaneous time of emergency? Given the visibility of "the everyday" in social and cultural theory today, and the claims being made for the disruptive potential of the everyday as an alternative to market-driven, industry-centered chronometric time, how, he asks, should we draw the temporal line between emergency and the everyday? While we typically conceive of these as opposite kinds of time calculation, Anderson argues that they must be put in relation, for they are frequently conjoined in claims that the contemporary condition of human life is life lived in uncertainty.

This is very different from John Maynard Keynes's 1930 prediction that by 2030 the wealth created by new technologies would bring about an era of universal leisure. In "Labor / Leisure," Aubrey Anable examines this assertion, probing the character of time as it is calculated in terms of work and play. Anable claims that, following the Industrial Revolution, the boundaries between labor time and leisure time were meaningfully blurred yet still distinct, but after 1960, these two terms have become, if not indistinguishable, at least vastly more temporally and experientially confused. Using digital games on mobile communications devices as an example, she argues that leisure time has shifted today from being a counterpart to labor time to being another modality of productive work.

Rounding out this section, Mark McGurl's "Real / Quality" takes up the idea of "work time," time as calculated according to workplace schedules and rhythms. He claims that the terms "real time" and "quality time" are not opposites but rather exist in broken or partial opposition, "as though they are the mutually reflecting shards of some prior, and presumably deeper, dialectical collision in the late capitalist life-world." Moreover, "real time" and "quality time" become gendered terms in the postindustrial workplace. The collision he identifies is between qualitative and quantitative or subjective and objective time, a relation placed under a new kind of stress in late capitalism.

Time as Culture: Mediating Time

To the degree that "time studies" refers to a capacious field of discourse in the humanities, time as culture is arguably the privileged locus of time studies today. Across recent cultural analyses of temporality, we see a concerted effort to show that the empty, homogenous time of modernity is actually full—full of impulses and rhythms that make the seeming uniformity of modern time a far more heterogeneous, and far more syncopated, experience than previous models had advanced. "History is the subject of a construction whose site is not homogenous, empty time," writes Benjamin in 1940, "but time *filled full* by now-time."[59] This takes a number of forms in time studies today: showing the lure of contingency in the reified temporalities of industrial modernity; revealing the alternative temporalities that texture the emptiness of modern times with, for example, queer, fantastic, and globalized cadences; unpacking contemporaneity itself so that it is no longer a purely spatial and geopolitical phenomenon, but an experience of temporal disjuncture and of intuiting the present as presence; and attending to how time functions formally, as a device that structures both works themselves and their arrangement as we access them as a horizon of experience.

The concept of time-as-culture drives us today toward a realization of multiplicity, a plurality of temporal differences within the contemporary. Modern times produce their own temporal others immanently, dragging in alternative rhythms and outmoded objects as the medium and material in which time becomes culture. All of the essays in this section pursue this idea and as a whole reflect the temporal plurality that thinking about time-as-culture glimpses askance. In "Aesthetic / Prosthetic," for example, Jesse Matz distinguishes between aesthetic time, or the ways in which aesthetic works create their own time, and prosthetic time, or the ways in which we work to externalize time. Taking an almost reparative approach, Matz is ultimately interested not in the bifurcation of aesthetic and prosthetic time, but how, when they converge in films such as *Source Code* and *Limitless*, a relationship develops between "temporal prosthesis and filmic art" that actually seeks today nothing less than to "cultivate a new humanity."

The literary arts are James Phelan's focus in "Analepsis / Prolepsis," keywords that are, as he explains, "rough synonyms for flashback and flashforward." As Phelan points out, we would be hard-pressed to imagine these temporal devices as contemporary inventions, given that we can locate them as early as Homer. Thus while Phelan makes the historical claim that contemporary literature has not enacted a total transformation of the flashforward and the flashback, he does argue that post-1960s fiction uses analeptic and proleptic temporalities to reveal new, contemporary anxieties about temporality and the politics of historical time.

In "Embodied / Disembodied," Michelle Stephens and Sandra Stephens reveal how post-1960s Conceptual art moves "from an embodied scopic regime of space, through a disembodied proprioceptive regime of time, to arrive, now, at a complex notion of the body as re-embodied, relational timespace." Considering the politics of colonial difference and postcolonial art, the more complex spatiotemporality of reembodiment comes to reside within—or upon—our skins in the artworks they consider, many of them the work of Sandra Stephens herself. A tactile temporality thus emerges on the surface of our bodies, dynamizing the spatial regime of embodiment with the kinesthetic and fleshy flow of time in ways sensitized to race, colonialism, and the body.

But what of sacred time? While this question could be the foundation of an entire essay collection in its own right, given the abundance and diversity of sacred times and sacred traditions worldwide, here we present a consideration of a sacred kind of time increasingly visible in postwar U.S. discourse and significant to contemporary Marxist and post-Marxist critical theory. In "Theological / Worldly," Stanley Hauerwas insists that Christianity is rooted in time, the ongoing product of "a revelation in history" rather than "a passage to the eternal." At the root of this insight is an other-worldly time, which Hauerwas takes to be the lived time of the Christian person—a temporality that works at the disjuncture of the time of God (theological time) and the time of modernity (worldly time). In this sense, Hauerwas is very much part of the movement in time studies today to show how full the presumed emptiness of what he calls worldly time can be.

Whereas Hauerwas turns to religion to reveal the plenitude of the historical present, in "Authentic / Artificial" Anthony Reed turns to race and diaspora to disrupt the seemingly consecutive temporal relationship of the authentic and the artificial. That which is authentic is often that which is most closely associated with "the folk," who are imagined to exist *before* modernity. That which is artificial is therefore that which comes *after* modernity. Tied to pernicious cultural hierarchies and political inequalities, this consecutive temporality, however, is in actuality a coeval time of reverberation and resemblance. Reed asks *when*, not where, the African diaspora can be located; through an account of the global circulation of hip hop, he suggests that this "when" may be "an impossible similarity rooted in shared time."

In "Batch / Interactive," it would seem that Nick Montfort would end up hewing mostly closely to a model of time that insists upon its abstraction and rationalization as empty and homogenous. For what interests Montfort are two modes of computer time: batch time and interactive time. Such temporalities might seem to be the ultimate embodiment of capitalism's empty, homogeneous time. But while batch time is organized around the computer itself, interactive time is organized around the human user of the computer. Thus what has emerged, Montfort argues, is a heterogeneous mix of computing temporalities in which human time becomes increasingly digital—as in the example of multitasking that he considers—even as the time of the computer is bent to the will of the human being who is using it.

In "Transmission / Influence," Rachel Haidu deconstructs notions of artistic influence along temporal lines. Her deconstruction is not proffered as an effort to do away with influence but as a means of redescribing and recuperating the concept for criticism today. To that end, she reconfigures what we usually call "influence"—the historical time of diachronic lineage we construct between two artists—as "transmission."

She argues that a transmissive temporality occurs in real time when one artist really looks at another's work in a synchronic moment, a present in which that artist comes into contact with some other artist's work (even, in some cases, his/her own work). Influence thus becomes a discursive and metacritical temporality.

The last essay of the volume enacts this notion of influence in hip hop montage. Paul D. Miller aka DJ Spooky, That Subliminal Kid asks us to consider the interrelations of sound and space as they intersect in mathematics, underscored in contemporary music and the writing about it. In this sense, the end of this collection of essays opens back out to the temporality of the immediate, of the spoken word, the artwork, and the multiplicity of times that characterize the contemporary moment.

NOTES

1. See, for instance, Geoffrey C. Bowker, *Memory Practices in the Sciences* (Cambridge, MA: MIT Press, 2005); Dipesh Chakrabarty, "The Climate of History: Four Theses," *Critical Inquiry* 25, no. 2 (2009): 197–222; Adam Frank, *About Time: Cosmology and Culture at the Twilight of the Big Bang* (New York: Free Press, 2011); Stephen Hawking, *A Brief History of Time*, updated and expanded 10th anniversary edition (New York: Bantam Books, 1996).

2. The following discussion cites numerous sources concerning contemporary time. When not fully cited in these endnotes, these sources appear in "Time Studies: A Bibliographical Reading List" at the end of this volume.

   As noted here, the essays in this book primarily are focused on a sociocultural experience of time by persons living in the present, though the essays reference many approaches to time apprehension. For example, "philosophy of time" investigates the nature of time: whether time passes, whether the past and future actually exist, when time begins, John McTaggart's famous distinction between A- and B-theories of time, and the nature of causation (see J. R. Lucas; Robin Le Poidevin and Murray MacBeath; Jeremy Butterfield; Adrian Bardon; Yuval Dolev). The phenomenology of time accounts for how we experience time as human beings and why some things appear to us as appropriate to the category of temporality. Such investigation might include the problem of asymmetry, or how the past, present, and future carve up time differently. Edmund Husserl, Martin Heidegger, and Jean-Paul Sartre are among those associated with this investigation, but modern and postmodern continental philosophy has also addressed different phenomenological concerns, including questions about the nature of memory and self-identity. (See Ned Markosian, "Time," *The Stanford Encyclopedia of Philosophy*, ed. Edward N. Zalta [Spring 2014], http://plato.stanford.edu/entries/time/#3D4Con, and, in the bibliography, David Wood; David Couzens Hoy.) Cognitive science researchers have asked questions concerning the brain and the neural mechanisms that account for our experience of temporal succession, persistence, and duration, while the *science of time* addresses how time works mathematically in quantum mechanics (e.g., in new mathematical formulations for timespace, new theories of infinite time, or a reconsideration of quantum theory's claim that the shortest duration for any possible event is about 10⁻⁴³ second). For different approaches to these two areas, see Tim Maudlin; Huw Price; Frank Arntzenius.

3   This is a standard operating principle for practicing historians. For an extended consideration of this point, see the final two chapters of Fredric Jameson, *Valences of the Dialectic* (London and New York: Verso, 2010), and Evan Calder Williams, *Combined and Uneven Apocalypse* (Winchester, UK: Zero Books, 2011).

4   David Scott, *Omens of Adversity: Tragedy, Time, Memory, Justice* (Durham, NC: Duke University Press, 2013), 1.

5   For dexterous work on this debate, see the forthcoming collection from Jason Gladstone, Andrew Hoberek, and Daniel Worden, eds., *Postmodern/Postwar— and After* (Iowa City: University of Iowa Press, 2016).

6   Roger Luckhurst and Peter Marks, "Hurry Up Please It's Time: Introducing the Contemporary," in *Literature and the Contemporary: Fictions and Theories of the Present*, ed. Roger Luckhurst and Peter Marks (New York: Longman, 1999), 3.

7   Steven Connor, "The Impossibility of the Present: or, From the Contemporary to the Contemporal," in *Literature and the Contemporary: Fictions and Theories of the Present*, ed. Roger Luckhurst and Peter Marks (New York: Longman, 1999), 30.

8   Terry Smith, "Introduction: The Contemporaneity Question," in *Antinomies of Art and Culture: Modernity, Postmodernity, Contemporaneity*, ed. Terry Smith, Okwui Enwezor, and Nancy Condee (Durham, NC, and London: Duke University Press, 2008), 8–9, italicized in original.

9   Site research by Joel Burges, *The Clock*, SFMOMA, San Francisco, California, Spring 2013. "Chronophilia" is a riff on Pam Lee's study *Chronophobia: On Time in the Art of the 1960s* (Cambridge, MA: MIT Press, 2006).

10 See Raymond Williams's useful survey of "culture," especially his definition of it variously "as 'the arts,' as 'a system of meanings and values,' or as a 'whole way of life'" in *Marxism and Literature* (New York: Oxford University Press, 1977), 13.

11 While Joel Burges spent time watching *The Clock*, the clips described have been chosen because they are currently readily available online should the reader want to view them. See "Christian Marclay, *The Clock*," https://www.youtube.com/watch?v=xp4EUryS6ac. The impracticality of this clock is noted by Peter Bradshaw, "Christian Marclay's *The Clock*: A Masterpiece of Our Times," *Guardian*, April 7, 2011, http://www.theguardian.com/film/filmblog/2011/apr/07/christian-marclay-the-clock.

12 See Frederic Jameson, *Postmodernism, or, the Cultural Logic of Late Capitalism* (Durham, NC: Duke University Press, 1991). Indeed, Jameson has been loyal to this idea through his many publications, and the sophistication of his claims is lost in this summary. See Jameson's own summary of his claims in "The Aes-

thetics of Singularity," *New Left Review* 92 (2015): 101–32. Many have contested Jameson's equation of a "spatial turn" with a new postmodern sense of time; see, for example, Arina Lungu, "Marx, Postmodernism, and Spatial Configurations in Jameson and Lefebvre," *CLCWeb: Comparative Literature and Culture* 10, no. 1 (2008), http://dx.doi.org/10.7771/1481-4374.1327.

13 See Umberto Eco, *Travels in Hyperreality: Essays*, trans. William Weaver (New York: Harcourt, 1986), originally published as *Il costume di casa* (*Faith in Fakes*) (Bompiani, 1973); Jean Baudrillard, *Simulacra and Simulation*, trans. Sheila Glaser (Ann Arbor: University of Michigan Press, 1994), originally published as *Simulacres et Simulation* (Éditions Galilée, 1981); Charles Jencks, *The Language of Postmodern Architecture* (New York: Rizzoli, 1977); Edward Soja, *Postmodern Geographies: The Reassertion of Space in Critical Social Theory* (London: Verso, 1989). Jencks valenced the "quotation of past styles" more positively than did Jameson.

14 David Harvey, *The Condition of Postmodernity* (Oxford: Basil Blackwell, 1989); John Frow, *Time and Commodity Culture: Essays in Cultural Theory and Postmodernity* (Oxford: Clarendon, 1997); Ursula K. Heise, *Chronoschisms: Time, Narrative, and Postmodernism* (Cambridge, UK: Cambridge University Press, 1997); Andreas Huyssen, *Twilight Memories: Marking Time in a Culture of Amnesia* (New York and London: Routledge, 1995).

15 Jameson defines it as "de-differentiation" and leveling of fields and categories—in the visual arts, for example, the leveling of differences between painting, video, performance, and music, leaving only an empty container called "the art installation" into which anything can be thrown when overseen by a "curator" who is the embodiment of the institutional context. See "The Aesthetics of Singularity," 107.

16 Jameson, "The Aesthetics of Singularity," 111.

17 In philosophy, the two ontologies of time are eternalism and presentism. As M. Joshua Mozerksky notes in "Presentism," in *The Oxford Handbook of Philosophy of Time*, ed. Craig Callender (Oxford: Oxford University Press, 2011), "[Eternalism] holds that all moments of time and their contents enjoy the same ontological status. Past and future moments, events and objects are just as real as the present ones; they are just not 'temporally here.' . . . [Presentism, on the other hand, is] the view that only the present entities are real. Eternalism and presentism are widely regarded as the best incarnations of the older competing view of time known as A- and B-theories, or tensed and tenseless theories" (16).

18 Moishe Postone, *Time, Labor, and Social Domination: A Reinterpretation of Marx's Critical Theory*, 201. In *Wheels, Clocks, and Rockets: A History of Technology* (1995; repr. New York: W. W. Norton, 2001), Donald Cardwell dates this turn to the eighteenth century.

19 For Benjamin's influential conceptualizing of "homogenous, empty time," see "On the Concept of History," in *Selected Writings, Volume 4: 1938–1940*, ed. Howard Eiland and Michael W. Jennings, trans. Edmund Jephcott et al. (Cambridge, MA: Belknap Press/Harvard University Press, 2003), 395.

20 Benedict Anderson, *Imagined Communities: Reflections on the Origins and Spread of Nationalism*, revised edition (New York: Verso, 2006), 9–38.
21 Peter Fritzsche, *Stranded in the Present: Modern Time and the Melancholy of History* (Cambridge, MA: Harvard University Press, 2004), 41.
22 Ian Baucom, *Specters of the Atlantic: Finance Capital, Slavery, and the Philosophy of History* (Durham, NC: Duke University Press, 2005), 316.
23 Bowker, *Memory Practices in the Sciences*, 35–74; Wolfgang Schivelbusch, *The Railway Journey: The Industrialization of Space and Time in the 19th Century* (Berkeley, CA: University of California Press, 1987).
24 Stephen Kern, *The Culture of Time and Space, 1880–1918*, 65–72. The quotation appears on page 67.
25 Mary Ann Doane, *The Emergence of Cinematic Time*, 104.
26 Fredric Jameson, "The End of Temporality"; Jameson, *Postmodernism*, 25.
27 Jürgen Habermas, "Modernity: An Incomplete Project," in *The Anti-Aesthetic: Essays on Postmodern Culture*, ed. Hal Foster (Port Townsend, WA: Bay Press, 1983), 3–15. This is contested by new interpretations of the Marxist dialectic, which claim that Marx's revisioning of Hegel understands Dasein to be capital itself, making the endpoint of history the synthesis leading to capital's own systemic self-realization. See Werner Bonefeld, *Critical Theory and the Critique of Political Economy: On Subversion and Negative Reason* (London: Bloomsbury Academic 2014). Our thanks to Harry Dahms for this reference.
28 See Jean-Luc Nancy, *The Inoperative Community* (Minneapolis: University of Minnesota Press, 1991); Gilles Deleuze and Félix Guattari, *A Thousand Plateaus: Capitalism and Schizophrenia*, trans. Brian Massumi (Minneapolis: University of Minnesota Press, 1991).
29 Terry Smith has argued, for example, that the only potentially stable thing about the multiplicious temporalities of the present is that, ironically, they may now be "eternal" (Smith, "Introduction," *Antinomies of Art and Culture*, 9).
30 The turn to "neoliberalism studies" is such an attempt to theorize capital while downplaying the Marxist conceptual apparatus, though many of the claims of this theory can be found (with different vocabulary) in postmodern studies. With perhaps more implications for time studies, Latourian actor-network theory, new materialist theories, and posthumanisms redefine materiality and its temporal modes of existence against both Enlightenment rationalism and older Marxist models.
31 Peter Osborne, *Anywhere or Not at All: Philosophy of Contemporary Art* (New York: Verso, 2013), 17. For Osborne, unlike Terry Smith, this *distinguishes* contemporaneity from modernity, which is a far "more structurally transitory category" (24, 16).
32 Connor, "The Impossibility of the Present," 31. In anthropology, Johannes Fabian (in *Time and the Other*, 1983) recognized such plurality of time but warned against "allochronism" or the privileging of one pole or the other, one kind of time over any other, and he advocates a more conceptual and dialogical approach to time by anthropologists.

33 David Couzens Hoy, *The Time of Our Lives: A Critical History of Temporality* (Cambridge, MA: MIT Press, 2009), 56–7. Hoy does not situate Heidegger vis-à-vis postmodernity.
34 On the turn to "planetarity studies," see *The Planetary Turn: Relationality and Geoaesthetics in the Twenty-First Century*, ed. Amy J. Elias and Christian Moraru (Evanston, IL: Northwestern University Press, 2015).
35 See Joel Burges's essay in this collection. The prophet of the "bad speed" of capitalism is of course Paul Virilio. See Virilio, *Speed and Politics*, trans. Mark Polizzotti (Los Angeles: Semiotext(e), 2006).
36 See Virilio, *Speed and Politics*.
37 Paul Virilio, with Bertrand Richard, *The Administration of Fear*, trans. Ames Hodges (Los Angeles: Semiotext(e), 2012), 16.
38 This is a paraphrase of Ian James, *Paul Virilio* (London and New York: Routledge, 2007), 30. See also Steve Redhead, *Paul Virilio: Theorist for an Accelerated Culture* (Toronto: University of Toronto Press, 2004). Virilio's theory is contested in digital media studies circles; see, e.g., Douglas Kellner, "Virilio, War, and Technology: Some Critical Reflections," *Illuminations: The Critical Theory Website*, http://www.uta.edu/huma/illuminations/kell29.htm.
39 *Slow Movement (website)*, http://www.slowmovement.com.
40 Lutz Koepnick, *On Slowness: Toward an Aesthetic of the Contemporary* (New York: Columbia University Press, 2014), 45. On the pleasures of speed and malignancies of velocity in the twentieth century, especially modernist and capitalist contexts, see Enda Duffy, *The Speed Handbook: Velocity, Pleasure, Modernism* (Durham, NC: Duke University Press, 2009), and Benjamin Noys, *Malign Velocities: Accelerationism and Capitalism* (Washington, DC: Zero Books, 2014).
41 Koepnick, *On Slowness*, 45, 48–49, 10.
42 Jonathan Crary, *24/7: Late Capitalism and the Ends of Sleep* (London and New York: Verso, 2013), 126.
43 Sarah Sharma, *In the Meantime: Temporality and Cultural Politics* (Durham, NC: Duke University Press, 2014).
44 Brian Massumi, *Semblance and Event: Activist Philosophy and the Occurrent Arts* (London and Cambridge, MA: MIT Press, 2011), 45.
45 Adam Jasper and Sianne Ngai, "Our Aesthetic Categories: An Interview with Sianne Ngai," *Cabinet* 43 (Fall 2011), http://www.cabinetmagazine.org/issues/43/jasper_ngai.php.
46 Lauren Berlant, *Cruel Optimism* (Durham, NC: Duke University Press, 2011), 4.
47 Kathleen Stewart, *Ordinary Affects* (Durham, NC: Duke University Press, 2007), 1, 4, 9.
48 Berlant, *Cruel Optimism* 4.
49 Siegfried Kracauer, "Time and History," *History and Theory*, 6, Beiheft 6 (1966), 65–88.
50 Berber Bevernage and Chris Lorenz, "Breaking up Time—Negotiating the Borders Between Present, Past and Future. An Introduction," in *Breaking up Time:*

Negotiating the Borders Between Present, Past and Future, ed. Chris Lorenz and Berber Bevernage (Göttingen: Vandenhoeck & Ruprecht, 2013), 8, quoting Marc Bloch, *Apologie pour l'histoire ou Métier d'historien* (Paris, 1997), 52, and Jacques Le Goff, *Histoire et mémoire* (Paris, 1988), 24; Norbert Elias, *Time: An Essay*, trans. Edmund Jephcott (Oxford: Blackwell, 1992), 36.

51 Lorenz and Bevernage, "Breaking Up Time," 9. The authors' notes provide an extremely useful listing sources on the relation of time and history. The historicizing of time concepts is exemplified in Reinhart Koselleck's project of "conceptual history," cited in numerous essays in this volume, as well as the French scholar François Hartog's theories about Western "regimes of historicity" (pastism, eschatologism, futurism, and presentism), each characterized by a mediated cultural relation to time. See Hartog, *Régimes d'historicité: Présentisme et expèriences du temps* (Paris: POINTS, 2003).

52 Email with Amy J. Elias, February 21, 2014.

53 Jörn Rüsen, "Introduction," in *Time and History: The Variety of Cultures*, ed. Jörn Rüsen (New York: Berghahn Books, 2007), 2, italics ours.

54 Rüsen, "Introduction," 2.

55 See Hayden White, "Preface," in *Metahistory: The Historical Imagination in Nineteenth-Century Europe* (Baltimore: Johns Hopkins University Press, 1973).

56 Amelia Groom, "Introduction//We're Five Hundred Years before the Man We Just Robbed Was Born," in *Time*, ed. Amelia Groom (London: Whitechapel Gallery/Cambridge, MA: MIT Press, 2013), 16.

57 On Walker's art, see for example *Kara Walker: My Complement, My Enemy, My Oppressor, My Love*, ed. Philippe Vergne, Sander Gilman, et al. (Minneapolis, MN: Walker Art Center, 2007); Joe Sacco, *The Great War: July 1, 1916: The First Day of the Battle of the Somme* (New York: W. W. Norton, 2013).

58 For a history of clocks and their social significance, see Alexis McCrossin, *Marking Modern Times: A History of Clocks, Watches, and Other Timekeepers in American Life* (Chicago: University of Chicago Press, 2013).

59 Benjamin, "On the Concept of History," 395. Not unrelatedly, but less messianically, Raymond Williams's tripartite scheme of the dominant, the residual, and the emergent has been important for the cultural analysis of time, with the residual rhythm often providing the critic with a vision of a resistant temporality. See Williams, *Marxism and Literature*, 124.

PART I

Time as History

Periodizing Time

1

Past / Future

AMY J. ELIAS

> The prognosis implies a diagnosis which introduces the past
> into the future.
> —Reinhart Koselleck, *Futures Past*

Much theoretical ink has been spilled about the "presentism" of post-WWII globalized societies and the loss of history that accompanies it. This presentism has been attributed to a traumatized Western collective consciousness confronting WWII as an "event" unprecedented in its history; to the time of the Spectacle that reduces the past to advertising slogans and depoliticized images of material desire; to finance capitalism's acceleration of time and eradication of spatial distance as it creates a technologized world economy; or to the speed of "real time" technology that makes impossible both deliberation and historical depth. It seems that we may be incarcerated in the present.[1] Yet while presentism often opposes "past" and "future" to "the present," the dialectical counter to time as diachronic history (past/future) is in truth not another kind of historical time (the synchronic present). The opponent is duration—timeless time, homogeneous time—whose synchronic partner is the Event. But there is duration, and there is duration. Understanding this leads one to realize that what is bemoaned (or celebrated) in most theories about presentism today is actually not a "present" at all, but rather marketplace duration, a dank obversion of Bergsonian temporal vitalism or Deleuzian rhizomatic flow. Our "now" is a technogenetic present, a cacophony of noise and color and movement; it is the construction and breaking apart of temporary mechanical assemblages, but in the hyperreal space of the world mall—no longer limited to shopping centers but flowing in the spaces of our cars, our earphones, and our computer screens in the unceasing movement of the electron.[2] Theories of "presentism" may attempt to define a situation in which, for the first time in modern history, duration has assumed a metaleptic character *inside* both planetary and historical time: humankind has created its own version of durational time inside (rather than outside) the box of historicity.[3]

Is techno-duration the only (a)historical space we have now, the terminal electrocution of past and present, or is this truly a metaleptic framing, a kind of time of capital *within* other kinds of possible time? The stakes of this question are high: if the latter, there may be a way back up the ladder of ontology and back out to planetary time. My own predilections lean toward metaleptic framing, so I would like here to explore how, in the twenty-first century, techno-duration characteristic of the world system may construct its own form of historical time, rather than rehearsing yet again how capital redefines linear history as perpetual present (an idea well trodden in twentieth-century postmodern theory).

My first idea is that in techno-duration, "past" and "future" have become redefined as "retrofuture" and "slipstream," contaminations and inversions of older notions of "past" and "future" time. These two terms signal neither progress nor decline—which would be developmental movement through historical time characteristic of modernity's historicism. Instead, "past" and "future" signal a doubled movement of simultaneous futurity and historicity that provokes an image of "moving stasis" compatible with techno-durational "presentism." My second thought, a form of melancholic realism toward which I really only gesture in this short discussion, is that in the twenty-first century, history has not "ended." There is no empirical evidence that we are caught interminably in the metaleptic techno-durational frame that demands a "past/future" relation as perpetual exchange, as the splinters in our minds keep reminding us.[4]

The Past as Retrofuture

The 1960s saw a significant shift in historical studies when historians such as Reinhard Wittram and Reinhart Koselleck "invented the concept of a *past future*, by which was meant a future that was not the future of the present but the future as it was conceived of at some time in the past."[5] Comparing pre-modern to modern time, Koselleck wrote that pre-modern Europe understood time eschatologically, in terms of biblical, apocalyptic history that demanded a "constant anticipation of the End of the World on the one hand and the continual deferment of the End on the other."[6] The agent of this deferral was the Roman Catholic Church and its secular allies.[7] But the Church's power eventually weakened, and by 1555 and the Religious Peace of Augsburg, Koselleck claims, politicians were concerned more with the temporal, secular future associated with civic peace than with the eternal future associated with personal salvation, and thus it became possible to open up "a new and unorthodox future" to the conceptual imagination.

Thus the period between 1500 and 1800 saw a radical change in how people understood the future: by the time of Robespierre "there has been an inversion in the horizon of expectations" concerning the historical future[8] and a "temporalization [*Verzeitlichung*] of history, at the end of which there is the peculiar form of acceleration which characterizes modernity."[9] Specifically, the eschatological *prophecy* is replaced by political calculation, embodied in the rational forecast or *prognosis*.[10] These are significantly different conceptions of future time. Prophecy exceeds human measurement, binds time in a moral universe, portrays current events as equivalent symbols of an already-known future, and in effect destroys time in a present determined by an apocalyptic future (and thus synchronic with it). In contrast, modern prognosis is dependent upon human calculation, opens time to a secular domain of probable and finite political possibilities, and defines events as unique and time as open-ended.[11] Prognostic history dovetails with a conception of Utopia newly created in the eighteenth century, moving from a tale of spatial/geographic travel to utopian *lands* to temporal travel into the *future*, when, in accordance with Enlightenment futurity, there will be a perfectibility of Man.[12] Yet this progress demands a future that "is characterized by two main features: first, the increasing speed with which it approaches us, and second, its unknown quality."[13]

While today's historiographers inherit the Enlightenment revolution in historical understanding and reject religious versions of eschatological prophecy, their notion of time has accelerated to such an extent that prognosis has also become discredited because it is unable keep up with the pace of technological innovation and societal mutation. We are dependent upon prognoses that are obsolete almost at the moment they are pronounced. Yet ironically, this may account for both our obsession with history, with the past, as well as our concomitant lack of belief in its ability to give us meaningful information about the future. This is why I have mentioned future time in this essay section devoted to past time: according to Koselleck, with the advent of the Enlightenment future (infinitely receding as the horizon of progress), the idea of *historia magistra vitae* ("history is the teacher of life") was undermined, but (paradoxically) as a result, the past became an object of observation and analysis in a historicist sense (an artifact rather than a living teaching), "history in and for itself."[14] In the nineteenth century, as the future became increasingly impermeable to prognosis, the past became mummified, a dead object to be observed and dissected. Once this happened, a second, postmodern stage of historical inquiry followed: metahistorical theorization of what kinds of futures the past created, a *history of past temporal concepts*. The past becomes viewed not in relation to the future, but from the perspective of a backward-looking present.[15] This is one backdrop to the melancholia of modernity but also a new engagement with the past generated by the linear sense of modern time taking hold in Europe after the French Revolution.[16]

We now live in the time of techno-duration, when the present dominates the past in just this way. As Hartog notes, a "regime of historicity" that some have called the postmodern and which he calls "presentism" follows Koselleck's modernity. It is characterized by omni-presence, an "invasion of the present into the realms of the past and future." Significantly, Hartog describes this dominating present as filled with memory and commemoration.[17] Reynolds has correlated this commemorative impulse to a "retromania" brought about through digital life: "In the analogue era, everyday life moved slowly . . . but the culture as a whole felt like it was surging forward. In the digital present, everyday life consists of hyper-acceleration and near-instantaneity . . . but on the macro-cultural level things feel static and stalled. We have this paradoxical combination of speed and standstill."[18] This combination is what I am calling "techno-duration," and in it, the present spreads out like tsunami waters over the past.

Techno-duration thus constructs a new kind of past, a "retrofuturism." "Retrofuturism," on the one hand, is a style emerging in the mid-twentieth century after the heyday of Art Deco and aligned now with what is termed "Googie architecture": a colorful, optimistic, "space age" 1950s and 1960s futuristic style (think *The Jetsons*).[19] The artists who produced artifacts of this kind (restaurants, hotels, theme parks, etc.) did not understand them to be "retrofuturistic" but, rather, "futuristic": the best example is always the 1939 New York World's Fair, which featured the World of Tomorrow, Futurama, and Trylon and Perisphere exhibits.[20] "Retrofuturism" is, then, a twenty-first-century historical perspective on the near past, a looking back upon these futuristic productions of the past that sees them as quaint utopian hopes of a future than never arrived. Today, such a retrofuturistic perspective should occur to visitors to the Magic Kingdom at Disneyland, where one can feel a palpable nostalgia for a 1950s vision of the technological future that is now itself obsolete.[21]

On the other hand, retrofuturism describes an aesthetic style that is produced *today* to imitate such past futuristic artifacts. This is commercialized retrofuturism, linked to what Fredric Jameson called the culture of pastiche. It has migrated to other aesthetic forms—TV shows, films, poster art, video games. This retrofuturism *as a contemporary style* may have varied sociopolitical aims. In *Retro: The Culture of Revival*, Elizabeth Guffey writes that "retro revivalism" after the 1970s actually separates the past conceived as "naïve" from the present, and Sharon Sharp asserts that retrofuturism can be the basis both of critique and a neo-conservative impulse.[22] The genre often resembles alternative-history narratives, and Henry Jenkins in fact uses "retrofuturism" to describe a post-1970s subgenre of science fiction set in the past, at a moment of utopian promise. Referencing Dean Motter's graphic novels, Jenkins writes that retrofuturism "allowed people to look backwards, examining older myths and fantasies against contemporary realities."[23] The attitude of critique certainly can be found in Michael Moorcock's *A Nomad of the Time Streams* series, subtitled "A Scientific Romance" and collating three time-travel novels published between 1971 and 1981. A self-described anarchist, Moorcock creates an upright British military commander, Captain Oswald Bastable, who time travels into the past to see different versions of the future. In each time voyage, readers are treated to overt commentary denouncing colonialism, sexism, and other social violences aligned specifically with a capitalist economic system.[24] A twenty-first-century literary example might be Thomas Pynchon's *Against the Day*, an encyclopedic but fantastic treatment of a historical moment (at the end of the nineteenth century) when the nature of the future was up for grabs.[25] Like Pynchon, neo-Marxist writer China Miéville has used steampunk to go retrofuturistic, particularly in the "salvage punk" novel *Railsea*. At the end of the novel, the protagonists find the "end of the line," the beginning of the railsea and hence the past origin of what was a failed utopian *capitalist* scheme: cover the entire earth with train rails to extend commerce and travel everywhere.[26] In film we see retrofuturism everywhere in postapocalyptic SF, from *Metropolis* (1927) to *Back to the Future* (1985) to *Twelve Monkeys* (1995), and in the visual arts one might point to many examples, such as Cyprien Gaillard's 2005 series of six etchings titled "Belief in the Age of Disbelief," which reworks seventeenth-century Dutch etchings by inserting modernist high-rises into the landscapes, thus transforming twentieth-century utopian architecture into a vision of picturesque ruins.[27]

Retrofuturism is the genre that exemplifies Fredric Jameson's claim that, traditionally, science fiction attempted to predict the future, but postmodern SF can now only dramatize our *inability* to imagine utopia.[28] Retrofuturism looks to visions of past utopian futures, and its lessons are the Ecclesiastean preachments of vanity and failure.[29] In both its commercialized and critical forms, it makes past historical optimism visible in the most cinematic of ways, but then, by showing the misguided nature of that optimism, it renders all forms of futurism or utopianism naïve and doomed to failure. Retrofuturism may be the most complicated and effective of counter-Enlightenment histories, the past allowed by today's techno-duration, for futurism is not merely rehearsed in fragmented form for market consumption but is resurrected in its (often anti-capitalist) utopian instantiations as a failed hope. Yet because retrofuturism is *the picture of the failure of progress as such*, it can have a doubled valence: on the one hand, the perfect kitsch object for market recirculation, and on the other (in versions based in critique, such as Miéville's), a repudiation of the very ideology of technological progress that bolsters that market today.

Retrofuturism thus is an *exemplum* in addition to a state of being and a possible vehicle of critique. As our conception of the historical past, it is a reminder that we are heroic, but deluded, dreamers in relation to (historical) time. This is its intersection with what I will call the slipstream future time of the techno-duration.

The Future as Slipstream

The historical present as techno-duration inherits the modern condition of being unable to predict the future in a constantly accelerating present. However, unlike the moderns, we have nearly abandoned the philosophy of progress characterizing the rise of post-Enlightenment historiography. As a result, we look to the future not with hope but with anxiety—even to the point of turning away our gaze from the coming apocalypse that we cannot seem to stem. We move forward in time expecting future shock and the unexpected blow, perplexed by or resigned to a world-event logic that eludes us.[30] Thus rather than moving forward in history with our eyes on the future, we move forward with our eyes on the past, desperately searching for some foothold and precedent that might give the movement of history meaning. We are not accelerating forward toward future time; we are, rather, caught in its slipstream.

Science fiction took up the term "slipstream" in the 1990s to designate "fiction of strangeness" straddling SF and fantasy—and, according to Bruce Sterling, a quintessentially contemporary (post-1989) mode of writing.[31] But deviating from Sterling and these other critics, I would like to transfer this term to a specifically historiographical context aligned with the term's use in aerodynamics. There, "slipstream" signals a wake behind a moving object in which air or water is moving at a velocity comparable to that of the moving object, relative to the fluid through which the object is moving. We might think of biking in the air slipstream of a semi-truck, or waterskiing in the wake of a moving boat. In metaphorical terms, "slipstream" thus signals a forward movement based on a kind of imperative momentum or force—a flow of force that catches something up in a forward motion against that thing's own natural orientation or speed.

In the time of techno-duration, if the past is retrofuturism, the future is slipstream. We look forward in time almost against our will, in the *wake* of a future understood to be antithetical to reason, out of control, or operating outside of human agency. In this sense, "slipstream" is the movement of Benjamin's Angel of History, now revealed as the present itself, drawn into the slipstream of time but oriented toward the past in a desperate search for origins, foundations, historical meaning.

Thus if in "retrofuturism-as-past," the present dominates over the past, in slipstream futurism, the present disappears and is absorbed into the interplay of past and future, for the present is always in motion, dragged blindly forward while hysterically or despondently scanning the landfill of the past for any clue to what lies before it.

As a type of *historical* narrativization, slipstream futurism is thus aligned with "figural fulfillment," a pre-modern model of history in which the future is experienced not as a horizon of hope and/or progress but as something that has already happened as an anticipated event. Erich Auerbach was one of the first to define this in terms of historical movement: "Figural interpretation establishes a connection between two events or persons, the first of which signifies not only itself but also the second, while the second encompasses or fulfills the first. The two poles of the figure are separate in time, but both, *being real events or figures, are within time, within the stream of historical life.*"[32] A figure is thus based in historical facts and is fundamentally different from allegory,[33] but through the historicity of figure-fulfillment, the past and the present point to one another *and to a future where both will be fulfilled or completed*. An event in the past is understood to continue into the present and point to its own completion in the future. Christianity's eschatological time is a prime example of this in pre-modernity: the appearance of the Christ is prefigured in the Old Testament prophecies, but the Christ also prefigures a future New Jerusalem, where His time is understood to be fulfilled and completed—a fulfillment that has *already happened* according to the eschatological mode. The figures, writes Auerbach, "point not only to the concrete future, but also to something that always has been and always will be; they point to something which is in need of interpretation, which will indeed be fulfilled in the concrete future, but which is at all times present, fulfilled in God's providence, which knows no difference of time."[34]

However, the figure-fulfillment model was rejected as decidedly antihistorical by the historcism of modernity, which understood history as a horizontal unfolding of individuated events/happenings. The Christian figure-fulfillment model of history seemed to be radically undermined in the eighteenth century with the onset of secular modernity. Yet as prominent a theorist as Hayden White has retained for modernity the figure-fulfillment model of history that Auerbach links to Christian eschatology: surprisingly, when we look closely at how modern history is written, we see that secular, disciplinary history (particularly in the nineteenth century) moves not from the past to the present, but from the present to the past, as historians single out a moment in history and actually resuscitate figure-fulfillment logic to explain it. The present becomes foreshadowed, foretold, pre-figured, in the events of the past, as the historical narrative takes the shape of a coherent story—both history and Aristotelian dramatic structure overtly or obliquely pointing to a future fulfillment or "ending."[35] This is the basis, in fact, of the cause-effect logic of periodization—the idea that a specific historical event sets in motion and explains following events on a macro scale.

In the time of postwar techno-duration, however, we move back to an almost eschatological sense of figure-fulfillment, as the future becomes ever more apocalyptic and metempirical. There are different theoretical variants of this, and they are not always congenial to one another. For example, in critical theory, theories of the Event proliferate after the 1990s, implying that the historical past has no connection to the future and that when change happens, it will do so in a cataclysmic epistemic, political, and/or ontological break, after which will be ushered in a new politics, a new episteme, a new world. This is a version of apocalypticism for the postsecular age, and it is not surprising that the radical Left espousing it has also looked to Christianity for its new enunciation.[36] These theories abandon the Enlightenment theory of stadialist progress upon which Marx's theories were partially grounded, and they assume that modern historical prognosis is always already obsolete in the face of an ever-accelerating future. The revolutionary Event thus becomes the figure that will be fulfilled, as the birth of Christ was a figural Event for the fulfillment of Christian history.[37]

Other current theoretical approaches often subtly embed figure-fulfillment logic into a past-future relation. According to Richard Grusin, for instance, we scan the past and the present today for possible clues to prognosticate the future, so that when the future happens, it happens as *déjà vu*. Drawing from theories of paranoia and affect, Grusin argues that in the twenty-first century, specifically after the 9/11 attacks on the World Trade Center, "premediation" is a specific effect of media, a kind of temporal distortion or flattening of time in which global media scan the recent past and immediate present in order to try to anticipate all possible future disaster scenarios. Because the media are part of the military-industrial complex, they do so in order both to address citizens' fear of terrorism and to keep the citizenry in a state of anxiety that precludes real political action. Thus the media construct "an affect of anticipation." Premediation is a kind of figural reading of the present to prepare for the disaster that has already happened, is already built into historical time.

In the slipstream future, as the past and the future set up a figure-fulfillment relation, the *present* disappears—we assume that the past and the future are connected by a prophetic thread that we cannot see, and paranoia becomes the temper and logic of the present.[38] For Koselleck, indeed, what gets increasingly lost in modern futurity is the present;[39] if modernity began to lose the present as "presence" in the intensity of its futurist gaze, it did so because it retained belief in the movement of history along the trajectory of progress. We post-moderns have lost this sense of progress, but we also lose the present as it disappears not into the future, but into a uneasy mixture of pre- and post-Enlightenment thinking about the relation between past and future time.

Combining the work of Koselleck and Auerbach, we might say that the time of techno-duration constructs a new kind of future based in secular eschatology, a weird wedding of older religious notions of eschatological history and the scientific time of modernity. Having lost our faith in modern prognosis, we *prophesy* about a future we see as inevitable, as having already happened, in a state of fearsome awe and paranoia but using modern techniques of mediated historical analysis. In the metaleptic frame of techno-durational time, however, we have only a negative version of religious eschatology: our fuliginous past is now the prefiguration only of an opaque, undefined apocalyptic future—the end of ecosystem, of the Anthropocene, of history itself in the world system. We operate in the mode of pre-modernity that understands the future as endtimes, and are likewise dragged forward into a future that seems already in place. We have thus reconfigured the historical past or near past so that it gives a deeply pessimistic lesson to us about the future. We have moved from prognosis back to prophecy, but our prophets are precisely those of techno-duration.

Perhaps that is a clue that there is a third space still available to think historical time.

Techno-Duration vs. the Real

Is the time of techno-duration the only history now available to us? Or can time studies help us to see that our past (as retrofuture) and our future (as slipstream) are the historical categories of a kind of time that is really metaleptically framed *inside or alongside* both planetary and historical time—a version of durational time inside (rather than outside) the box of historicity, itself inside the system of the planetary? This would follow from Koselleck's own work, in which he posited "historical layers of time" operating in the present and moving at different rates, a "multilayeredness of historical courses of time."[40]

In *In the World Interior of Capital*, Peter Sloterdijk distinguishes Walter Benjamin's arcade capitalism (which implied some kind of "outside" to the capitalist system to which we might escape) from a twenty-first-century "interiorized" moment of globalization imaged in the "Glass Palace": "Whatever happens today within the domain of spending power takes place in the framework of a generalized 'indoor reality,'" a world palace of the privileged "in a crystallized world system" built on the principle of therapeutic comfort as control and where "everything is subject to the compulsion of movement" rather than "indwelling" and placedness.[41] This is techno-duration, a time aligned with the global space of the airtight and encompassing world system. Significantly, in this space, history disappears: Sloterdijk identifies history with the modern age of exploration, and when in the twentieth century we understood the globe to be a closed, bounded space, we exited history to exist in a perpetual "foamed" present of enclosure, density, and control.[42] This might be the negative fulfillment of Koselleck's desire that "prognostic certainty ought to increase again if it becomes possible to incorporate more delaying effects into the future . . . that become calculable as soon as the economic and institutional framework that conditions of our actions becomes more stable."[43]

But the globe is also a planet with a geologic time of its own, and the resources upon which the Glass Palace depends are not infinite. In *The Seeds of Time*, Fredric Jameson states that "the thinking of totality itself . . . has the palpable benefit of forcing us to conceive of at least the possibility of other alternate systems."[44] And today, as Naomi Klein and others have argued in the activist arena, the alternative temporal system is a geologic and environmental one as well as a social one that can easily open a sinkhole beneath the arcade, revealing it to be not a world but a concept.[45] Certainly, new studies in eco-criticism as well as posthumanism and systems theories point out how myopic is a concept of time built upon anthropocentric models and limited social frames.[46] To undo the knot of time woven by techno-duration, and to see past, present, and future in a new relation, we need not revert either to naïve fantasy or, in Mark McGurl's works, to a "pragmatic voluntarism (as though we simply choose our relation to time) whose limits any critical posthumanism would want to explore."[47] We might agree with Jameson that our dilemma is "a situation that endows the waiting with a kind of breathlessness, as we listen for the missing next tick of the clock, the absent first step of renewed praxis," yet with Dipesh Chakrabarty agree that, practically, historiography needs to develop a wider analysis of capital *and the planet* if it is to understand what "future" may mean, next.[48] To climb up the ladder of metaleptic ontology constructed by techno-duration, or to have that climb forced upon us—it is not yet clear if even these are the only alternatives that the future holds.

NOTES

1 See Dominick LaCapra, *History and Memory after Auschwitz* (Ithaca, NY: Cornell University Press, 1998); Guy Debord, *The Society of the Spectacle* (Detroit, MI: Black and Red, 2000); François Hartog, *Régimes d'historicité: Présentisme et expèriences du temps* (Paris: POINTS, 2003). Hans Ulrich Gumbrecht, in *After 1945: Latency as Origin of the Present* (Stanford, CA: Stanford University Press, 2013), has argued that the post-1945 period is a stoppage of history/time. One also thinks of Francis Fukuyama's *The End of History and the Last Man* (New York: Free Press, 1992), as well as the heated public conversation surrounding it.

2 "Technogensis" originates in Bernard Steigler's investigation into the merging of technology and capitalism in his three-volume work *Technics and Time* (Stanford, CA: Stanford University Press, 1994–2001). Paul Virilio in *Lost Dimension*, trans. Daniel Mosehberg (1983; Los Angeles: Semiotext(e), 2012), articulates my own sense of spatial loss.

3 Metalepsis in narrative theory is a violation of the separation between syntactically defined levels of narrative diegesis and/or an ontological transgression of narrative universes, often figured as "stories within stories"; see Gérard Genette, *Narrative Discourse: An Essay in Method*, trans. Jane E. Lewin (1972; Ithaca, NY: Cornell University Press, 1980). Duration has been affiliated with and against global capital in different ways too numerous to cite here; of course the term extends back to the work of Henri Bergson, *Time and Free Will: An Essay on the*

Immediate Data of Consciousness, trans. F. L. Pogson (London: George Allen & Company, 1910; repr. New York: Dover, 2001).
4 See Andy Wachowski and Lana Wachowski, *The Matrix*, directed by Andy Wachowski and Lana Wachowski (1999; Warner Home Video, 1999), DVD. Such a sentiment oddly underlies Fredric Jameson's cognitive mapping in *Postmodernism, or, the Cultural Logic of Late Capitalism* (Durham, NC: Duke University Press, 1991) and of Lauren Berlant's notion of the "affectsphere" in *Cruel Optimism* (Durham, NC: Duke University Press, 2011). I take the term "melancholic realism" from Ian Baucom, *Specters of the Atlantic: Finance Capital, Slavery, and the Philosophy of History* (Durham, NC, and London: Duke University Press, 2005).
5 See Lucian Hölscher, citing the proceedings of the 26th German Historikertag in Berlin, 1964, in "Mysteries of Historical Order: Ruptures, Simultaneity and the Relationship of the Past, the Present and the Future," in *Breaking up Time: Negotiating the Borders between Present, Past and Future*, ed. Chris Lorenz and Berber Bevernage (Göttingen and Bristol, CT: Vanderhoeck & Ruprecht, 2013), 149.
6 Reinhart Koselleck, *Futures Past: On the Semantics of Historical Time*, trans. Keith Tribe (New York: Columbia University Press, 2004), 11.
7 Koselleck, *Futures Past*, 13.
8 Koselleck, *Futures Past*, 12.
9 Koselleck, *Futures Past*, 11.
10 Koselleck, *Futures Past*, 18.
11 Koselleck, *Futures Past*, 19.
12 Reinhart Koselleck, "The Temporalization of Utopia," trans. Todd Samuel Presner, in *The Practice of Conceptual History: Timing History, Spacing Concepts* (Stanford, CA: Stanford University Press, 2002), 85.
13 Koselleck, *Futures Past*, 22.
14 Koselleck, *Futures Past*, 60. "*Historia magistra vitae est*" is attributed to Cicero.
15 For a useful discussion of Koselleck's notion of time, see Helge Jordheim, "Against Periodization: Koselleck's Theory of Multiple Temporalities," *History and Theory* 51 (May 2012): 151–71.
16 For this theory of modernity and modern time sense, see Peter Fritzsche, *Stranded in the Present: Modern Time and the Melancholy of History* (Cambridge, MA: Harvard University Press, 2010).
17 On commemoration, see Paul Ricoeur, *History, Memory, Forgetting*, trans. Kathleen Blamey and David Pellauer (Chicago: University of Chicago Press, 2004).
18 Simon Reynolds, *Retromania: Pop Culture's Addiction to Its Own Past* (New York: Faber and Faber, 2011), 427.
19 See Alan Hess, *Googie: Fifties Coffee Shop Architecture* (San Francisco: Chronicle Books, 1986).
20 See the University of Virginia's online American Studies project on the 1939 World's Fair: John C. Barans, "Welcome to Tomorrow," *America in the 1930s* (May 1998; last updated September 1, 2009, by abh9h@virginia.edu), http://xroads.virginia.edu/~1930s/display/39wf/frame.htm.

21 Reynolds, *Retromania*, 368–72, discusses "nostalgia for the future" at Disneyland. Scott Bukatman calls the "retro-futures" of the Disney theme parks "meganostalgia" in *Matters of Gravity: Special Effects and Supermen in the 20th Century* (Durham, NC: Duke University Press, 2003), 31.
22 Elizabeth Guffey, *Retro: The Culture of Revival* (London: Reaktion Books, 2006); Sharon Sharp, "Nostalgia for the Future: Retrofuturism in *Enterprise*," *Science Fiction Film and Television* 4, no. 1 (2011): 27.
23 Henry Jenkins, "'The Tomorrow That Never Was': Retrofuturism in the Comics of Dean Motter" (Part 1), *Confessions of an Aca-Fan: The Official Weblog of Henry Jenkins* (June 18, 2007), http://henryjenkins.org/2007/06/the_tomorrow_that_never_was_re.html.
24 Michael Moorcock, *A Nomad of the Time Streams* (Clarkston, GA: White Wolf, 1995).
25 Thomas Pynchon, *Against the Day* (New York: Penguin, 2006). For a more extended discussion of this novel in relation to retrofuturistic steampunk, see Amy J. Elias, "Cyberpunk, Steampunk, Teslapunk, Dieselpunk, Salvagepunk: Metahistorical Romance and/vs the Technological Sublime," in *Metahistorical Narratives and Scientific Metafictions: A Critical Insight into the Twentieth-Century Poetics*, ed. Giuseppe Episcopo (Naples: Edizioni Cronopio, 2015), 201–20.
26 China Miéville, *Railsea* (New York: Del Rey, 2012).
27 Cyprien Gaillard, "Belief in an Age of Disbelief," 2005, discussed with illustrations in *Yesterday Will be Better: Mit der Erinnerung in die Zukunft/Taking Memory into the Future*, ed. Madeleine Schuppli, Claudia Jolles, Felicity Lunn, and Philippe Pirotte (Aarau: Kerber Verlag/Aargauer Kunsthaus, 2010), 108–11.
28 Jenkins references Jameson's article "Progress versus Utopia; Or, Can We Imagine the Future?," *Science Fiction Studies* 9, no. 2 (1982): 147–58, but Jameson developed this idea in more depth in *Archaeologies of the Future: The Desire Called Utopia and Other Science Fictions* (New York: Verso, 2007).
29 Tyrus Miller sees promise in the "retroavantgarde" in *Time-Images: Alternative Temporalities in Twentieth-Century Theory, Literature, and Art* (Newcastle upon Tyne: Cambridge Scholars Publishing, 2009). In contrast, Terry Smith has labeled a turn to recycling older styles in the visual arts "retromania," which he writes is symptomatic of late capitalism's evisceration of creativity; see *What is Contemporary Art?* (Chicago: University of Chicago Press, 2009), 250.
30 See Immanuel Wallerstein, *World Systems Analysis: An Introduction* (Durham, NC: Duke University Press, 2004).
31 See Bruce Sterling's and Lawrence Pearson's master list of slipstream books, originally published in the fanzine *Nova Express* (1999) and available online as of February 2014 at http://home.roadrunner.com/~lperson1/slip.html. For a good genre overview, see Paweł Frelik, "Of Slipstream and Others: SF and Genre Boundary Discourses," *Science Fiction Studies* 38 (2011): 20–44. N. Katherine Hayles and Nicolas Gessler have correlated the genre to ontological instability in "The

Slipstream of Mixed Reality: Unstable Ontologies and Semiotic Markers in *The Thirteeth Floor, Dark City* and *Mulholland Drive*," PMLA 119, no. 3 (2004): 482-99.
32 Erich Auerbach, "Figura," in *Scenes from the Drama of European Literature: Six Essays* (New York: Meridian Books, 1959), 53-54, italics mine.
33 Auerbach, "Figura," 57.
34 Auerbach, "Figura," 58.
35 See Hayden White, *Figural Realism: Studies in the Mimesis Effect* (Baltimore: Johns Hopkins University Press, 1999).
36 See, for example, Alain Badiou, *Saint Paul: The Foundation of Universalism* (Stanford, CA: Stanford University Press, 2003), or Slavoj Žižek, *The Fragile Absolute, or Why Is the Christian Legacy Worth Fighting For?*, 2d ed. (New York: Verso, 2009).
37 Antonis Liakos uses Auerbach similarly in relation to end-of-history theories and the posthumanism of Agamben; see "The End of History as the Liminality of the Human Condition: From Kojeve to Agamben," in *Crafting Humans: From Genesis to Eugenics and Beyond*, ed. Marius Turda (Göttingen and Taipei: V&R unipress/National Taiwan University Press, 2013), 63-70.
38 Grusin's theory of premediation can be found in Richard Grusin, *Premediation: Affect and Mediality after 9/11* (London: Palgrave Macmillan, 2010). For a different take on prophetic threads connecting past and present, see Amy J. Elias, "Paranoia, Negative Theology, and Inductive Style," *Soundings: An Interdisciplinary Journal* 86, nos. 3-4 (2003): 281-313.
39 Koselleck, *Futures Past*, 22.
40 Reinhart Koselleck, "The Unknown Future and the Art of Prognosis," trans. Todd Presner, in *The Practice of Conceptual History: Timing History, Spacing Concepts* (Stanford, CA: Stanford University Press, 2002), 143. See also John Zammito, "Koselleck's Philosophy of Historical Time(s) and the Practice of History," *History and Theory* 43 (2004): 124-35. Jacques Rancière has also written about multiple temporalities; see *On the Shores of Politics*, trans. Liz Heron (London and New York: Verso, 2007).
41 Peter Sloterdijk, *In the World Interior of Capital*, trans. Wieland Hoban (Cambridge, UK: Polity, 2013), 248.
42 The degree to which Sloterdijk predicates his definition of *history* on masculinist models of conquest and Glass Palace on metaphors of stifling domestic life is worth investigation.
43 Koselleck, *The Practice of Conceptual History*, 147. See also Perry Anderson, *A Zone of Engagement* (London and New York: Verso, 1992), chapter 13.
44 Fredric Jameson, *The Seeds of Time* (New York: Columbia University Press, 1994), 70.
45 See Naomi Klein, *This Changes Everything: Capitalism vs. the Climate* (New York: Simon and Schuster, 2015). Amir Eshel, in *Futurity: Contemporary Literature and the Quest for the Past* (Chicago: University of Chicago Press, 2013), argues that in fact contemporary literature shows that it is still possible to imagine the future

and that "the only future we seem to lack is the kind envisioned by the grand social utopias of the nineteenth and twentieth centuries" (16).

46 See, for example, Bruno Latour, *We Have Never Been Modern* (Cambridge, MA: Harvard University Press, 1991).

47 See Mark McGurl, "The Posthuman Comedy," *Critical Inquiry* 38, no. 3 (Spring 2012), 540. Little research to date has been done concerning posthuman time as a temporal category or way of understanding time itself, though many studies imply such a focus by linking posthumanism to spatial models (hyperobjects, the planet ecology, planetarity, bodies in ontological relation, etc.). Thus posthuman time looks much like space-time within the science of time.

48 See Dipesh Chakrabarty, "The Climate of History: Four Theses," *Critical Inquiry* 35, no. 2 (2009): 197–222. Charkrabarty cites Giovanni Arrighi's *Adam Smith in Beijing: Lineages of the Twenty-First Century* (London: Verso, 2007) to note that, in this late work, Arrighi "is much more concerned with the question of ecological limits to capitalism" (200). For further discussion of this idea, see Amy J. Elias and Christian Moraru, eds., *The Planetary Turn: Relationality and Geoaesthetics in the 21st Century* (Evanston, IL: Northwestern University Press, 2015).

2

Extinction / Adaptation

URSULA K. HEISE

Extinction and adaptation are key concerns in the understanding of temporality and history at the turn of the third millennium. They describe processes in the biological world, but they also function now as narrative metaphors for different kinds of cultural engagement with ongoing processes of modernization and globalization.

Extinction

Extinction and adaptation are, according to Darwinian theory, normal components of evolutionary processes that have taken place during all of the 3.5 billion years of biological life on Earth. Genetic changes that arise randomly in biological organisms create handicaps or advantages for certain individuals and populations as they interact with a complex network of other species and configurations of soil, water, climate, and vectors of disease. Dynamic processes of ecological change lead to the increase of some plant or animal species and the decrease or extinction of others at what biologists refer to as the "background rate"; some of the extinct species may be succeeded by differently adapted daughter species.

    This conception of evolution confers no exceptional status on the contemporary age. But in the 1970s and 1980s, paleobiologists such as David Raup and Jack Sepkoski drew attention to the importance of mass extinctions in shaping evolutionary change. Periods of mass extinction function according to a different logic than ordinary evolution in that they drive large numbers of species to extinction regardless of their adaptations.[1] The best known of the five mass extinction events known to science occurred sixty-five million years ago, when a meteorite hit the Earth and led to the demise of the dinosaurs as well as 80% of species then existing—not a consequence of bad genes but bad luck, as Raup emphasizes. For the reptiles, that is—good luck, by contrast, for mammals, whose subsequent evolution, including that of *homo sapiens*, was enabled by the disaster.

Biologists such as Norman Myers, Paul and Anne Ehrlich, and E. O. Wilson have argued that such an abrupt evolutionary reversal might be occurring again now, in what some claim is the sixth mass extinction, but the first one triggered by human activity.[2] Current rates of species extinction, driven by habitat loss, introduced species, human population growth, pollution, and overharvesting,[3] are estimated to exceed the normal background rate by fifty to five hundred times,[4] although the precise numbers are the subject of intense debate.[5] Precise assessments of extinction rates are handicapped by the bias of current research toward birds, mammals, and amphibians, whereas invertebrates, plants, and micro-organisms have been studied in far less detail.[6] Such assessments are also complicated by our uncertainty about the total number of species on the planet.[7] But the general trend toward species loss is not in dispute, and species extinction is commonly used by scientists as an indicator of a general reduction in biodiversity, a term that was coined in 1984 and refers to biological diversity ranging all the way from genes to populations, species, and ecosystems.[8]

Along with climate change and widespread pollution, the current mass extinction of species often functions as a shorthand for global ecological crisis in environmentalist discourse as well as in the public media. Especially since the 1980s, dozens of websites, hundreds of books and documentary films, and thousands of photographs have brought either the fate of individual endangered species or the larger panorama of biodiversity loss to public awareness in a variety of countries. Public attention—and not infrequently, conservation funding—tends to focus on what has half seriously and half jokingly come to be called "charismatic megafauna": those species that humans tend to find attractive and culturally significant (typically animals rather than plants, vertebrates rather than invertebrates, mammals or birds above other taxa). Primates, bears, wolves, pandas, tigers, rhinos, whales, raptors, or parrots tend to garner attention along with strikingly colorful frog or beautiful butterfly species, whereas plants, fungi, and micro-organisms are largely invisible in extinction discussions. The genres of elegy and tragedy, with their familiar tropes of decline, loss, nostalgia, mourning, and melancholia, often shape portrayals of species endangerment and seek to mobilize readers' or viewers' affect and empathy.[9] Frequently, in such accounts, the loss or endangerment of a particular species comes to stand in symbolically for other losses a cultural community or nation believes itself to have suffered during its modernization or colonization, whether these be traditional ways of life, a closer connection to the natural world, ownership or access to land and natural resources, or a more harmonious social community.[10]

The narrative that structures such accounts is not just the story of the decline of nature that has informed proto-environmentalist and environmentalist discourses as they have arisen in resistance to successive waves of industrialization and modernization since the early nineteenth century in Europe, North America, and Australia, and in resistance to imperialism, colonialism, and foreign resource appropriations in parts of Africa, Latin America, and Asia. It is also a narrative that more generally understands the present as a time of lost diversity and abundance, of scarcity, and of limits.

That these concepts refer to more than ecological conditions emerges clearly, for example, in analyses of language extinctions, which are often discussed in terms that parallel those surrounding biological extinctions. Many of the world's approximately six thousand languages are no longer spoken, or in a few cases have given rise to successor languages. Pidgins and trade languages arise out of specific historical moments and zones of contact, and they disappear when the contact ceases or shifts geographically. Extinction of indigenous languages often occurs because the cultural communities that used them disappear or are forced to adopt a different language. This kind of disappearance can sometimes be materially connected to biological extinction, for example, when the ecological environment in which a particular community once thrived is destroyed. But quite frequently, the reasons for linguistic extinction are cultural and political rather than ecological, and not all cultures disappear along with their languages, just as not all languages remain extinct. For these reasons, many linguists are skeptical of close analogies or causal associations between biological and linguistic extinctions.[11] Yet the extinction narrative in its larger connotations does tend to include these analogies.

If language extinctions resemble species extinctions, it is not so much in their causes and consequences as in the cultural significance that is attributed to them. The extinction of some languages is perceived as a greater loss than that of others, in a process somewhat analogous to the preference for charismatic megafauna in biological endangerment—the loss of indigenous languages associated with deep histories, distinct sets of cultural practices, and an aura of authenticity tend to be mourned more deeply than the disappearance of trade languages, pidgins, or creoles that arose in circumstances of more recent histories, colonial encounters, or cultural hybridization.[12] From the linguist's point of view, this is an arbitrary distinction, but it makes sense in the framework of the narrative construction of modernization as loss or decline. In this story template, the diversities that modernization processes diminish tend to be valued more than the diversities that modernization generates. Extinction functions as the central trope that signals these losses.

Some of the environmental writers who focus on the decline of biodiversity as a sign of global ecological crisis go one step further to forecast the extinction of humankind itself as a consequence of its environmental destruction. The paleobiologist Richard Leakey and his co-author Roger Lewin compare humans' impact on the Earth to the asteroid that caused the extinction of the dinosaurs, and they contend that future destruction might ultimately include human life itself: "The world's biological diversity will plummet, including the productivity on which human survival depends. The future of human civilization therefore becomes threatened."[13] As a consequence, "we, *Homo sapiens*, may also be among the living dead."[14] The narrative according to which humans' knowledge of and ability to manipulate the material world ultimately triggers their self-destruction is familiar from portrayals of nuclear culture between the 1940s and the 1980s, here repurposed for the context of biodiversity loss. Considering that human population growth figures among the major causes of species extinction and is poised to continue until at least 2050, the prospect of humanity's self-destruction as a consequence of decreasing biodiversity is not considered likely by most demographers and biologists. But forecasts of future harm to humans themselves can function as powerful arguments for the conservation of other species. Significantly, the narrative of modernity's ultimately destructive and self-defeating logic is what, in fact, lends plausibility to the scenario of human extinction.

Biodiversity loss, climate change, and toxification—three major current ecological crises—are all associated in environmentalist thought and writing with a particular kind of temporality, a dual and seemingly contradictory emphasis on slowness *and* speed. On one hand, many environmentalist activists argue, these planetary ecological crises unfold too slowly to be readily perceived and understood by the public at large. From Paul and Anne Ehrlich's famous metaphorization of species extinction as rivets that are popped from a plane one after the other until the plane becomes incapable of flight, to Rob Nixon's recent emphasis on the "slow violence" that environmental degradation inflicts on the world's poor, environmental writers have emphasized the gradual unfolding of crises that are hard to address in the fast-paced socioeconomic, technological, and political cycles of contemporary societies. On the other hand, environmentalists—often the same ones—tend to urge rapid action to forestall future catastrophe, and to emphasize that time windows for prevention are likely to close in the very near future, a temporal trope that runs from the Meadows' *Limits to Growth* (1972) to Al Gore's *An Inconvenient Truth* (2006). Global ecological crisis today is typically associated with this temporal tension—too slow to see and understand, and yet too fast by the normal standards of ecological change, and quite possibly too rapid for current political decision-making processes.

Adaptation

In Darwinian theory, extinction and adaptation form part of the same evolutionary processes—as better adapted species arise, less adapted ones die out. Indeed, in one understanding of Darwin that was not uncommon in the nineteenth century, extinction is to be welcomed as it leads to better adapted species and more functional biological systems. This view is less acceptable now that ecology has shown that species do not evolve against a background of static conditions, but in the context of ecosystems that change dynamically with weather fluctuations, natural disasters, and the arrival or disappearance of disease vectors, among other dimensions. In such dynamic systems, extinction and adaptation present themselves as components of evolutionary change rather than progress. Adaptation, which has often been misinterpreted as an at least partially intentional process, really refers to combinations of the randomness of genetic mutation with processes of natural selection. Random differences between individuals of a species can sometimes result in a benefit or handicap that allows some individuals to procreate at a greater rate than others, so that their genetic information comes to shape future generations. Richard Dawkins's theory of the "selfish gene," a reading of Darwinian theory in combination with twentieth-century genetic research that particularly emphasizes the nonintentionality of such processes, characterizes plant, animal, and human bodies as mere "robots" by means of which genes perpetuate themselves.[15] This implies less a determinism whereby individual organisms are unable to act against their genes—as Dawkins himself points out, this would make contraception logically impossible—than the idea that genes set default templates, so to speak, for the behavior of organisms and thereby govern adaptation.

Cultural understandings of extinction and adaptation do not always follow such scientific arguments closely. If scientific logic portrays extinction and adaptation as parts of the same processes, cultural logic more often sees them as opposites. Extinction equals death in many kinds of public narrative; we tend to be startled by the suggestion that Latin is spoken across vast swathes of Europe today in the form of French, Spanish, Italian, and Romanian (among other languages), or that our parks are populated by flighted mini-dinosaurs that we call starlings and pigeons. Extinction, in other words, typically highlights the disappearance of earlier forms in cultural contexts, whereas adaptation is associated with narratives of self-perpetuation through changeable forms. In contexts of cultural modernization or of colonization, the invocation of extinction and narratives about the "last of" a particular group or species are usually designed to evoke surprise and regret over disappearance. In Werner Herzog's *Wo die grünen Ameisen träumen* (*Where the Green Ants Dream*, 1984), for example, a film about the encroachments of a mining company on land in the Australian outback that an aboriginal community claims as its own, a climactic trial scene about land rights features a lengthy, impassioned speech by an aboriginal elder in his own language. When the white judge asks for a translation, he is told that no one in the room, including the other aborigines, knows or can translate what the elder said, as he is the last and only speaker of his language; whatever legal, historical or moral claim he might have spelled out is thereby doomed to vanish from the record.[16]

By contrast with such stories of temporal rupture, adaptation narratives foreground a temporal continuity that is achieved by both positive and negative modes of change—they include the ingenuity of the trickster, cultural reappropriation, hybridization, and reinvention, but also amnesia, inauthenticity, corruption, and betrayal of older traditions and cultural frameworks. Indra Sinha's novel *Animal's People* (2007), for example, manifests this duality through a picaresque narrator-protagonist so crippled by the chemical explosion at Bhopal (renamed Khaufpur in the novel) that he can walk only on all fours, but he superbly uses his handicap as well as his gifts of observation and language to analyze and manipulate the postcolonial society around him and to help bring about political change.[17] This kind of adaptation is often associated with its biological counterpart, but it differs fundamentally from the biological process in that it combines material and political pressures with individual intentionality and agency.

The ambiguities of survival over time that the term "adaptation" brings with it in cultural, economic, and political contexts frequently inflect debates about current environmental crises—not just about industrial accidents such as those in Chernobyl, Bhopal, or the Gulf of Mexico, but also about climate change and species loss. In the 1990s, debates over global warming, especially in North America, focused on the urgent necessity to curb the emission of greenhouse gases. "Mitigation" was the watchword of the decade, whereas scientists and activists who suggested, timidly at first, that "adaptation" might also be important to explore were often fiercely rebutted with the charge that they were distracting attention from the principal task of reducing greenhouse gases, and conceding the battle to the enemy by portraying climate change as inevitable. Over the course of the 2000s, as glaciers were found to have already substantially receded and populations near the Equator began to suffer the consequences of the warming climate, the emphasis shifted. Climate change could indeed no longer be prevented or even substantially diminished over the short term—whatever mitigation measures are undertaken in the present will now most substantially affect neither the current generation of humans nor its children, but only its grandchildren. In this context, adaptation assumed a new respectability as environmentalists in developed nations recognized that lending support for adaptation in already irrevocably altered ecosystems had to form part of their ethical responsibilities. Climate change, in other words, is no longer a disaster looming in the future but a present and unfolding reality in which adaptation is, especially for the most affected regions, no longer a matter of choice.

In a different way, adaptation also plays a crucial role in thinking about and managing biodiversity. Most moderately educated readers, even those with a rudimentary exposure to evolutionary science, will be aware that adaptation can generate diversity, as Darwin's well-known discovery of finch species with variously adapted bills on different islands of the Galapagos Archipelago showed. Karen Tei Yamashita has brilliantly translated such processes into magical realism. Her novel *Through the Arc of the Rainforest* presents an abandoned junkyard in the Amazonian jungle where adaptation has given rise to a whole new set of species,

> an area which resembled an enormous parking lot, filled with aircraft and vehicles of every sort of description. The planes and cars had been abandoned for several decades, and the undergrowth and overgrowth of the criss-crossing lianas had completely engulfed everything. . . . What was most interesting about the discovery of the rain forest parking lot was the way in which nature had moved to accommodate and make use of it. The entomologists were shocked to discover that their rare butterfly only nested in the vinyl seats of Fords and Chevrolets and that their exquisite reddish coloring was actually due to a steady diet of hydrated ferric oxide, or rusty water. There was also discovered a new species of mice, with prehensile tails, that burrowed in the exhaust pipes of all the vehicles. These mice had developed suction caps on their feet that allowed them to crawl up the slippery sides and bottoms of the aircraft and cars. . . . [A] new breed of bird, a cross between a vulture and a condor . . . nested on propellers and pounced on the mice as they scurried out of exhaust pipes. Finally, there was a new form of air plant, or epiphyte, which attached itself to the decaying vehicles. . . . Meantime, . . . human life was adapting . . . in ways as unexpected as those found in the rain forest parking lot and as expected as the great decaying and rejuvenating ecology of the Amazon Forest itself.[18]

Yamashita exaggerates real processes of adaptation just enough in this passage to turn them into metaphors for the transformations wrought on the Brazilian rainforest by modernization, global capitalism, and pollution, transformations that her novel portrays as both destructive and generative.

Yamashita no doubt modeled part of her fictional rainforest scenario on the historical precedent of the peppered moth in England, whose coloration changed from predominantly light to predominantly dark as many trees in its habitats were covered with soot as a consequence of industrial production.[19] But humans' intentional and unintentional impacts on evolutionary adaptation make themselves felt in many contexts, as the biologist Stephen Palumbi has shown; from changes in the average size of certain fish species caused by humans' harvesting of the largest specimens before they are able to reproduce, to the emergence of herbicide-resistant weeds and antibiotic-resistant bacteria in agriculture and medicine, the human presence itself has become an environment to which other species adapt.[20] Are such incidents in fact extinction events in which the earlier form of a species disappears, superseded by a new form, or adaptations by which the old species ensures its survival? The answer depends on the temporal perspective one brings to bear on such scenarios and differential valuations of the past or of ongoing change, as natural adaptation interlaces with humans' manipulations of their environments.

De-extinction / Re-animation

Environmentalists eager to move beyond narratives of decline toward more future-oriented perspectives have in recent years suggested the possibility of "de-extinction": revival of extinct species by means of DNA recovered from fossil or museum specimens, and genetic assistance to species currently endangered in the wild.[21] Current biotechnology makes the prospect of reviving extinct species at least plausible, even though practical challenges persist—attempts to reconstruct the *bucardo* (*Capra pyrenaica pyrenaica*), an extinct subspecies of mountain goat native to the Spanish Pyrenees, has yielded what Lothar Frenz has called a "seven-minute renaissance," a cloned *bucardo* kid that lived all of seven minutes before its lungs collapsed, thereby bestowing on its subspecies the dubious distinction of having gone extinct not just once but twice.[22]

Beyond the sheer technical challenges in reconstructing extinct species from remnant DNA, other questions beset the idea of reversing extinction events that had until recently been considered irreversible.

Given that many species have perished because their habitat disappeared, where would the reconstructed specimens live? Would they ever move beyond sample populations kept in zoos as rare and expensive collector items? If reintroducing them into the wild turned out to be impossible, they would make no real ecological difference, and the question then becomes why exactly they should be revived in the first place—what ethical claim would their technological resurrection respond to? And even more fundamentally: *would* they in fact be the same species as those gone extinct? The woolly mammoth, for example, one of the "candidate species" for resurrection,[23] would presumably be reconstructed from a combination of fossil mammoth with contemporary elephant DNA, creating a species with a genome unlike that of the original woolly mammoth.[24] This mix of genes in biotechnologically resurrected species was anticipated in Steven Spielberg's film *Jurassic Park*, which featured dinosaurs reconstituted with the help of frog DNA.

De-extinction, therefore, raises complex temporal as well as ethical questions. It might seem plausible to de-extinct a recently vanished species such as the passenger pigeon, whose last individual died at the Cincinnati Zoo in 1914. But candidate species such as the dodo, which was last sighted on the island of Mauritius in the 1660s, make us revert back to a far more remote past, and the woolly mammoth, which went extinct circa 2000 BCE, seems wholly disconnected from our present. More than a return to the past, such revivals would generate an invented, human-created biological world without any historical precedent.

Considering such futuristic ecological possibilities, it comes as no surprise that authors of science fiction and speculative fiction have explored the possibilities of technologically assisted adaptation and de-extinction. In Kōbō Abe's 第四間氷期 (*Dai yon kan pyōki* [*Inter Ice Age 4*], 1959), Bruce Sterling's *Schismatrix* (1985), and Orson Scott Card's *Ender* series (including *Ender's Game* [1985], *Speaker for the Dead* [1986], *Xenocide* [1991], and *Children of the Mind* [1996], among other volumes), biotechnology becomes evolution's new tool for creating innovative ecologies on and beyond planet Earth.[25] In Kim Stanley Robinson's *2312* (2012), humans have ingeniously terraformed planets, moons, and satellites across the solar system and have reengineered their own bodies, which now range from froglike to pygmylike, incorporate animal genes and abilities, and oscillate across a wide spectrum of gender identities.[26] In the meantime, Earth continues to be riddled with ecological problems from climate change and seawater rise to species extinction, and social crises from poverty to power struggles. The new (post)humans from Mercury, Venus, Titan, and Io seek to aid the Earth's immiserated societies and ailing ecosystems, though their efforts are often frustrated by those societies' large populations, long histories, and complex social structures. One of their projects, called the "Reanimation," consists of the reintroduction of species extinct on Earth from "terraria" where they have been bred on other planets. The animals are released by the thousands in a kind of rain, in aerogel-filled bubbles that float down to the Earth's surface, accompanied by activists (such as the Mercury-born Swan Er Hong) who are charged with helping the animals after their landing:

> Swan looked around, trying to see everywhere at once: sky all strewn with clear seeds, which from any distance were visible only as their contents, so that she drifted eastward and down with thousands of flying wolves, bears, reindeer, mountain lions. There she saw a fox pair; a clutch of rabbits; a bobcat or lynx; a bundle of lemmings; a heron, flying hard inside its bubble. It looked like a dream, but she knew it was real, and the same right now all over Earth: into the seas splashed dolphins and whales, tuna and sharks. Mammals, birds, fish, reptiles, amphibians: all the lost creatures were in the sky at once, in every country, every watershed. Many of the creatures had been absent from Earth for two or three centuries. Now all back, all at once. Swan came down in the midst of a cluster of animals.[27]

Considering the hard-headed, science-based utopianism of Robinson's earlier novels, this is an unusually surrealist scene, reminiscent of René Magritte's *Giaconda* painting with its rain of bowler-hatted in men in dark coats. Robinson chooses a deliberately implausible solution to the real problem of biodiversity loss, emphasizing that the nature of the future will not be "natural" in the sense of any return to an originary past, but will be a readaptation, in contrast to normal evolutionary adaptation deliberately carried out by humans so as to create ecosystems of their own design. The cultural and indeed political origins of this future nature are further emphasized through Robinson's portrayal of some of the protests and struggles that accompany the Reanimation, which is not initially welcomed by all of Earth's inhabitants.

One conceptual step further than Robinson's *2312*, German author Dietmar Dath's speculative novel *Die Abschaffung der Arten* (2008; translated as *The Abolition of Species*) portrays a future society on planet Earth, part of which is ruled by a lion king with wolf and dragonfly advisors.²⁸ This sentient-animal scenario would seem like a quaint revival of both the medieval German animal fable of Reineke Fuchs and of twentieth-century Disney films if it were not for the fact that another part of the world is more futuristically occupied by the computer-descended "Ceramicans." And when the reader learns that the lion king's mate was at one point of her life a swarm of insects rather than a lioness, the usual concept of "species" itself, in either its biological or allegorical functions, turns out to be a relic of the past. The "Gente," as the inhabitants of this posthuman world call themselves in a dual allusion to gene technology and the Spanish word for "people," are in reality genetically reengineered animals or humans. "[D]er Name 'Pferd' zum Beispiel bezeichnete natürlich nicht dieselbe Sorte Wesen wie vor der Befreiung, sondern der neue Pferdekopf wies so gut wie jedes andere Haupt jedes anderen Geschöpfs, das Sprache hatte, Hominidenzüge auf," the narrator explains ("The name 'horse,' for example, did of course not refer to the same sort of creature as before the Liberation, but the new horse's head, just like the heads of any other creature, displayed hominid features"). Even more explicitly, he observes: "Die Unterscheidungen zwischen den echten Spezies aber waren, da jedes Geschöpf nur mehr nach seiner je eigensten Art schlug und nahezu alle mit allen andern Nachkommen zeugen konnten, ebenso sinnlos geworden wie die Unterscheidungen zwischen den Menschenrassen" ("Because each creature molded itself after its very own species and almost all individuals could produce offspring with all others, the distinctions between genuine species had become as meaningless as distinctions between human races.")²⁹ The Liberation of animals from human domination, which occurred approximately five hundred years before the beginning of the novel's plot, turns out to be at the same time the liberation from the constraints of biological species itself.

The civilization of posthuman animals on Earth is vanquished by the Ceramicans, but not before the animals export some seeds for future offspring to Venus and Mars, where they start colonies. Two descendants from these colonies return to Earth toward the end of the novel, and one of them reflects on the extinction of the old kind of humans that eventually gave rise to their own kind: "'Die Menschen mußten sterben, damit die Menschheit eine Chance hat. Denn das waren die Gente ja: die erste realisierte Menschheit'" ("Humans had to die so that humankind would have a chance. For that's what the Gente were: the first fully realized humankind"). Dath's vision of a future beyond species takes up the evolutionary combination of extinction and adaptation to reflect on the porous boundaries between humans, nonhuman species, and inanimate environments. Full humanity is only achieved beyond humans' self-definition as a species, as humans (re)become animals—an invitation to rethink the temporal directionality of all the stories that we tell about our own evolution through both adaptations and extinctions. Whereas extinctions and adaptations tend to mark turning points in the historical trajectories of species, languages, and cultural communities that are usually conceived of as sequential and mostly linear, Dath's vision invites European and American readers in particular to move beyond currently popular end time narratives—the end of nature, of diversity, of choice, and perhaps even of humankind itself—to a reimagination of the future that has not yet been consumed by the present.

NOTES

1 See David M. Raup, *Extinction: Bad Genes or Bad Luck?* (New York: W. W. Norton, 1991); David Jablonski, "Mass Extinctions: New Answers, New Questions," in *The Last Extinction*, eds. Les Kaufman and Kenneth Mallory, 2d ed. (Cambridge, MA: MIT Press, 1993), 47–68; and David Sepkoski, *Rereading the Fossil Record: The Growth of Paleobiology as an Evolutionary Discipline* (Chicago: University of Chicago Press, 2012). For a cultural history of the concern over extinction in the United States, see Mark V. Barrow, Jr., *Nature's Ghosts: Confronting Extinction from the Age of Jefferson to the Age of Ecology* (Chicago: University of Chicago Press, 2009).
2 Norman Myers, *The Sinking Ark: A New Look at the Problem of Disappearing Species* (Oxford: Pergamon, 1979); Paul and Anne Ehrlich, *Extinction: The Causes and Consequences of the Disappearance of Species* (New York: Ballantine, 1981); and Edward O. Wilson, *The Future of Life* (New York: Vintage, 2002).
3 Wilson, *The Future of Life*, 50.
4 Jonathan E. M. Baillie, Craig Hilton-Taylor, and Simon N. Stuart, eds. *IUCN Red List of Threatened Species: A Global Species Assessment* (Gland, Switzerland: IUCN, 2004).
5 See Mark J. Costello, Robert M. May, and Nigel E. Stork, "Can We Name Earth's Species before They Go Extinct?," *Science* 339 (January 25, 2013): 413–16; William F. Laurance, "The Race to Name Earth's Species," *Science* 339 (March 15, 2013):

1275; Camilo Mora, Audrey Rollo, and Derek P. Tittensor, "Comment on 'Can We Name Earth's Species before They Go Extinct?,'" *Science* 341 (July 19, 2013): 237-c.; and Mark J. Costello, Robert M. May, and Nigel E. Stork, "Response to Comments on 'Can We Name Earth's Species before They Go Extinct?,'" *Science* 341 (July 19, 2013): 237-d.

6   See Craig Hilton-Taylor, Caroline M. Pollock, Janice S. Chanson, et al., "State of the World's Species," in *Wildlife in a Changing World: An Analysis of the 2008 IUCN Red List of Threatened Species*, ed. Jean-Christophe Vié, Craig Hilton-Taylor, and Simon N. Stuart (Gland, Switzerland: IUCN, 2008).

7   As discussed by Robert May, "How Many Species?," in *The Fragile Environment: The Darwin College Lectures*, ed. Laurie Friday and Ronald Laskey (Cambridge, UK: Cambridge University Press, 1989), 61–81; and Costello, May, and Stork, "Can We Name Earth's Species," 413–16.

8   "Biodiversity" is itself a concept that combines scientific assessments with value judgments, as has been pointed out by David Takacs, *The Idea of Biodiversity: Philosophies of Paradise* (Baltimore: Johns Hopkins University Press, 1996); James Maclaurin and Kim Sterelny, *What Is Biodiversity?* (Chicago: University of Chicago Press, 2008); and Donald S. Maier, *What's So Good about Biodiversity? A Call for Better Reasoning about Nature's Value* (Dordrecht: Springer, 2012). For initial definitions of the term, see Edward O. Wilson and Frances M. Peter, eds., *Biodiversity: Papers from the 1st National Forum on Biodiversity, September 1986, Washington, D.C.* (Washington, DC: National Academy Press, 1996).

9   For discussion of these tropes, see Catriona Mortimer-Sandilands, "Melancholy Natures, Queer Ecologies," in *Queer Ecologies: Sex, Nature, Politics, Desire*, ed. Catriona Mortimer-Sandilands and Bruce Erickson (Bloomington: Indiana University Press, 2010), 331–58, and Ursula K. Heise, "Lost Dogs, Last Birds, and Listed Species: Cultures of Extinction," *Configurations* 18 (2010): 49–72.

10 Heise, "Lost Dogs," 60–69.

11 For analyses of the causes and consequences of language extinction, see David Crystal, *Language Death* (Cambridge, UK: Cambridge University Press, 2000); Lenore A. Grenoble and Lindsay J. Whaley, eds., *Endangered Languages: Language Loss and Community Response* (Cambridge, UK: Cambridge University Press, 1998); and Daniel Nettle and Suzanne Romaine, *Vanishing Voices: The Extinction of the World's Languages* (Oxford: Oxford University Press, 2000).

12 K. David Harrison, *When Languages Die: The Extinction of the World's Languages and the Erosion of Human Knowledge* (New York: Oxford University Press, 2007); Paul B. Garrett, "Dying Young: Pidgins, Creoles, and Other Contact Languages as Endangered Languages," in *The Anthropology of Extinction: Essays on Culture and Species Death*, ed. Genese Marie Sodikoff (Bloomington: Indiana University Press, 2012), 143–62.

13 Richard Leakey and Roger Lewin, *The Sixth Extinction: Patterns of Life and the Future of Humankind* (New York: Anchor, 1995), 239.

14 Leakey and Lewin, *The Sixth Extinction*, 245. Cf. the discussion in Michael Boulter, *Extinction: Evolution and the Ends of Man* (New York: Columbia University Press, 2002), 185–91. While for some readers the phrase "living dead" may conjure visions of zombies and other popular gothic figures, "living dead" in the biological realm refers to species that still exist, but not in sufficient abundance to survive over the long term.
15 Richard Dawkins, *The Selfish Gene*, 30th anniversary ed. (1976; Oxford: Oxford University Press, 2006).
16 *Where the Green Ants Dream*, screenplay by Bob Ellis and Werner Herzog, directed by Werner Herzog (Pro-ject Filmproduktion, 1984); released also as *Wo die grünen Ameisen träumen* (1984).
17 Indra Sinha, *Animal's People* (New York: Simon & Schuster, 2007). For insightful interpretations of this novel, see chapter 6 of Upamanyu Pablo Mukherjee, *Postcolonial Environments: Nature, Culture and the Contemporary Indian Novel in English* (Houndmills: Palgrave Macmillan, 2010), and chapter 1 of Rob Nixon, *Slow Violence and the Environmentalism of the Poor* (Cambridge, MA: Harvard University Press, 2011).
18 Karen Tei Yamashita, *Through the Arc of the Rainforest* (Minneapolis: Coffee House Press, 1990), 99–101.
19 Michael Majerus, *Melanism: Evolution in Action* (New York: Oxford University Press, 1998).
20 Stephen R. Palumbi, *The Evolution Explosion: How Humans Cause Rapid Evolutionary Change* (New York: W. W. Norton, 2001).
21 See an advocate of such views, Stewart Brand, at the Long Now Foundation website: http://longnow.org/revive/.
22 Lothar Frenz, *Lonesome George oder Das Verschwinden der Arten* (Berlin: Rowohlt, 2010), 61–67.
23 According to the Long Now Foundation, http://longnow.org/revive/candidates/.
24 I am grateful to the philosophers Ronald Sandler and James Maclaurin for discussing these issues in depth in their presentations at the "Thinking Extinction" symposium at Laurentian University in November 2013.
25 Kōbō Abe, 第四間氷期 (*Dai yon kan pyōki*) (*Tokyo*: Shinchōsha, 1959; *Inter Ice Age 4*, trans. E. Dale Saunders, New York: Alfred A. Knopf, 1970); Bruce Sterling, *Schismatrix* (Westminster, MA: Arbor House Publishing, 1985); Orson Scott Card, *Ender's Game* (New York: Tor Books, 1985), *Speaker for the Dead* (New York: Tor Books, 1986), *Xenocide* (New York: Tor Books, 1991), and *Children of the Mind* (New York: Tor Books, 1996).
26 Kim Stanley Robinson, *2312* (New York: Orbit, 2012).
27 Robinson, *2312*, 453–54.
28 Dietmar Dath, *Die Abschaffung der Arten* (Frankfurt: Suhrkamp Verlag, 2008; *The Abolition of Species*, trans. Samuel P. Willcocks, London: Seagull Books, 2013). The translations in this essay are my own.
29 Dath, *Die Abschaffung der Arten*, 34.

3

Modern / Altermodern

DAVID JAMES

Criticism has reached a moment when the distinctions between modernity and contemporaneity have never been more debated; when the case for seeing periodization as a professional restriction and intellectual impediment is gaining traction; when postmodernism as a critical category, an epoch of aesthetic production, and a cultural pathology of our late-capitalist condition is considered to be passed; and when artistic modernism itself is being regarded as unfinished and irrepressible, a keyword that resurfaces afresh to capture a range of experimental practices across the visual, textual, acoustic, and plastic arts. At such a disciplinary moment as this—and notwithstanding the precept that *modernity* legitimately refers to a *longue dureé* stretching from the Renaissance to the present—what are the challenges and opportunities of approaching the temporality of contemporary aesthetic modes through the lens of the *modern* and its various pre-modifiers (post-, re-, alter-)? Unwieldy though it appears, such a question is the backdrop to what follows, as I engage with the way debates about art and contemporaneity have informed discussions about the kinds of periodizing work modern(ism) now performs. In so doing, I distinguish *modernism* as a historically identifiable (though by no means historically limited) constellation of artistic ambitions that generate aesthetic novelty, negation, and revolution from *modernity* as a transhistorical designator of social and technological advancement that can be applied across broader swathes of time and space.

To some extent, all periodizing categories lead a double life: they can function as designators for *when* art happens and as evaluative adjectives for distinguishing *how* art works at specific historical junctures. But the problem we face in working specifically with modernism and modernity involves their potential for mutual contamination as overlapping models for imagining history, as the hypothetical flexibility of modernism at the level of argumentation is achieved at the expense of its empirical validity and applicability at the level of artistic practice. We now have the advantage of being able to look back at how modernism and postmodernism have been utilized not only to identify phases in artistic production—the 1850s to the 1940s (Baudelaire to Beckett) and 1945 to 1989 (from the bombing of Hiroshima to the fall of the Berlin Wall), respectively, to hazard some typical if arbitrary spans—but also to name paradigms of cultural transition across eras marked by the rise of finance capitalism and by irrepressible globalization. However, I want to work through the methodological, disciplinary, and theoretical implications of how it "seems today," in Jean-Michel Rabaté's words, "that modernism has absorbed most of the twentieth century, that it goes back deep into the nineteenth century and that it has moreover swallowed postmodernism."[1]

One purpose of surveying those implications here will be to think about the correlations between era and vocabulary, epoch and analysis. For the issue arises as to whether modernism (as a moment and movement, as a catalyst for various politico-aesthetic manifestos, as a kind of disciplinary rubric for certain professional commitments and investments) can still provide an appropriate interpretive lexicon for apprehending the arts of our time. Given modernism's critical liberation from temporal parameters confined to the early twentieth century and from spatial parameters confined to metropolitan Europe, are there ways of grasping modernism today (in terms of its periodizing activities and as an adjective for certain kinds of aesthetic activities) that can disrupt linear models of historical progression in the arts, that emerge from transcultural dynamics of aesthetic revaluation rather than solely from a desire for militant artistic dissent—and that no longer, in other words, enlist the "shock troops of the modern," in Amy J. Elias's words, "marching into the utopian future and away from the desiccated, conservative, politically and/or aesthetically compromised past"?[2]

Complicating this question is the truth that to posit a keyword for an age is to court anachronism. As interpretive operations rather than empirical facts, keywords announce their own arbitrary imposition upon contingent and internally complex historical circumstances; as periodizing tools, they often seem out of joint with the very cultural formations whose currency they supposedly frame. It is the double bind one faces in categorizing any era, of course. But nowhere is it more evident than when anatomizing that shape-shifting phenomenon, "the contemporary." Our unfolding times have inspired a gamut of all-embracing terms—*cosmodernism, digimodernism, the new sincerity, post-irony, the Anthropocene*—even though, as soon as they're announced, many of these abstractions feel misaligned with the *Zeitgeist* they long to capture. So why should any category stick, even when it's outstripped by its referents? Why do terms linger when temporalities morph? Shouldn't the issue of post/modernism's continued relevance have been safely wrapped up long ago, overtaken by the moving horizon of contemporary artistic production? How do we account for the curious persistence and institutional acceptance of such labels in theory, and what does this say about our expectations of their pertinence in practice? Keywords have a knack of bedding in. And it's in this sense that a "time," as novelist Jonathan Lethem reminds us, "is marked not so much by ideas that are argued about as by ideas that are taken for granted," because the "character of an era hangs upon what needs no defense."[3]

Modern / Modernism

However much modernism is taken for granted, that fact alone hasn't detracted from its surviving appeal—not least for debates about how we critically organize and reinforce scholarship that specializes in the moving target of contemporary culture. If the "mentality of aesthetic modernity," as Habermas called it, was "characterized by a set of attitudes which developed around a transformed consciousness of time," then modernism today seems to have transcended time altogether—pulling its anchor free of fixed historical (and indeed geographical) locations, despite its association with the seismic upheavals of early-1900s Europe.[4] Modernism has reemerged from, if indeed it was ever fully submerged in, the explanatory systems associated with late-twentieth-century postmodernism, fully reprised as the term of choice for identifying how artistic advancement occurs in response to sociocultural change. Dissolving the temporal borders altogether, Susan Stanford Friedman has argued that we should cultivate our "recognition that the 'periods' of modernism are multiple and that modernism is alive and thriving," so as to explore how modernist art occurs "wherever the historical convergence of radical rupture takes place."[5] This logic of originality-as-rupture ignores one of the challenges we face in reading modernism as an unfinished project: to find ways of doing justice to art's contemporary responses to a modernist past that don't just reimport traits of the modern itself—like rupture, innovation, or fragmentation—as criteria for evaluation. For art historians, this perpetual rehashing of modernism is a largely theoretical exercise, whose bearing on what's happening in aesthetic practice is difficult to gauge. "Many emerging artists," reflects Terry Smith, "sense that Modernism—no matter how often and subtly it is Remodernized—is past its use-by date." This new generation, he continues, moreover "regard[s] 'Postmodern' as an outmoded term, a temporary placeholder that is no longer adequate to describe conditions that, they believe, have changed fundamentally." Because they have "inherited" rather than lived through "the successes and shortcomings of the 1960s and 1970s—from anticolonialism to feminism—and now seek to relate these lessons to the even greater challenges of living in the conditions of contemporaneity," artists forging postmillennial careers require different categories of address, categories that unyoke them from a now-antiquated postmodernism without merely situating them as modernism's legatees or as renovators of its residual promise.[6]

Justified though these reservations about the currency of modernism and its derivations are, to rely on "contemporaneity" to do so much critical work—the two-handed job of classification *and* periodization—seems equally perilous. Smith himself admits as much in his earlier study *What Is Contemporary Art?* (2009), where he declares, in a panoramic flourish, that "[w]ith the passing of modernity, the evaporation of the postmodern and the rise of fundamentalisms, with the eruptions of an overstressed planet and the diminution of imaginable futures, contemporaneity seems to be all that we have."[7] Or almost all that we have, since it also appears that "every kind of past has returned to haunt the present, making it even stranger to itself."[8] Without wanting to suggest that the present's relation to modernism is best understood in such spectral terms, the premise here of its self-renewing return has gained more traction for theorists in literary studies and art history alike, who have sought to rethink modernism today while refusing to dismiss the impact of postmodernism's aesthetic, cultural, and ideological legacies.[9]

Post-Modern / Postmodernism

If one trend in modernist scholarship today has been toward the expansion of its temporal and cultural perimeters, another has concerned the reassessment of modernism's postwar successor—revealing, in turn, how much the "new modernist studies" itself might owe to postmodernism's interrogation of unifying grand narratives and of Eurocentric conceptions of historical development. A banner for an unrivalled arsenal of philosophical, narratological, and geopolitical vocabularies, postmodernism now attracts some methodological hesitation rather than immediate reaffirmation. This is not to detract from the success of postmodernism as a cultural and formal keyword for conceptualizing our times, whether intellectually or artistically. Postmodern discourse has usefully combined sociological query about globalized cultures with new understandings of aesthetic self-reflexivity across visual, plastic, textual, and acoustic arts.[10] In contrast to the modernist ambition to represent and recuperate momentary experiences of time from the physical pace, psychic unsettlement, and mediating regimes of modernity, postmodern artists and theorists have taught us to doubt such reparative impulses. They celebrate instead the compression of space and time as it casts lived experience into states of contingency, displacement, and unpredictability, in ways that thoroughly resist the compensatory logic of modernist form. Skeptical of the quest for an ontological sense of order underlying our phenomenally disparate and accelerating sense of time, postmodernism ironizes the belief in "aesthetic redress," as Jesse Matz notes in this volume, for "[a]lmost by definition, postmodernism rejects the salvific aspirations often essential to the modernist arts." Such is the "lesson," as Linda Hutcheon once described it, "of postmodernism's complex relation to modernism: its retention of modernism's initial oppositional impulses, both ideological and aesthetic, and its equally strong rejection of its founding notion of formalist autonomy."[11]

And yet, nowadays in literary and cultural criticism the postmodern no longer gets an epistemic free ride. Of all the iconic thinkers of late-twentieth-century society and the postmodern "condition," Fredric Jameson offered a decade ago significant qualifications to the global paradigm he did so much to engineer as a cultural barometer of time. For Jameson, "what is often called postmodernism or postmodernity will simply document yet a further internal break and the production of yet another, even later, still essentially modernist moment."[12] Upsetting an evolutionary account of the postwar era as a steady departure from and self-conscious scrutiny of modernist culture, Jameson homes in on the phenomenon of "late modernism" as a "specifically historical period concept."[13] To Jameson, this is a preferable term for grasping the ideological dimensions of modernity, one that also allows us to spot the conceptual contradiction of postmodernity, "unable" as it is "to divest itself of the supreme value of innovation."[14] For "well beyond the life span of modernism itself," notes Jameson, "which in the late 1950s and 1960s seemed to touch a kind of limit and to have exhausted all available and conceivable novelties"—in essence "something of a caricature of capitalism itself, as the ultimate limit of a saturated world market becomes thinkable"—postmodernism reveals that, "despite its systematic and thoroughgoing rejection of all the features it could identify with high modernism," it nevertheless "seems utterly unable to divest itself of this final requirement of originality" that lay at the heart of the modernist enterprise.[15]

When it comes to the postwar arts, though, the problem is not only postmodernism's inadvertent affiliation with an era defined by modernist aspirations, but also its association with a limited set of cultural centers. Without due interrogation, postmodern time becomes synonymous with Western time; yet the danger with simply exporting the postmodern, as a condition either for cultural change or for artistic potentiality since 1950, is that it turns into a homogenizing—thereby *a*temporal—ascription. As Brian McHale observes, reflecting on his own foundational work on postmodern literary history, the "unexamined assumption that postmodernist fiction was really one and the same thing everywhere, even when it emerged far from its Euro-American 'homeland,' seems indefensible today."[16] Countering this sort of assumption, Wang Ning has mapped the range and variety of Asian postmodernisms, viewing them not in retrospect—confined to a group of postwar decades—so much as arising in response to immediate times. "In China," writes Ning, "postmodernist literature is characterized by avant-garde experimentation with narrative techniques, the new realist fiction filling in the gap between traditional realism and the postmodern doctrines, the consumerist popular culture and the deconstructive critical practice."[17] As this all-embracing catalogue of features demonstrates, it's unclear as to why Ning ultimately needs *postmodern* as a term for this disparate range of cultural practices, generic reformations, and critical trajectories. Stretching wide here against the background of the supposed "waning of modernist literature,"[18] postmodernism conveniently wraps around most objects irrespective of shape or substance. Providing one has this kind of dehistoricized version of modernism to hand, the postmodern can of course be made relevant once more. To assert that in Japan "postmodernism is mostly practiced by avant-garde writers who challenge outdated modernist doctrines" is to invite the question of whether postmodernism retains literary-historical purchase today—not because it offers an enabling language of valuation and critique, but because it can be touted as inherently progressive and formally liberating in contrast to the perceived drawbacks of modernism's "doctrines."[19]

Is the postmodern, then, flexible enough for our current times, even if its terms of reference and cultural reach can be duly globalized without being generalized? For McHale, there are still "good reasons for skepticism about the transnational scope of postmodernism" when we observe how "it is the pluri-cultural postcolonial experience that appears genuinely global, not postmodernism, which settles into a narrower, more provincial niche."[20] That we now "find ourselves," in McHale's words, "in a genuinely *dialogic* moment"[21] suggests that the postmodern—with its taste for ludic indeterminacy and destabilizing forms of representational self-examination—may not be conceptually or affectively in sync with the experience of our times, not least when writers emerging now as "post-postmodernists tend to view unstable reality suspiciously."[22] Perhaps postmodernism's apparent "evaporation," to recall Smith's phrase, marks a terminological watershed, clearing the way for testing modernism's pliability as a keyword once more—and without necessarily dispensing with the socioeconomic understanding of postmodernity as a collective situation. Indeed, there's nothing necessarily contradictory about discovering across cultures alternative modernisms flourishing in response to the lived effects of advanced capitalism that, for materialist theorists like Jameson and David Harvey, have always been at the heart of postmodernity as a global condition. Critical thought about these adjusted and rejuvenated modernisms has taken many forms; new formulations serve various methodological and disciplinary ends. So-called *altermodern* reconfigurations buy into the critical ramifications of reviving modernism's utility as a dual-pronged concept—epochal yet aesthetically mobile, periodizable yet viably transhistorical—for addressing the temporality of artistic practice today.

Altermodern or Other Modern?

Alternative conceptions of modernism have been deployed since the turn of the century not so much to recuperate marginalized works as modern*ist* as to describe what artistic and cultural advancement might look like in light of reconfigured, pluralized models of modernity. As Dilip Parameshwar Gaonkar observes, "the very idea of alternative modernities has its origin in the persistent and sometimes violent questioning of the present precisely because the present announces itself as the modern at every national and cultural site today."[23] To question what the present actually means at the level of social and intimate experience, so diversely and unevenly across cultures, is also to pause before we embrace certain *alter*modernist formations as paradigmatic expressions of alternative modernities. Among the most utopian new frames for thinking about art in an age of mass communication and global migration has come from Nicolas Bourriaud. Curating the fourth Tate Triennial at Tate Britain in 2009, Bourriaud proposed *altermodernism* to capture "a constellation of ideas linked by the emerging and ultimately irresistible will to create a form of modernism for the twenty-first century."[24] Bourriaud reminds us that "today, temporalities intersect and weave a complex network stripped of a centre."[25] Lest we be tempted to sense something distinctly "postmodern" about this coreless temporal web, Bourriaud claims that "[n]umerous contemporary artistic practices indicate . . . that we are on the verge of a leap, out of the postmodern period and the (essentialist) multicultural model from which it is indivisible, a leap that would give rise to a synthesis between modernism and postcolonialism."[26] Bourriaud describes this synthesis as *alter*modern, because it demands a "vision of human history as constituted of multiple temporalities, disdaining the nostalgia for the avant-garde and indeed for any era—a positive vision of chaos and complexity."[27] Bourriaud tackles thorny distinctions, distinguishing his paradigm from any post/modern affinities, for altermodern "is neither a petrified kind of time advancing in loops (postmodernism) nor a linear vision of history (modernism), but a positive experience of disorientation through an art-form exploring all dimensions of the present, tracing lines in all directions of time and space."[28]

Critics have noted, however, the amorphousness of altermodernism as a model for identifying (let alone grouping) aesthetic practices, as well as the extent to which the very temper and characteristics of *altermodernity* offer more a recapitulation of, than an alternative to, the cultural past. For altermodernism may in fact enact a tactical recolonization of the artistic antecedents it strives to distance itself from. As Okwui Enwezor has noted, "what the altermodern proposes is a rephrasing of prior arguments," for "have not the practices of art always been predicated on trajectories and detours, on dynamic flows and modes of production and dissemination?"[29] Similarly, Terry Smith observes that it "is not entirely clear" whether or not "the spirit identified" in altermodernism "is a new kind of Modernism, a reversion to the Modernism first defined by Charles Baudelaire in Paris in the 1850s, or indeed whether it is Modernist at all."[30] Indeed, according to the occasion, Bourriaud ultimately displays a kind of love-hate attitude toward the very concept of *the modern* as a pliable, transhistorical epithet. Depending on the modern in one instance, dismissing it in the next, Bourriaud seems torn in his conceptual allegiances. Nonetheless, he reassures us that in our postmillennial era, "it is possible to reclaim the concept of modernity without for an instant feeling that it's a throwback."[31] At the same time, however, he declares—in a startling generalization—that modernism is "[a] thing of the past, with its outdated humanism and universalism, the colonial machinery of the West."[32] As a keyword, in turn, modern*ism* becomes here a victim of a process of reassessment for which it is partly desired and partly renounced, freighted either way with philosophical and world-historical harms against which a monolithic notion of modernity is then derogated.

Elsewhere, in literary studies, the discussion of modernism's salience for approaching the formal potential and political efficacy of contemporary writing has been more nuanced and methodologically robust. Arguments about modernism's geographical expansion have also disrupted linear accounts of cultural evolution, coinciding with a broader set of contentions about the periodicity of aesthetic forms that play devil's advocate with the legitimacy of periodization itself. Though displaying different sympathies toward the future of periodization, both Eric Hayot and Katie Trumpener have highlighted how the inherent assumptions behind period models have intellectual and institutional consequences, given that periods impose temporal organizations upon some of the most fundamental structures that determine how we view the scope and specialization of humanities research.[33] In Hayot's view, periodization's pandemic is a product of our own making, because the academy's apparent inability to see beyond period divisions amounts to a "collective failure of imagination and will on the part of the literary profession."[34] Equally, for Trumpener, "most literary history looks surprisingly tame in its approach to time."[35] Periods can be elongated, as they often are, or redescribed according to new politico-aesthetic peaks and troughs. But even with the help of such "concepts," writes Trumpener, like "the long century and the annus mirabilis," we tend to end up with "a prolonged, largely static or coherent sense of period alternating with packed, dense, glittering moments." Moreover, to accept that a century can double-up as an artistic period around which academic careers are cultivated and maintained is to overlook how the very "idea of the long century depends on the century as an arbitrary, abstract counting unit—or, in some cases, as a succession of royal reigns—expanded only slightly by a gentle pushing back of boundaries."[36] The lesson here? Periodization is always first and foremost *procedural*: a process of historical redescription with provisional rather than definitive outcomes and applications.

Given this, whither modernism as a concept for our times? If it still performs the work of periodization, its formal malleability and temporal transferability also reminds us that periodizing categories remain both denotative and connotative in relation to historical time; at once pinpointing and gesturing, they frame specific decades of cultural advancement while also remaining curiously unrestricted to particular epochal traits. Should we, then, embrace the intersection of *modernism* with *the contemporary*, or carefully police their differentiation for critical discourse and creative practice alike? And if a keyword fulfils, as modernism certainly does, that double duty as a weakly periodized moment and a strongly evaluative optic, does it remain hermeneutically profitable when it becomes so temporally transportable?

One answer would be to regard modernism not simply as an inheritance but as an aesthetic horizon or promise that contemporary art has yet to reach or fulfil. I'm thinking less here of Habermas's sense of modernity's "unfinished project," than of what modernism can still do as a reflexive and self-reforming *poeisis* that arises in different cultures and in response to different historical pressures. John Berger put it this way, when writing in 1969 on cubism: "it is part of history. But a curiously unfinished part," a localizable occasion for innovation that was nonetheless proleptic, promissory, "defining desires which are still unmet."[37] More recently Neil Lazarus has drawn attention to "the *ongoing* critical dimension of modernist literary practice" for postcolonial writing that "resists the accommodationism of what has been canonised as modernism." Distinguishing these contemporary modernisms, as we might find them in J. M. Coetzee, Kazuo Ishiguro, or Gabriel García Márquez, from those vaunted experimenters of the 1920s and 1930s, Lazarus argues that postcolonial modernists' "project" consists of "disconsolation" in a mode that "protests and criticises."[38] Using this model of continuance, one could chart the way modernism—as an ethos, not simply a bygone epoch—bequeaths to contemporary culture formal possibilities without eliding the particularity of artistic strategies separated across time, and without regarding innovative art now merely as a derivation of, or passive homage to, modernism's historical monumentality.

Extensions to modernism's temporality have thus been matched by the cultural enlargement and diversifications of its territory. This move carries its own risky idealisms, of course. Even while endorsing a model of "global polycentrism" that can recognize how "every modernism is derivative of cultural forms it adapts from elsewhere," Susan Stanford Friedman has nonetheless acknowledged that "[w]ith its emphasis on fluidity, multidirectionality, and reciprocal exchange, this approach can slide into a utopian discourse of happy hybridity, forgetting the role of power asymmetries constituted through empires, imperial hegemonies, and local stratifications."[39] Friedman's cautionary note is foremost directed at the recent transnational redrawing of modernism's cartographic boundaries. But it is equally applicable to the job of extending the temporal borders of modern art so as to disrupt the Eurochronology that binds modernist aesthetics to "forms of historical time," in Hayot's words, "privileged by modernity at large."[40] Gaonker reinforces this concern when reminding us that we "can provincialize Western modernity only by thinking through and against its self-understandings which are frequently cast in universalist idioms."[41] Moreover, it's the very notion of time itself "as that which precedes but anticipates 'modernization'" that, as Vasant Kaiwar and Sucheta Mazumdar point out, has served not only to "frame critical facets of modernity" but also to posit "different and quite incompatible totalizing schemes," given modernity's own "uneven geography" of development in global capitalism.[42] Such are the challenges we face in formulating altermodern ways to "think past the basic structures that keep European patterns of development at the center of history."[43] This is why it becomes crucial to formulate other models of time for registering patterns of artistic activity that can reflexively acknowledge the conceptual and institutional values we attribute to those patterns when tracing significant cultural change or unprecedented aesthetic emergence. By the same token, just as historical simultaneity doesn't necessarily equal congruence between aesthetic practices across cultures, apparent formal affinities between modern and contemporary art don't automatically signal genuine points of "reciprocal exchange" or developmental advancement across time.[44]

The "time dimension" of our research, as Susan Gillman calls it, therefore remains integral to how we articulate what we professionally and intellectually do. As such, the "literary-historical unit of the century, along with its various subdivisions into decades and turns, is hard to escape."[45] This is not an excuse to leave the conventions of periodization unexamined or assume that it is one of those habituated aspects of intellectual behavior that are more or less insuperable. Rather, precisely because of its continued and not easily dispensable role in disciplinary self-fashioning and self-recognition, periodization needs at least to be entertained, especially at a time when modernism has become revived as a concept untethered to historically bracketed moments of radical artistic reorientation. A keyword whose value is now correlated with its very resistance to periodicity, modernism has acquired the capacity (as we heard Rabaté declare earlier) to suffuse critical discourse on contemporary culture. Modernism conceptualized as formal ambition has surpassed—or at least appears no longer accountable to—modernism conceptualized as past event.

That periodizing terms can still be equated with temporal coordinates for identifying certain cultural revolutions reminds us (as Jonathan Lethem might have warned) how prone one can be to taking modernism for granted. There are two slants to the act of taking something for granted, of course: a concept can become overfamiliar and fail to command the attention it deserves; but also a concept can simply be unconsidered, taken at face value to be valid, relevant, and true to our times. Overfamiliarity and unquestioned pertinence—combined causes that leave modernism susceptible to terminological malady. Poignantly, Jameson accepted that "it might be better to admit that the notions that cluster around the word 'modern' are as unavoidable as they are unacceptable."[46] Yet the habit of taking the modern for granted may also be as voluntary as it is inescapable. That is to say, it's consoling to characterize an era with a vocabulary that seemingly requires no defense.

Realignments, resuscitations, and enlargements of modernism suggest that there might be a certain comfort to be had in amending rather than altering "the modern." Perpetuating modernism offers interpretive solace as it distracts from the far more unpredictable task of forging other glossaries for artistic contemporaneity. While we're on the lookout for aesthetically singular preoccupations that escape the explanatory convenience of modal comparison, emit no visible signs of affinity, and disrupt arcs of transhistorical influence, is the importation of variegated moderns into the present the next-best substitute on offer? Modernism's famous "disconsolation," to recall Lazarus's phrase, ironically turns out to be a form of succor for critical discourse, as myriad alternatives—*re-*, *digi-*, *cos-*, and *alter-* itself—only redouble our compulsion to reassess how modernism can endure as undoubtedly different and yet in other ways unadulterated. Thus, one closing paradox: the comforting familiarity of modernism's hallmark structural and epistemological difficulty is precisely what allows us to distinguish formal novelty now, as one can find critical reassurance in the fact that the identifiable and institutionally accredited radicality of modernism thrives when serving as a litmus test for artistic experimentation in new times and places. Granted, there's nothing wrong with embracing this paradox, especially if it brings us closer to how art today affronts temporal or generic schemes of categorization and challenges jaded politico-aesthetic systems of explanation that work seamlessly in theory but inelegantly in practice.

In both degree and kind more capacious and mobile than ever, then, modernism is an unfinished argument as much as a principle of organization. Its sheer variety compensates for its immanent authority and omnipresence across discussions of twentieth- and twenty-first-century culture. Such compensations invite us to accept rather than resent modernism's prevalence and to work actively to prevent its stagnation as a basis for characterizing important artistic intentions and effects. The more elements and justifications that modernism aggregates for its contemporaneity, though, the more likely we are to concede that its abstractness as a keyword is an indispensable part of its epithetic function, even if that abstraction eventually blunts the force of how and what it describes. This concession can be yet another consolation, of course, mitigating for now the problem of whether we ought to be hypothesizing other ways of addressing significant dispositions in contemporary cultural production. Although our vocabularies today are arguably more heterogeneous and sophisticated than ever, they're no less susceptible to the lure of terms from earlier times, especially if a term seems charismatic enough to be institutionally reconsecrated as urgent—notably when conversations about the contemporary are also moderated by the added methodological urgency of successive "turns." So if the collective disciplinary momentum is on the path to seeking out a truly alternative lexicon of historical organization, value judgment, and aesthetic taxonomy (be that "altermodern" or some other modern we have yet to conceive), it will do so knowing full well that the solace afforded by durable, mutually agreed keywords of inquiry and professional solidarity is hard to refuse.

NOTES

1 Jean-Michel Rabaté, "Introduction," in *A Handbook of Modernism Studies*, ed. Rabaté (Oxford: Wiley-Blackwell, 2013), 11.
2 Amy J. Elias, "The Dialogical Avant-Garde: Relational Aesthetics and Time Ecologies in *Only Revolutions* and *TOC*," *Contemporary Literature* 53, no. 4 (2012): 738.
3 Jonathan Letham, "The Ecstasy of Influence: A Plagiarism," reprinted in *The Ecstasy of Influence: Nonfictions, etc.* (London: Vintage, 2013), 101.
4 Jürgen Habermas, "Modernity: An Unfinished Project," repr. in *Habermas and the Unfinished Project of Modernity*, ed. Maurizio Passerin d'Entrèves and Seyla Benhabib (Cambridge, UK: Polity, 1996), 40.
5 Susan Stanford Friedman, "Periodizing Modernism: Postcolonial Modernities and the Space/Time Borders of Modernist Studies," *Modernism/Modernity* 13, no. 3 (2006): 439.
6 Terry Smith, *Contemporary Art: World Currents* (London: Lawrence King, 2011), 11.

7   Terry Smith, *What Is Contemporary Art?* (Chicago: University of Chicago Press, 2009), 255.
8   Smith, *What Is Contemporary Art?*, 255.
9   See, for example, Pelagia Goulimari, ed., *Postmodernism: What Moment?* (Manchester, UK: Manchester University Press, 2011); Ihab Hassan's groundbreaking *The Postmodern Turn: Essays in Postmodern Theory and Culture* (Columbus: Ohio State University Press, 1987); and Linda Hutcheon, *A Poetics of Postmodernism: History, Theory, Fiction* (London: Routledge, 1989), especially chapter 3: "Limiting the Postmodern: The Paradoxical Aftermath of Modernism."
10 Consider in this context the revisionary orientation of essays in the special issue of *Twentieth-Century Literature* edited by Jason Gladstone and Daniel Worden, "Postmodernism, Then," 57, nos. 3–4 (Fall 2011), and Brian McHale's reassessment of the implicit Euro-American concentration of earlier work on postmodernism in "Afterword: Reconstructing Postmodernism," *Narrative* 21, no. 3 (2013): 357–64.
11 Linda Hutcheon, *The Politics of Postmodernism* (London: Routledge, 1989), 26.
12 Fredric Jameson, *A Singular Modernity* (London: Verso, 2002), 151.
13 Jameson, *A Singular Modernity*, 13.
14 Jameson, *A Singular Modernity*, 5.
15 Jameson, *A Singular Modernity*, 151–52.
16 McHale, "Afterword," 360.
17 Wang Ning, "Introduction: Historicizing Postmodernist Fiction," *Narrative* 21, no. 3 (2013): 267.
18 Ning, "Introduction," 267.
19 Ning, "Introduction," 267.
20 McHale, "Afterword," 359, 362.
21 McHale, "Afterword," 363.
22 Robert L. McLaughlin, "After the Revolution: U.S. Postmodernism in the Twenty-First Century," *Narrative* 21, no. 3 (2013): 289.
23 Dilip Parameshwar Gaonker, "On Alternative Modernities," in *Alternative Modernities*, ed. Gaonker (Durham: Duke University Press, 2001), 14.
24 Nicolas Bourriaud, "Altermodern," in *Altermodern: Tate Triennial* (London: Tate Publishing, 2009), 12.
25 Bourriaud, "Altermodern," 12.
26 Bourriaud, "Altermodern," 13.
27 Bourriaud, "Altermodern," 13.
28 Bourriaud, "Altermodern," 13.
29 Okwui Enwezor, "Modernity and Postcolonial Ambivalence," in *Altermodern: Tate Triennial* (London: Tate Publishing, 2009), 31, 32.
30 Terry Smith, *Contemporary Art: World Currents*, 322.
31 Nicolas Bourriaud, *The Radicant* (New York: Lukas & Sternberg, 2009), 15.
32 Bourriaud, *The Radicant*, 12.

33 See Eric Hayot, *On Literary Worlds* (New York: Oxford University Press, 2012), and Katie Trumpener, "In the Grid: Period and Experience," *PMLA* 127, no. 2 (2012): 349–55.
34 Hayot, *On Literary Worlds*, 149.
35 Trumpener, "In the Grid," 354.
36 Trumpener, "In the Grid," 354.
37 John Berger, *Selected Essays and Articles: The Look of Things*, ed. Nikos Stangos (London: Penguin, 1971), 135, 162.
38 Neil Lazarus, *The Postcolonial Unconscious* (Cambridge, UK: Cambridge University Press, 2011), 30, 31.
39 Susan Stanford Friedman, "World Modernisms, World Literature, and Comparativity," in *The Oxford Handbook of Global Modernisms*, eds. Mark Wollaeger with Matt Eatough (New York: Oxford University Press, 2012), 513, 515.
40 Hayot, *On Literary Worlds*, 6.
41 Gaonker, "On Alternative Modernities," 15.
42 Vasant Kaiwar and Sucheta Mazumdar, "Introduction," in *Antinomies of Modernity: Essays on Race, Orient, Nation*, ed. Kaiwar and Mazumdar (Durham, NC: Duke University Press, 2003), 2.
43 Hayot, *On Literary Worlds*, 6.
44 The question of whether modernist practices can be transcribed not only across conventional temporal partitions but also across ethnic and cultural thresholds has been addressed by revisionist accounts of how postcolonial writers rework modernist protocols; see Urmila Seshagiri, "Modernist Ashes, Postcolonial Phoenix: Jean Rhys and the Evolution of the English Novel in the Twentieth Century," *Modernism/Modernity* 13, no. 3 (2006): 487–505.
45 Susan Gillman, "Oceans of *Longues Durées*," *PMLA* 127, no. 2 (2012): 330.
46 Jameson, *A Singular Modernity*, 13. Habermas, along similar lines, states that "[m]odernism represents a great seductive force, promoting the dominance of the principle of unrestrained self-realization" ("Modernity," 42).

4

Obsolescence / Innovation

JOEL BURGES

Although obsolescence is central to capitalist economies from at least the 1940s to the present, it is not peculiar to the post-1945 period. The same is true of its twin, innovation, the precursor of which is invention.[1] Innovation and obsolescence, at their most general, refer to twinned processes of technological change that play out across most periods of human history. We typically think of these twins as a smoothly forward-moving sequence of technical advances that have improved functionality and efficiency (innovation), displacing and depreciating the use-value of that which came before with the improvements that they create (obsolescence). But in truth, such advances are rarely a "rout [in which] 'old-fashioned, inferior' technologies [are] swiftly replaced by 'new and better' ones in a linear process." They are instead the outcome of "protracted competitions, often lasting decades, centuries, or longer" in which a variety of apparatuses may—or may not—"die out."[2]

Nonetheless, the twentieth and twenty-first centuries have seen these processes unfold at accelerated temporal and historical rates and increasingly far-reaching spatial and social expanses.[3] For example, with regular software updates and new generations of smart phones a regular part of daily life for a third of the planet at this point, indeed, a part of life the rhythms of which form a technical echo of those diurnal and seasonal tempos that once determined quotidian existence centuries ago, digital computing exemplifies how innovation and obsolescence have become ever more central to the structural dynamics of late capitalism. The post-WWII period is characterized by redundant overproduction in which innovation and obsolescence assume key economic as well as technological roles in the advanced capitalist economies.[4]

But what do these twinned techno-economic processes have to do with time? With increasing intensity during the post-1945 "long present," innovation and obsolescence spark the emergence of what I call a "horizon of temporal sensation" shaped by the commodified coming and going of technologies and products, along with the practices and qualities with which they are associated—from smart phones and software updates to the need to find a pay phone and the clacking of a typewriter.[5] This is a horizon in which a half-felt, half-thought experience of historical time unfolds by way of our ongoing sensitivities to what is today known in marketing circles as the "product life cycle." In the novel *Pattern Recognition*, for example, William Gibson's character Cayce Pollard paradigmatically articulates this horizon in her acutely temporal and historical "allergy" to the currency of products and logos as they cycle through their "lives."[6]

The "product life cycle" is a concept that refers to how "the growth of sales of a product follows a systematic path from initial innovation through a series of stages: early development, growth, maturity, and obsolescence," the last being the "inevitable" point at which demand for the product "slackens."[7] In what follows, I describe our sensitivities to this cycle of innovation and obsolescence as a rhythmic experience of historical time, nuancing prior Marxist models of that experience by conceiving of its rhythm as an affective cadence and sensory tempo inseparable from technologically driven consumption. That rhythm is defined by a counterpoint of temporal sensations by which (1) we anticipate the new "generation" of a product always coming on the horizon of consumption, a reified object of desire that pulses like a glimmering mirage sent from the future only just perceptible in the present, at the same time that (2) we know that whatever we have bought or will buy is already obsolete, a contemptible relic of the recent past that will soon embarrass us with its datedness.[8] What these sensations of anticipation-desire and contempt-embarrassment embody is a temporal horizon in which the experience of historical time is inseparable from innovation and obsolescence, which rhythmically alternate between future and past in a contrapuntal cycle that repeatedly and recursively turns these times into one another within the current moment.

Focusing us primarily on the sphere of consumption, however, this rhythm alone is insufficient to account for what makes innovation and obsolescence produce a horizon of authentically historical experience. To account for that, we must grasp how this rhythm is tied to a deeper one in the realm of production: the rhythm by which human beings are rendered obsolete due to technological innovations in industrial productivity intended to sustain profits. Put epigraphically, every time an iPod appears, a human worker disappears.[9] The contradiction that arises out of the relationship between these two rhythms clarifies how innovation and obsolescence in the consumer sphere mediate a horizon of temporal sensation in which an actively experiential relationship to history is genuinely at work, a horizon that is even indicative of a politics of obsolescence immanent to the rhythms of the present. The "crisis in historicity" long ago diagnosed as a condition of our present may, then, be the crisis of falling *into* history—of running up against the very limits of capital, and wanting to act against them, but not always clearly seeing the path forward to transformational protest.[10]

Obsolescence

In "What Is Good for Goldman Sachs Is Good for America: The Origins of the Current Crisis," Robert Brenner employs a series of dead metaphors in accounting for the economic crisis that began in 2007 and 2008. He speaks of the post-1970s economy's "incapacity to drive itself forward *on its own steam*." Despite a mid-1990s period in which "the economy seemed to be *functioning on all cylinders, harvesting the fruits* of the significant increase in the rate of profit during the previous decade," for Brenner the truth of post-1970s economic history erupted with a "housing market that *ran out of steam* in 2006."[11] While "harvesting the fruits" was the idiomatic product of the seasonal cycles of ancient agricultural labor, images of steam- and cylinder-power were the colloquial fruit of the Industrial Revolution. What makes these dead metaphors is how their referential relationship to an originating semantic context has lost its vitality in current usages. Here that loss turns on references we no longer fully hear to both a long obsolete mode of production and an obsolescent phase in the capitalist mode itself. What they stress in the context of Brenner's important essay, even if this is not his point, is how fundamental one force of production has been to the long downturn usually dated to 1973: obsolescence. Although they evoke much earlier times, what Brenner's dead metaphors crystallize is the historic role that obsolescence has assumed in the structural dynamics of advanced capitalism.[12]

This role predates the post-1973 period. It begins in the 1920s and then solidifies in the 1940s and 1950s such that, by 1960, "obsolescence" had become an idiomatic part of the vocabulary of the long present.[13] Crucial to this process was the rise of planned obsolescence as a business practice and industrial philosophy that rationalized obsolescence as a force in the production and consumption of commodities, and thus in the reproduction of capitalist life itself.[14] In the 1920s, for example, obsolescence was conceptualized as an investment instead of a risk. A 1928 article entitled "Investing in Obsolescence" argued that obsolescence should not be understood as "undesirable" or as a "loss," but rather as "purely an incident of progress and advance," as a moment of "improvement and evolution." As the writer explained, "[T]he successful operation of the old equipment, before obsolescence, was the effect of an investment in the progress of a mechanical art of which any particular machine is only a temporary expression—that's what we mean by 'Investing in Obsolescence.'"[15]

The notion of investing in obsolescence, of making the most out of the *temporariness* of the mechanical arts, comes into its own in the 1940s and 1950s, when planned obsolescence matured within capitalist economies, especially that of the United States. But where the author of "Investing in Obsolescence" was referring to fixed capital such as the machines on which goods are made, the author of "A Theory of Purposeful Obsolescence" was in 1947 focused on the superannuation of the goods themselves as a growing feature of the economy. "The key to the concept," economic scholar Paul Gregory wrote in that essay, is "deliberateness."[16] According to Gregory, this deliberateness was visible across an array of mid-twentieth-century industries and goods, from razor blades and flashlight bulbs to food, drugs, and books. Purposeful obsolescence was variously "physical" and "psychological."[17] The former occurs when "manufacturers produce goods with a shorter physical life than the industry is capable of producing under existing conditions," intensifying the temporariness of the use-value of a product materially, while the latter occurs when "manufacturers or sellers induce the public to replace goods which retain substantial physical usefulness," intensifying the temporariness of the use-value of a product psychologically.[18] "The line is hard to draw," observed Gregory, but the Model A Ford was a great improvement over the Model T, a television set is a different product from an ordinary radio, and the owners of these obsolete objects gain by buying the new ones. But with purposeful obsolescence, the "old" article loses value (at least psychologically) and the owner must then buy—or is induced to buy—the "new" or the "latest" model, which is seldom better and sometimes worse.[19]

To Gregory, falling into obsolescence was something like going out of fashion. And he was right. Today the adjective *obsolete* is nearly synonymous with the accusation *outmoded*, the former no longer as clearly an indication that a new use-value has been invented (as in Gregory's contrast between a television and a radio) or an old use-value has been improved (as in his example of the Model A).

In 1947, Gregory was observing a phenomenon that would catch fire in the 1950s and 1960s, though a backlash would set in during the later decade in part due to the 1960 publication of Vance Packard's bestselling *The Waste Makers*, a book that helped to make "planned obsolescence" a household word.[20] Gregory perceived the historicity and temporality that planned obsolescence epitomizes; the intensifying temporariness of goods that it encourages makes this business philosophy and industrial practice the mid-twentieth-century codification of Baudelaire's mid-nineteenth-century description of modernity as a time of contingency and ephemerality.[21] More specifically, however, there is a regressive impulse at work in planned obsolescence, because some products manufactured with its "theory" in mind are intentionally out of sync with state-of-the-art manufacturing processes and standards. In many cases, the product is past date before it is made, and will soon be passé regardless. As industrial designer George Nelson put it in 1967, "To a designer, anything that is, is obsolete."[22]

Today, the history of digital computing provides vivid and obvious examples of obsolescence and innovation. In the 1950s, however, it was the automobile industry that most fully embraced the temporary temporality and regressive historicity of planned obsolescence. This embrace began in the 1920s, when the industry went into a slump because most consumers with the buying power to own a car already did, driving automobile makers toward planned obsolescence.[23] In response, car manufacturers imported the shorter cycles of design associated with fashion,[24] leading to a rhythm that is still familiar to us: the three-year styling change that "would eventually define the lifespan of all so-called durable goods in America. Between these major styling changes, annual face lifts rearranged minor features, such as chrome work. But even these minor moves created the illusion of progress and hastened the appearance of datedness that obsolescence required."[25] By the 1950s, the annual style change had become the standardized rhythm for car production and consumption, converting whatever had been current five years before to datedness. Between 1948 and 1959, for instance, General Motors periodically superannuated prior models of the Cadillac by steadily exaggerating its tail fin and adding an "an average of forty-four pounds of surplus chrome," mostly "luxury" gadgets.[26]

Between our cars and our computers, then, it becomes clear how in the postwar period, obsolescence assumed the structural role it has continued to play in post-1973 capitalism. We also begin to see how obsolescence increasingly gave rise to a techno-economic rhythm that produces not only goods, but also an experience of historical time that involves the objective production of a past that the historical consumer subjectively perceives. The narrator of Nicholson Baker's 1985 novel, *The Mezzanine*, captures this brilliantly:

> But other things, like gas pumps, ice cube trays, transit buses, or milk containers, have undergone disorienting changes, and the only way we can understand the proportion and range and effect of those changes, which constitute the often undocumented daily texture of our lives (a rough, gravelly texture, like the shoulder of a road, which normally passes too fast for microscopy), is to sample early images of the objects in whatever form they take in kid-memory—and once you invoke those kid-memories, you have to live with their constant tendency to screw up your fragmentary historiography with violas of lost emotion. I drink milk very rarely now. . . . But I continue to admire the milk carton, and I believe that the change from milk delivered to the door in bottles to milk bought at the supermarket in cardboard containers with peaked roofs was a significant change for people roughly my age—younger and you would have allied yourself completely with the novelty as your starting point and felt no loss; older and you would have already exhausted your faculties of regret on earlier minor transitions and shrugged at this change.[27]

The obsolescence of delivered milk takes on a generational temporality for this cerebral and nostalgic narrator, and the notably "disorienting changes" he describes (which he will counterintuitively but accurately call a "tradition"[28]) are akin to those that the annual automotive style change rhythmically enacts, especially as they accumulate, say, in the baroque development of the Cadillac into a commercial work of art in the 1950s.[29] Whether they occur annually or generationally, such rhythmic changes in the make and model of commodities continually build up the past within the passé, engendering moments of temporality for the historical consumer.

Thus for all the ease with which we view such changes as *lacking* temporality and historicity, the coming and going of goods provides a rhythm by which we perceive historical difference, what Baker's narrator dubs a "fragmented historiography." In that "historiography," obsolete stuff—a previous generation's milk container, the 1959 tail fin, the most recent operating system for one's Macintosh—appears as "an accumulation of historical time," as a contemporary horizon of temporal sensation made up of so many "valorizable chunks of time and fragmented leftovers" that has, I shall assert below, a politics of historical time embedded within it.[30]

Innovation

But first, innovation. Theodor W. Adorno was always careful to distinguish the truly modern, the authentically new, and the historically advanced—all those future-oriented phrases we associate with innovation—from that into which they devolve in late capitalism: an up-to-date Zeitgeist. In a postwar work, *Aesthetic Theory*, Adorno writes,

> Art is modern when, by its mode of experience and as the expression of the crisis of experience, it absorbs what industrialization has developed under the given relations of production. This involves a negative canon, a set of prohibitions against what the modern has disavowed in experience and technique; and such determinate negation is virtually the canon of what is to be done. That this modernity is more than a vague Zeitgeist or being cleverly up to date depends on the liberation of the forces of production. Modern art is equally determined socially by the conflict with

the conditions of production and inner-aesthetically by the exclusion of exhausted and obsolete procedures.[31]

Somewhat earlier in *Aesthetic Theory*, he states his point in terms of innovation: "If a possibility for innovation is exhausted, if innovation is mechanically pursued in a direction that has already been tried, the direction of innovation must be changed and sought in another dimension."[32] The reminder that Adorno issues to us about innovation is that it was thoroughly imbricated with the historical avant-gardes of the early twentieth century. But as is suggested by the distinction that Adorno was still drawing in the 1960s between the modernity of a modern art that truly breaks with the past, and the modernity of an up-to-date Zeitgeist that only simulates such a break, aesthetic innovation in the twentieth century continually ran the risk either of seeking out a falsely extrinsic relationship to, or of engendering an unduly affirmative relationship with, innovations such as the progressively more elaborate tail fin of postwar automobiles.[33] After Adorno's death, in fact, there may be no more innovation of the kind he so persistently theorized—and desired. The history of the later twentieth century is one in which innovation loses its negativity as it mutates from the radical slogan of the avant-garde to the entrepreneurial mythology of corporate culture, its rupturing futurity superseded by that culture's belief that "innovation means radically changing the system but also maintains that everyone should work for change *within* the system."[34]

Between these two historical moments is the one I have been describing, in which the postwar rise of planned obsolescence instantiated an experience of historical time that remains with us today. This is an experience in which the past is rhythmically produced by way of the deliberate dating of goods both materially and perceptually. But what truly makes such past production integral to a productive rhythm is how obsolescence and innovation constitute a cycle in which whatever is now past was once future, not so much blurring the distinction between them as rendering them a counterpoint of times that recursively turn into one another within a horizon of temporal sensation. In the mid-twentieth-century discourse around planned obsolescence, innovation was linked to the sensation of anticipation-desire in the consumer sphere, making the projecting and producing of the future crucial for industry. Castenholz observed, for instance, that "obsolescence is a future occurrence, which represents an advantage gained in the future by future operations."[35] And in 1960, Brooks Stevens, a major defender of planned obsolescence, wrote: "The designer must anticipate future demand. The average consumer product requires from six to nine months in tooling, and in the automobile field the lead time may be eighteen months. The industrial designer is called upon to be working from a year to two ahead of that particular article which may be enjoying booming acceptance at the moment."[36] What was clearly emerging with the rise of planned obsolescence was the product life cycle as we know it, with obsolescence and innovation not only becoming ever more contrapuntal, but the futurity of the latter constituting an anticipatory impulse to both make and "own something a little newer, a little better, a little sooner than necessary."[37] That contrapuntal cycle thus signals a horizon of temporal sensation that renders the future "past" by way of a more rapidly recurring series of microinnovations ("a *little* sooner"), the rhythms of which have in the long present marked our experience of historical time due to their redundantly overproductive growth since 1945.

Taken together, then, obsolescence and innovation would appear to be dynamic forces in the steady *attenuation* of historical time to a series of microtemporalities that eradicate the difference of past and future from the long present altogether.[38] But they are also dynamic forces in the steady *accumulation* of historical time in that period, however much we may not like the shape that accumulation takes. What there is to dislike about that accumulation is, to draw on Moishe Postone's important work, the efficiency with which capitalism reproduces itself as "a perpetual present, an apparently eternal necessity" by way of its social forms, especially value.[39] This ever-present reproduction occurs regardless of the "ongoing directional movement" and "flow of history" that the capitalist mode of production concretely induces through a state of constant flux, for example, through technological transformations that would seem to propel and compel us to render capitalism obsolete in favor of a different future by simultaneously decreasing the need for the capitalist division of labor and increasing the production of material wealth.[40] But in the end, they don't—or we don't. Postone describes this situation as a "treadmill pattern." The immanent historical dynamic by which the trajectory of production introduces a concrete experience of change into contemporary life that, while pointing to a transformed future beyond capitalism, dialectically entails "the constant translation of historical time into the framework of the present, thereby reinforcing that present" and temporally reconstituting the social formation of capital for good.[41]

There is nonetheless a material sense in which historical time remains untranslated, concretely accumulating in the obsolete stuff that the cycle of innovation and obsolescence produces. Consider Kevin Jerome Everson's independent film *Century* (2012).[42] The "cast" is an old Buick being demolished over the course of a seven-minute long take. Everson, however, has stated that the real cast of *Century* is actually the African American workers whose daily logic of sensation was formed—at the gestural and epidermal levels of "hand and touch"—in the stamping plant, shut down in 2010, in the town where Everson grew up:

> I wanted to make something where it didn't have any people in it but was all about people. And all about black folk. My cousins made those cars. They worked at the Fisher Body General Motors Plant in Mansfield, Ohio, which was a stamping plant, making doors, side panels, hoods. They were forming it. Not putting it together. They're made by automation but there's some kind of hand and touch in it, so to speak. I was thinking sculptural. I wanted to do something where I could transform the object too.[43]

In making a film about unmaking a Buick Century, Everson evokes the spectral gestures, sensory economy, and automated rhythms of the now obsolete laborers who "sculpted" the car in the first place. The seven-minute-long take in this respect quite literally gives us the time to sense and recall, even if we cannot actually see, a superannuating horizon of what anthropologist André Leroi-Gourhan would call "human technicity."[44] By focusing us on the car's demolition in and over a sustained length of uninterrupted time punctuated not by editorial cutting, but by the irregular rhythm of mechanical crushing, *Century* labors to draw us into a diachronic encounter with the wholesale obsolescence of an entire techno-economic way of life in Mansfield, Ohio, one that has rendered commodities and workers alike relics of the past. What *Century* crystallizes is how obsolete stuff can become the material in which accumulations of historical time persist. To reverse Adorno's thought, those accumulations endure "inner-aesthetically" in Everson's film. However, *Century* depends not on the "exclusion," but rather on the *inclusion* "of exhausted and obsolete procedures" as it draws on prior rhythms of industrial productivity and human labor.

In that accumulative aesthetic of historical time, what *Century* further crystallizes is how the rhythm of fragmented historiography that Baker's narrator senses in his encounters with milk containers is structurally bound to a deeper temporality, arguably more closely tied to obsolescence than innovation. As *Century* shows, obsolescence is not only central to a techno-economic rhythm by which products relentlessly succeed one another in the sphere of consumption. It is also a less immediately sensible rhythm by which technologies regularly replace workers with automated processes in, as Karl Marx famously called it, "the hidden abode of production," or render them obsolete by way of the anarchy of uneven technological development in the late-capitalist totality.[45] The latter rhythm pulses within the former, our encounters with the product life cycle, particularly at those times when some good we loved declines into datedness, engendering moments in which we experience the anxiety of our own obsolescence in this deeper and more total sense. This anxiety is, in Adorno's words, "a fear of falling behind the *Zeitgeist*, of being cast on the refuse-heap of discarded subjectivity," not unlike an old Buick slated for demolition.[46] There is no surer realization of this fear than when a laborer loses her job to a machine (a process that has moved from being a Luddite worry to a late-capitalist reality). At that point, she is not only out of work but out of sync as well, unable to keep up with the overproductive rhythm of obsolescence and innovation that generates so much obsolete stuff. In fact, she herself is now an instance of that stuff as it appears, not in the realm of consumption, but rather in the sphere of production. In short, you know your historical time has come when you are both out of sync and out of work, unable to purchase the goods manufactured on the machine that outmoded you—even though you're paradoxically still expected to. This paradox situates you in a contradictory position, one that confronts you with the historical limits of capital; consigned to the techno-economic past, you are nonetheless still expected to buy its future.

In creating a confrontation with the limits of capital by way of this contradiction, however, an actively experiential relationship to history materializes, both anxiously and angrily, in the lived experience and cultural forms of late capitalism. In 1975, for example, as part of a much larger strike, pressmen at the *Washington Post* protested their obsolescence by a computerized press, an act of automation intended, according to the *Post* publisher, to "eliminate archaic union practices."[47] Holding their foreman hostage, they took apart the machine that was taking their jobs, and were eventually legally indicted for doing so. The specter of these latter-day Luddites lives on in mass culture today. One thinks, for example, of the figure of Turbo from the 2012 animated film about video game characters *Wreck-It Ralph*.[48] Aptly named after technological speed itself, Turbo refuses to accept his obsolescence when a more advanced video game hits the market and he loses his fans. Pegged for removal from the arcade, Turbo hacks into and hides out in the game *Sugar Rush* under the guise of the character King Candy, attempting to survive his superannuation into a future from which the film ultimately eradicates him anyway. *Wreck-It Ralph* kills Turbo off, teaching its audience that a politics of obsolescence in which a worker takes action against his techno-economic demise is a grotesquerie to be exorcised from the system. Much as both of these stories seem in the end to confirm the power of capital to reconstitute itself as a perpetual present, they also reveal that obsolete stuff, especially when that stuff includes human beings laboring to stay employed and alive, contains within itself a politics of historical time that actively and authentically engages the limits of the long present as we know it.

NOTES

1. On invention, especially the "duty to invent," see Lewis Mumford, *Technics and Civilization* (Chicago:University of Chicago Press, 1963), 52–55.
2. Michael Brian Schiffer, "The Explanation of Long-Term Technological Change," in *Anthropological Perspectives on Technology*, ed. Schiffer (Albuquerque: University of New Mexico Press, 2001), 216, 217.
3. James W. Cortada, "How New Technologies Spread: Lessons from Computing Technologies," *Technology and Culture* 54, no. 2 (2013): 229.
4. For background to my claim here, see Robert Brenner, *The Economics of Global Turbulence: The Advanced Capitalist Economies from Long Boom to Long Downturn, 1945–2005* (New York: Verso, 2006), 1–10, 27–40.
5. My use of the term "horizon" here is derived from Miriam Hansen's dazzling account of Oskar Negt and Alexander Kluge's discussion of the public sphere as

a "horizon of experience" in her foreword to Negt and Kluge, *Public Sphere and Experience: Toward an Analysis of the Bourgeois and Proletarian Public Sphere*, trans. Peter Labanyi, Jamie Owen Daniel, and Assenka Oksiloff (Minneapolis: University of Minnesota Press, 1993): ix-ili.

6 William Gibson, *Pattern Recognition* (New York: Berkley Books, 2003), 17–18.
7 Peter Dicken, *Global Shift: Reshaping the Global Economic Map of the Twenty-First Century*, 4th ed. (London: Sage Publications, 2003), 104–5.
8 See Joel Burges, "Adorno's Mimeograph: The Uses of Obsolescence in *Minima Moralia*," *New German Critique* 40, no. 1 (2013): 65–91.
9 Clearly this aphoristic sentence is not literally true, but a rhetorical condensation of a set of technological forces that produce unemployment in capitalism. This aphorism is also meant to allude to one of Karl Marx's claims: "Capital itself is the moving contradiction, [in] that it presses to reduce labor time to a minimum, while it posits labor time, on the other side, as the sole measure and source of wealth" (*Grundrisse: Foundations of the Critique of Political Economy (Rough Draft)*, trans. Martin Nicolaus [New York: Penguin Books in association with *New Left Review*, 1973], 706). For more empirical work on this contradiction in the post-1945 period, and how it relates to technological unemployment, see Stanley Aronowitz and William DiFazio, *The Jobless Future*, 2d ed. (Minneapolis: University of Minnesota Press, 2010); Martin Ford, *Rise of the Robots: Technology and the Threat of a Jobless Future* (New York: Basic Books, 2015); Carol Benedict Fry and Michael A. Osborne, "The Future of Employment: How Susceptible Are Jobs to Computerization?" Engineering Sciences Department and the Oxford Martin Program on the Impacts of Future Technology, University of Oxford (September 17, 2013), http://www.oxfordmartin.ox.ac.uk/publications/view/1314; David F. Noble, *Forces of Production: A Social History of Automation* (New York: Oxford University Press, 1984).
10 Fredric Jameson, *Postmodernism, or, The Cultural Logic of Late Capitalism* (Durham, NC: Duke University Press, 1991), 25.
11 Robert Brenner, "What Is Good for Goldman Sachs Is Good for America: The Origins of the Current Crisis," Center for Social Theory and Comparative History, UCLA (April 18, 2009), http://www.sscnet.ucla.edu/issr/cstch/papers/BrennerCrisisTodayOctober2009.pdf): 3, 25, 4, my emphasis.
12 For a powerful account of the long downturn that stresses overproduction, giving context to my claims about obsolescence, see Brenner, *The Economics of Global Turbulence*.
13 For the reader interested in further accounts of the history of planned obsolescence, see these excellent resources: Daniel M. Abramson, "Obsolescence: Notes towards a History," in *Building Systems: Design, Technology, and Society*, ed. Kiel Moe and Ryan E. Smith (New York: Routledge, 2012), 159–70, and Giles Slade, *Made to Break: Technology and Obsolescence in America* (Cambridge, MA: Harvard University Press, 2006).

14 This is what makes obsolescence a productive force. See Raymond Williams, "Productive Forces," in *Marxism and Literature* (New York: Oxford University Press, 1977), 90–94.
15 W. B. Castenholz, "Investing in Obsolescence," *Accounting Review* 3, no. 3 (1928): 269, 270.
16 Paul Gregory, "A Theory of Purposeful Obsolescence," *Southern Economic Journal* 14, no. 1 (1947): 24.
17 Gregory, "A Theory of Purposeful Obsolescence," 24, 25, 33, 24.
18 Gregory, "A Theory of Purposeful Obsolescence," 24.
19 Gregory, "A Theory of Purposeful Obsolescence," 37.
20 Vance Packard, *The Waste Makers* (1960; Brooklyn, NY: Ig Publishing, 2011).
21 See Charles Baudelaire, *The Painter of Modern Life and Other Essays*, trans. and ed. Jonathan Mayne (London: Phaidon Press, 1995).
22 George Nelson, "Obsolescence," *Perspecta* 11 (1967): 173.
23 See Slade, *Made to Break*, 30–36, and Karal Ann Marling, *As Seen on TV: The Visual Culture of Everyday Life in the 1950s* (Cambridge, MA: Harvard University Press, 1994), 136.
24 Marling, *As Seen on TV*, 136.
25 Slade, *Made to Break*, 45.
26 Marling, *As Seen on TV*, 139, 140–141.
27 Nicholson Baker, *The Mezzanine* (New York: Vintage Books, 1988), 41–2.
28 Baker, *The Mezzanine*, 42.
29 In the 1950s, "[t]he car became an armature on which to mount a whole panoply of expressive shapes. In time, the car transcended its prosaic function altogether and became a piece of figurative sculpture, a powerful work of art" (Marling, *As Seen on TV*, 140).
30 I borrow the compelling phrase "accumulation of historical time" from Moishe Postone, *Time, Labor, and Social Domination: A Reinterpretation of Marx's Critical Theory* (New York: Cambridge University Press, 1993), 299. For Negt and Kluge, writes Hansen, a "key problem" in the long present is "the temporal matrix of the horizon of experience," that is, how "individual and collective learning cycles interact under the regime of an industrial-capitalist temporality that divides time into a mere succession of valorizable chunks of time and fragmented leftovers" (*Public Sphere and Experience*, xxxv).
31 Theodor W. Adorno, *Aesthetic Theory*, ed. and trans. Robert Hullot-Kenner (Minneapolis: University of Minnesota Press, 1997), 34.
32 Adorno, *Aesthetic Theory*, 22.
33 Beyond Adorno, see Peter Bürger, *Theory of the Avant-Garde*, trans. Michael Shaw (Minneapolis: University of Minnesota Press, 1984), and Paul Mann, *The Theory-Death of the Avant-Garde* (Bloomington: Indiana University Press, 1991).
34 Alan Liu, *The Laws of Cool: Knowledge Work and the Culture of Information* (Chicago: University of Chicago Press, 2004), 125.
35 Castenholz, "Investing in Obsolescence," 272.

36 "'Planned Obsolescence'—Is It Fair? Yes! Says Brooks Stevens, No! Says Walter Dorwin Teague," *Rotarian* (February 1960): 12.
37 Quoted in Glenn Adamson, ed., *Industrial Strength Design: How Brooks Stevens Shaped Your World* (Cambridge, MA: MIT Press, 2003), 204.
38 Fredric Jameson, "The End of Temporality," *Critical Inquiry* 29, no. 4 (2003): 702–5.
39 Postone, *Time, Labor, and Social Domination*, 299.
40 Postone, *Time, Labor, and Social Domination*, 293, 301–6, 339. This is a complex point that cannot be adequately unpacked here, but as Postone writes, "production in capitalism does not develop in a linear fashion. The dialectical dynamic [of abstract present time and concrete historical time] does, however, give rise to the possibility that production based on historical time can be constituted separately from production based on abstract present time—and that the alienated interaction of past and present, characteristic of capitalism, can be overcome" (*Time, Labor, and Social Domination*, 301).
41 Postone, *Time, Labor, and Social Domination*, 300.
42 *Century*, directed by Kevin Jerome Everson (2012). Film; 0:07 min.
43 Terri Francis, "Interview: Kevin Jerome Everson—Contemporary American Artist and 'Black Filmmaker,'" *Shadow and Act: On Cinema of the African Diaspora*, January 21, 2014, http://blogs.indiewire.com/shadowandact/interview-kevin-jerome-everson-contemporary-american-artist-and-black-filmmaker. The American Film Institute website for its 2013 festival refers to the Buick as the "cast" of *Century* (http://afifest.afi.com/sections/CENTURY). I borrow the phrase "logic of sensation" from Gilles Deleuze, *Francis Bacon: The Logic of Sensation* (Minneapolis: University of Minnesota Press, 2003).
44 On human technicity, see André Leroi-Gourhan, *Gesture and Speech* (Cambridge, MA: MIT Press, 1993), especially the section "Memory and Rhythms," 219–66.
45 Karl Marx, *Capital: A Critique of Political Economy, Volume 1*, trans. Ben Fowkes (New York: Penguin Books, 1990), 279; Brenner, *Economics of Global Turbulence*, 27–40.
46 Theodor W. Adorno, *Minima Moralia: Reflections on a Damaged Life*, trans. E.F.N. Jephcott (New York: Verso, 1974), 221. For a discussion of anxieties over obsolescence in a rather different context, see Kathleen Fitzpatrick, *The Anxiety of Obsolescence: The American Novel in the Age of Television* (Nashville, TN: Vanderbilt University Press, 2006).
47 As quoted in Cal Winslow, "Overview: The Rebellion from Below, 1965–81," in *Rebel Rank and File: Labor Militancy and Revolt from Below in the Long 1970s*, ed. Aaron Brenner, Robert Brenner, and Cal Winslow (New York: Verso, 2010), 23. See as well Gavin Mueller, "The Rise of the Machines," *Jacobin: A Magazine of Culture and Polemic* 10 (April 2013), https://www.jacobinmag.com/2013/04/the-rise-of-the-machines/.
48 Richard Moore et al., *Wreck-It Ralph*, directed by Rich Moore (Walt Disney Animation Studios, 2012). Film.

5

Anticipation / Unexpected

MARK CURRIE

"Anticipation" and "unexpected" provide us with headings under which to organize and think about the valences of keywords that have taken hold recently in humanities discourses. I am thinking about words such as these: invention, inventiveness, advent, *arrivant*, event, eruption, irruption, emergence, singularity, unforeseeability, unpredictability, uncertainty, the untimely, and the messianic. Most of these terms have straightforward denotations in everyday life, yet at the same time they have specialized contexts in philosophy and theory. We know, for example, what an event is, and yet we recognize that "we" and "us" also designate specialist communities in sentences such as these: "Indeed there are those of us who are inclined to think that unexpectability conditions the very structure of the event. Would an event that can be anticipated . . . actually be an event in the full sense of the word?"[1] The "we" of everyday speech communities is displaced here to a specialized group of event philosophers, who share in the assumption that "an event worthy of the name cannot be foretold."[2] There is something about this blend of everyday with technical meanings that propels emergent words.

In the list of terms above, the phrase "emergent words" has a double meaning. These are words that emerge, but they are also words that name emergence itself. They are signs of change, but they also take change itself as their topic. The double temporality of new arrivals that speak of arrival and novelty perhaps derives from a doubling that belongs to all thinking about what is often called "epochal temporality," or periodizations that characterize an epoch in terms of its comprehension or experience of time. Epochal temporality is doubly temporal: first it periodizes, and then it takes temporal experience as the basis for the periodization. Our contemporary "emergent words" offer a convoluted subset: emergent words seem to say that what is emerging in the present is a preoccupation with what is emerging in the present, or perhaps, if we alter the tense, that futurity is the future.

It would be a mistake to assume that this double time of epochal thinking disturbs the sleep of only a few poststructuralist philosophers. In fact it is a much more widespread response to a condition that many refer to as "uncertainty," and understand, particularly since the global financial meltdown of 2008, as a crisis of prediction. We live, according to one of our most pervasive contemporary self-analyses, in a world less certain, less knowable, and less foreseeable than the one we inhabited in the past. There is something about the very notions of "world" and "globe," the very collectivity of the "epoch," that seems to be at stake in this meltdown, as if the universal logic of capital once created for us a shared and foreseeable future, dependable investments, happy retirements—and that now these certainties have vanished. The confidence of the global has been refigured as universal unknowability, and to this condition, so this analysis goes, we must all adapt. The notion of the unforeseeable future has therefore acquired a social meaning far broader than the one that underlies the philosophy of futurity in Paris, and we see its signs all over the new world.

More specifically, we see the doubleness of emergent terms everywhere in the self-representations of the new financial regime. The largest bank in the world, HSBC, represents the changing ambitions of corporate finance particularly clearly. Having traded for many years on a slogan that disavows the impersonality of global capital—"The world's local bank"—HSBC is one of many financial corporations to have launched campaigns in recent years that shift focus to the condition of global uncertainty. One example of the campaign is the advertisement for HSBC that hung above Terminal 2 at JFK airport in 2013, which made the following two statements on either side of an image of a compass: (1) "In the future, investors will need to be explorers," and (2) "Tomorrow will be nothing like today." Here is the future imaged as the unknown, as the unforeseeable, and as a break from the present. Financial corporations, for obvious reasons, are particularly inclined to affiliate themselves to the new condition of the unexpected, and inclined moreover to blend the unpredictable future of investments with connotations of the creative unpredictability of the corporation itself. We can see this in a campaign like one recently launched in the UK for First Direct Bank, which draws on these more positive meanings of unexpectedness with madcap slogans—"We are the crab that walks forwards"—to support the campaign's overall self-depiction as "The Unexpected Bank." If the negation of sameness took an anti-global stance in the earlier HSBC campaign, we see here a temporalization of sameness as repetition, and the emergence of unpredictability as a positive value in the place of individuality or locality, as if one kind of unpredictability (creativity, novelty, surprise) might help the corporation to respond to the other (uncertainty). Faced with the opacity of the future, the banking corporation becomes a creative guide in adverse, because unknowable, conditions.

The deployment of unforeseeable futurity in banking draws on the same notion of emergent "double time" as epochal thinking in general, but if its emergent words designate emergence, they do so specifically in order to transform the threat of uncertainty into positive values like creativity or openness to the future. It might also be said that, in a different context, the most positive connotations of the new cohort of futural concepts derive from a contemporary view of knowledge itself, most obviously exemplified by the emergence in theoretical physics of concepts of uncertainty, randomness, chaos, and hyperchaos, of complexity in biology, and of the unknowable in mathematics. Our celebration of the unforeseeable finds credibility in the fact that advanced systematic thinking has abandoned its most extravagant projects centered on the description of a clockwork universe and has come to reconcile itself with the unpredictable and the aleatory condition of matter.

How are these connotations, positive and negative, of the *unexpected* related to the notion of *anticipation*? This seems a straightforward question, since the unexpected, whether positive or negative, must be the negation of expectation, or the failure of anticipation. It is the arrival of what we did not see coming. It is that to which our expectations were oblivious. But in a less obvious way, the category of the unexpected is actually produced by anticipation, expectation, and prediction. The unexpected event is more prominent in a world that is methodical and accurate and expert in its predictions, and the degree of unexpectedness is proportional to the extent to which we are accustomed to seeing things coming. The unexpected storm, for example, acquires a significance in a world of increasingly sophisticated weather forecasting that it did not have when weather was generally unforeseeable, and all storms were unexpected. The science of prediction itself foregrounds and gives emphasis to whatever is unforeseen, which stands out against a backdrop of regular, rule-governed, or repetitive behavior. Paradoxical though it may be, we might be justified in regarding these emerging ideas about a world less predictable than it used to be as ideas actually produced by the general environment of reliable and accurate prediction rather than by simple failures of prediction. This is not simply a structuralist logic that points out the normally imperceptible relation between a concept and its dark structural other, the conceptual inseparability of the contingent and the necessary. It is instead a recognition of the heightened experience of surprise produced by the labor of its annihilation. The modern species of surprise at the unexpected is the product and not the negation of sophisticated expectation. For this reason it is always necessary to regard the unexpected not only as a term of futurity, but also one of belatedness, of realization of what, in the past, we did not know. The unexpected arrives from the future, but it also springs an unexplained past, or a gap in what we knew about the world.

Anticipation

Most accounts of anticipation in the modern tradition of philosophical thinking about time (a tradition normally thought of as beginning in the work of Saint Augustine) understand it to designate a part of the present with a future orientation. In philosophy after Augustine, the temporal present is understood to be a synthesis, made up of what has been, what is to come, and what is. But if these are the constituent parts of the present moment, there is something ontologically uneven about them. If moments are made of traces of what has been and traces of what is to come, it must be acknowledged that these two types of trace have a quite different existential and cognitive status. If we think about presence in this way, as a synthesis, we run immediately into what is often called the ontological asymmetry of time, which accords actuality to what has been and virtuality to what lies ahead. In other words, mental acts of recollection differ fundamentally from acts of projection into the future, and similarly, linguistic utterances that refer to what has been are capable of what grammarians call the factive mode, where references to what is to come are made of the linguistic markers of virtuality. Future tenses in English, for example, are made up of modal expressions that encode degrees of certainty and uncertainty in the utterance, for the obvious reason that future time references refer to events that have not yet taken place and cannot be known in a factive mode. Regardless of all the uncertainties that may be in play in recollection, recovery and representation of the past, there is something concrete about pastness that gives it a substantiality to which futurity can never lay claim.

The belief that the past and the future exist on a different ontological footing is one of the great presuppositions in thinking about time, but it is not the gravest of problems for this line of thinking about the moment as a synthesis. Augustine was troubled particularly by the nature of presence in the present moment—with how any kind of corporeality or cohesion could be established logically for it. For Augustine, though there may be reasons to distinguish the past from the future, neither exists in the present, and what is present is what exists. What this means is that the notion of a present moment as a synthesis is contradictory in the most fundamental way, in that presence is constituted by what it is not, and what exists contains components of what does not. More sobering still, if the component of presence in the present moment is thought about separately from the traces of past and future that it contains, it seems to vanish before the eye. If the present has any duration, for Augustine, it is divisible: it can be divided into the bits of it that have been and the bits of it that are still to come. In the act of division, bits of the synthetic present are essentially retrieved from presence and consigned instead to the very nonexistence that characterizes the past and the future. This new problem, often referred to as the *vanishing present*, is vexing in the sense that time loses its foundational concept, and presence itself seems to refer to nothing other than the traces of what has been and what is to come.

If we leap forward to a more modern context, the present's tendency to vanish is also at the center of Derrida's critique of Husserl's account of internal time consciousness. If Husserl's version of the Augustinian synthesis understands the present as a crossed structure of protentions and retentions, Derrida's critique insists that protentions and retentions provocatively comprise rather than surround the concept of presence: "And deconstructing the simplicity of presence does not amount to accounting for the horizons of potential presence, indeed for a 'dialectic' of protention and retention that one would install in the heart of presence instead of surrounding it with it."[3] For Derrida, in *Of Grammatology*, this installation of a dialectic at the heart of presence is just another way of describing the "form of the living present," and of conserving the fundamental homogeneity and successivity of time. What Derrida advances as the *trace structure* of time admits a more fundamental illogicality, which recognizes that Husserl's phenomenological model is constituted as a warp of language and logic.

The question of anticipation brings into view many of these logical problems in the phenomenological approach to time. As most philosophers of time are wont to do, Augustine and Husserl both reach for an analogy, a model of time that will make clear the synthetic structure of the moment, and both, in slightly different ways, find this model in a notion of *inscription* or *writing*. Augustine famously settles on the recitation of a psalm to illustrate the problems of presence, while Husserl chooses the act of listening to a melody. For Augustine, the living present is what he calls the faculty of attention, which is the part of the psalm that the recital currently attends to, however much that part is marked by the retention of what has already passed and the anticipation of what is still to come.[4] Similarly, for Husserl, the present of a melody is the period for which sounds still resonate in the ear, however much those sounds draw on the notes that have faded or anticipate those to come.[5] In both of these examples there is a clear connotation of the scripted nature of the future—the psalm and the melody are known in advance, inscribed in writing, and susceptible to accurate anticipation.

The most obvious reason that this matters is that inscription, or writing, as a model of time, imports into the conceptualization of time a property that does not obviously pertain in lived experience. In writing, the future is already there. It has a physical trace. In writing, the future is neither open nor contingent, since there is nothing that can be done to alter what is written. Even if we object that, for the psalm and the melody, there is some element of performance, some possibility of deviation from a script that restores openness to the future, the unexpected plays no part in either example. Both depend upon their models for the shared proposition of a future that is already there. In this way, anticipation draws out something illogical in the phenomenology of time, not only because it participates in the logical problem of the vanishing present, but also because it presents an illogicality that offends against the ontological asymmetry of time.

It makes sense, after all, to talk about a trace of the past in the present, but it is apparently nonsensical to think that the future can leave its mark in any comparable way on the present. Here, for example, is Charles Arrowby, the narrator of Iris Murdoch's *The Sea, the Sea*, reflecting on the madness of a future projection that has the same substance as a recollection:

> The idea of killing Ben had not entirely left my mind. It was as if, contrary to reason and more calm reflection, a deep trace had been left in my mind, like a memory trace, only this was concerned with the future. It was a sort of "intention trace," or what might exist in the mind of someone who could "remember" the future as we remember the past. I am aware that this scarcely makes sense, but what I felt here was neither a rational intention nor a premonition nor even a prediction. It was just a sort of mental scar which I had received and had to reckon with. I refrained as yet from planning. I vaguely envisaged the moment of "battering through" as a scene of legitimate self-defence. And I searched for a blunt instrument.[6]

What scarcely makes sense in this passage is that the intention to murder could leave a mark comparable to a memory in the mind. It is, in fact, a recognizable criminal psychosis to believe that the future has this kind of unalterable substantiality. It is tempting to make sense of the passage as a commentary on the metaphor of inscription that we have been discussing, or on the basic conditions of written narrative, in which the future does indeed have the same substantial trace as the past, in the form of writing. If, as Arrowby suggests here, it is contrary to reason to think that the openness of the future and the completion of memory can merge in this way, the statement also forms an unwitting commentary on the notion of writing. For understood as an analogy for temporal experience, the commentary models exactly this commingling of necessity and contingency, futurity and completion, or virtuality and completion that he regards as a psychological irrationality.

Writing, then, seems to offer an analogy that gives expression to the idea that the moment is a synthesis of what has been and what is to come, and yet it also contains something impossible from the point of view of temporal becoming, namely the merging of prospection and retrospection. According to the analogy between time and writing, a present moment is constituted by traces of past and future, by protention and retention, by recollection and prospect, but because the future is as yet unknown within the moment, the apprehension of the trace of the future is deferred. It waits on retrospect. It is unknowable until retrospection arrives to supply the positive content that could not be perceived in that moment's occurrence. It is striking that most who have taken up the question of the trace structure of time succeed in doing so only by reducing the synthesis of the moment to a relation between the present and the past, bracketing traces of the future under the heading of what cannot be known. Though Husserl understands the present moment as a dialectic of retention and protention, it is in fact retention that receives all the attention in his work. Contemporary philosophers who have followed his work on time have tended to do the same, the most notable study in this tradition being Paul Ricoeur's *Memory, History, Forgetting*, a work that approaches the trace structure of time by focusing it on the question of memory, bracketing the protention altogether. The trace, for Ricoeur, is thus the "temporal mark of the before."[7] He never ventures to explore the trace of the future or the impossibility and deferral that anticipation introduces into the present moment.

This is perhaps less surprising in a work focused on memory and history than it is in a study of writing in relation to the phenomenology of time, such as Roman Ingarden's *The Cognition of the Literary Work of Art*. Ingarden's discussion is an extended reflection on the idea that the present moment of reading is analogous to the experiential present in the activity of memory that fills the present with traces of what is already read. Though he declares an interest in the "temporal phases of the approaching future," in practice these are bracketed and ignored on the basis of what he calls the "phenomenal vagueness" of anticipated events, whose "qualitative determination is completely different from what it seemed in mere anticipation."[8] As for the unanticipated future, Ingarden declares that surprises are "not without importance for the aesthetic apprehension of the literary art work," but retreats altogether from any discussion of them.[9]

It is useful to think of Derrida's work as a refusal of this bracketing or suppression of the protention, or the subtraction of anticipation from an account of the moment. For Derrida, the trace is a structure much more challenging to traditional temporal concepts than these followers of Husserl have recognized: "The concept of *present, past* and *future*, everything in the concepts of time and history which implies evidence of them—the metaphysical concept of time in general—cannot adequately describe the structure of the trace."[10] The concept of the trace carries something of the irrationality that we saw in Arrowby's "memory of the future," which the key terms in the metaphysical concept of time simply cannot describe. It is not immediately apparent, in this discussion in *Of Grammatology*, why the metaphysical concepts of past, present and future are so incompatible with the structure of the trace, but it is clear in this statement that the concept of time and the structure of the trace cannot be thought of as the same thing. Derrida in fact gives so many accounts of the structure of the trace that it is always dangerous to settle on any one of them, but it is safe to say (1) that it responds to a general neglect of the pole of anticipation in thinking about the present moment as a synthesis of anticipation and retrospection; (2) that it revels in the element of impossibility or deferral of thinking about anticipation; and (3) that it makes explicit the contradiction between an already inscribed future and the open future of human experience that inheres in the analogy of time and writing.

The Unexpected

The emergence of a new strand of thinking about the future, and especially the future that cannot be anticipated, can be thought of as a rejection of the whole notion that time finds an adequate analogy in writing. Catherine Malabou argues this case: that we need to get away from the *gramma* of grammatology and the graphism of the trace on the grounds that writing can no longer provide a governing image for the production of thought.[11] The importance of sweeping writing aside in this way underlies many of the new philosophical efforts to displace Derrida's notion of the trace with a more developed account of the unexpected. We see this in the work of Quentin Meillassoux, perhaps the key figure in the movement, currently the cutting edge of the new French philosophy known as Speculative Realism. This perspective in ontological philosophy seeks not only to displace writing as an influential analogy for temporal experience but also to break more fundamentally from what Meillassoux calls "correlationism," or the post-Kantian view that any absolute knowledge of things in themselves is unavailable to us, and that we have access only to the appearances, thoughts and linguistic forms through which things in themselves are apprehended.

To gloss this argument, I am going to borrow a distinction more commonly deployed in game theory, probability, and economics, and which is of the utmost importance to the contemporary invocation of the unexpected. Game theory likes to distinguish between *aleatory variability* and *epistemic uncertainty*. Aleatory variability is the inherent randomness that governs the roll of a dice: outcomes are uncertain but limited to a determinate set of possibilities, which are known in advance, and for which the probability of a given outcome can be calculated. Epistemic uncertainty, on the other hand, is unforeseeability that derives from lack of knowledge. Whereas the uncertainty of the latter category could be reduced by collecting more knowledge or constructing a more complete picture of the factors that determine an outcome, nothing will reduce the speculative unpredictability of a roll of the dice. Perhaps the most interesting thing about this distinction is how difficult it is to uphold. It is tempting to think that we could, in theory, obtain knowledge that would reduce the uncertainty of aleatory randomness—for example, by producing a complete picture of the physical forces and conditions exerted upon the dice. Financial corporations, however, find the distinction between kinds of uncertainty useful, as the basis on which to decide whether to seek information that will reduce uncertainty or submit to the inherent randomness of aleatory possibility.

In game theory, a third possibility is often introduced: *ontological uncertainty*. This describes a condition in which outcomes are neither governed by a closed set of possibilities nor rendered less uncertain by knowledge. Ontological uncertainty is a condition of absolute unpredictability, in which events are not only unforeseen, but are in principle unforeseeable. The standard example here is the throwing of a 7 with a single dice—an outcome that did not exist in the set of pre-understood possibilities, so that, unlike numbers 1 to 6, its eventuation is in principle unforeseeable, and not merely unforeseen. Again it is easy to object that these kinds of uncertainty are not categorically separable, or that it is necessary to resort to the view, as game theory tends to do, that the distinctions have a heuristic value, even if they cannot be defended.

The distinctions are certainly useful for an understanding of Meillassoux's position on the importance of the unexpected. His argument in *After Finitude*, for example, places its emphasis on the absolute necessity of contingency—that events always, necessarily, could have happened otherwise—and this means that time can bring forth events that are exactly not instances of aleatory variability, or of epistemic uncertainty. If aleatory variability is the randomness that prevents us from knowing which of a closed set of possibilities will eventuate, and epistemic uncertainty is the randomness that results from a lack of knowledge, Meillassoux's principal interest lies in ontological uncertainty—in the advent of something logically unforeseeable, since no amount of knowledge, or any pre-existing possibility, could have signaled it in advance. He writes,

> If we maintain that becoming is not only capable of bringing forth cases on the basis of a pre-given universe of cases, we must then understand that it follows that such cases irrupt, properly speaking, *from nothing*, since no structure contains them as eternal possibilities before their emergence: *we thus make irruption* ex nihilo *the very concept of a temporality delivered to its pure immanence.*[12]

This irruption *ex nihilo*, for Meillassoux, is more than just an instance of the unexpected. It sheds light on the nature of time in general, and strikes at the heart of causation: "In every radical novelty, time makes manifest that it does not actualize a germ of the past but that it brings forth a virtuality which did not pre-exist in any way, in a totality inaccessible to time, its own advent."[13] The importance of the distinction between aleatory and ontological uncertainty, therefore, is not just that it recognizes a special kind of event, the kind contained by no structure in advance, but that this kind of singularity, this irruption from nothing, breaks with thinking about time as the realization of potentialities:

> time is not the putting-in-movement of possibles, as the throw is the putting in movement of the faces of the die: time creates the possible in the very moment it comes to pass, it brings forth the possible as it does the real, it inserts itself in the throw of the die, to bring forth a seventh case, in principle unforeseeable, which breaks with the fixity of potentialities.[14]

I claimed earlier that Derrida's account of the trace had something of the irrational and even the impossible at its core. Though Meillassoux belongs to the generation of philosophers determined to distance themselves from writing as the governing metaphor of all thought, we see in this notion of the irruption *ex nihilo* the same emphasis on the impossible and the irrationality of a future that brings forth, in the very moment it comes to pass, the possibility from which it is presumed to follow. This is the very same structure that runs through Derrida's work, and which begins life in his description of the "strange logic" of the supplement that disrupts Husserl's thinking about time from within. Derrida describes the structure in *Speech and Phenomena* as a "possibility" that "produces that to which it is said to be added on."[15] That "possibility" carries both Meillassoux's sense of a pre-existing potentiality and the futural meaning of something that may happen—it is both the origin and the outcome, collapsing in a single moment of production. We see it repeatedly in Derrida's later work, which focuses more rigorously on ideas of emergence, irruption and event, and it is clearly present in this description of what he calls the "event-machine":

> Will this be possible for us? Will we one day be able, and in a single gesture, to join the thinking of the event to the thinking of the machine? Will we be able to think, what is called thinking, at one and the same time, *both* what is happening (we call that the event) *and* the calculable programming of an automatic repetition (we call that a machine)?[16]

Here are two apparently incompatible ways of thinking about time. The possibility of thinking in both ways at the same moment is something that lies ahead, but remains, for the moment, an impossibility:

> That is why I ventured to say that this thinking could belong only to the future—and even that it makes the future possible. An event does not come about unless its irruption interrupts the course of the possible and, as the impossible itself, surprises any foreseeability. But such a supermonster of eventness would be, this time, for the first time, also produced by the event.[17]

As for Meillassoux, the unexpected event is for Derrida an irruption that breaks with a notion of time as the realization of potentialities, but it also regards this impossible thinking of the event-machine, indeed the "impossible itself," as the thing that makes the future possible.

Thus the terms "anticipation" and "unexpected" divide the future in two, establishing the difference between the effort and the failure to foresee it. But wherever we delve into the thinking of futurity, we find that the unexpected is, in turn, divided into two, to the degree that we can distinguish between different types of unforeseeability: the inherent randomness of aleatory variability, and the absolute unforeseeability of ontological uncertainty. In this second division there is a kind of dialectic between the notion of possibility (anticipation, time as the realization of known potentialities) and impossibility (the unexpected, time as the irruption of what is in principle unforeseeable). According to Derrida, our language is at this moment inadequate to describe the interaction of possibility and impossibility, and we are forced, in the meantime, to talk of the apparently impossible, retroactive production of the possibilities that are indicated by the emergent event.

It is not only in a deviant concept, like Arrowby's "memory of the future," that impossibility shows itself. The philosophy of time is in fact structured around the division of incompatible concepts, and every description of time seems to be inhabited by its dark structural other: anticipation by the unexpected, aleatory variability by epistemic uncertainty, temporal becoming by retrospection, and novelty by repetition. This is the condition that we live out when we think about epochal temporality, and find that any account we can offer of a historically specific experience of time, or any contemporary regime of temporal words. If it is true that a cohort of futural words pertaining to emergence, novelty, unforeseeability, and irruption have arrived in the humanities, it must also be true that they carry within them, as their secret sharer, all the words pertaining to repetition that they seek to depose. The relation between Meillassoux and Derrida is in fact a local manifestation of a more general interdependence that our emergent words have on the cultural theory of postmodernism, which (however absurdly) admitted no novelty and routinely associated our cultural forms with repetition. I am thinking of analytical concepts like these: recycling, recontextualization, rewriting, the blocked future, the perpetual present, the end of history, and the death of originality. In our epochal thinking we seem to have moved from a futureless present into a world of radical novelty and chaotic randomness. In the new world we may purport to celebrate the unexpected, but in doing so, our epochal periodizations merely live out apparently incompatible, and yet cognitively entangled, positions in the philosophy of time.

NOTES

1. Jacques Derrida, "My Chances/*Mes Chances*: A Rendezvous with Some Epicurean Stereophonies," in *Taking Chances: Derrida, Psychoanalysis and Literature*, ed. Joseph H. Smith and William Kerrigan (Baltimore: Johns Hopkins University Press, 1988), 5.
2. Derrida, "My Chances," 5–6.
3. Jacques Derrida, *Of Grammatology*, trans. Gayatri Chakravorty Spivak (Baltimore and London: Johns Hopkins University Press, 1976), 67.
4. See Augustine, *Confessions*, trans. R. S. Pine-Coffin (Harmondsworth: Penguin Books, 1961).
5. See Edmund Husserl, *The Phenomenology of Internal Time-Consciousness*, trans. James Churchill (Bloomington: Indiana University Press, 1964).
6. Iris Murdoch, *The Sea, the Sea* (London: Vintage Books, 1999), 471.
7. Paul Ricoeur, *Memory, History, Forgetting*, trans. Kathleen Blamey and David Pellauer (Chicago: University of Chicago Press, 2006), 58.
8. Roman Ingarden, *The Cognition of the Literary Work of Art* (Evanston, IL: Northwestern University Press, 1973), 106, 103.
9. Ingarden, *The Cognition of the Literary Work of Art*, 103.
10. Derrida, *Of Grammatology*, 67.
11. See Catherine Malabou, *Plasticity at the Dusk of Writing: Dialectic, Destruction, Deconstruction*, trans. Carolyn Shread (New York and London: Columbia University Press, 2009).
12. Quentin Meillassoux, "Potentiality and Virtuality," in *The Speculative Turn: Continental Materialism and Realism*, ed. Levi Bryant, Nick Srnicek, and Graham Harman (Melbourne: re.press, 2011), 232.
13. Meillassoux, "Potentiality and Virtuality," 232.
14. Quentin Meillassoux, *After Finitude: An Essay on the Necessity of Contingency*, trans. Ray Brassier (New York and London: Continuum Books, 2008), 78.
15. Jacques Derrida, *Speech and Phenomena: And Other Essays on Husserl's Theory of Signs*, trans. David B. Allison (Evanston, IL: Northwestern University Press, 1973), 89.
16. Jacques Derrida, *Without Alibi*, ed. Peggy Kamuf (Stanford, CA: Stanford University Press, 2002), 72.
17. Derrida, *Without Alibi*, 273.

PART II

Time as Calculation

Measuring Time

6

Clock / Lived

JIMENA CANALES

Almost nobody noticed. On New Year's Day 1972, every second of every official clock was "shortened" by a small but significant amount. The velocity of planet earth was slowing, and scientists needed clock time to reflect this delay. By the last decades of the twentieth century, atomic clocks had improved so much that the scientific community in charge of timekeeping felt confident to base time measurements on them. In consequence, they authorized a change in the length of every subsequent second till kingdom come.

The solution appeared simple enough, and by all accounts the new system matched better with "time itself"—except for one problem. We were used to telling time according to the earth's movement through the solar system. The length of the day and night and the regularity of the seasons depended on the earth's rotational and orbital velocity. By marking time by reference to atoms, instead of to astronomy, we risked losing the tight connection of time to daylight and to the seasons. In thirty-seven thousand years from that New Year's Day 1972, we would be a day out of phase with the calendar. Eventually we would celebrate the dawn of the new year in the middle of summer. Perhaps the calendar could be readjusted. After all, Europe had adopted the Gregorian calendar only in the sixteenth century. But another solution appeared to be much simpler. Just skip over a certain length of time that was too short for anyone to notice. So it was decided that, every year, scientists would deftly and quietly "leap" over a single second.[1]

The idea of changing the calendar or adjusting time units was not new. In the aftermath of the French Revolution, for example, scientists and political reformers concocted a brilliant plan to convert hours, minutes, and seconds to the decimal system: they built clocks based on the number ten instead of the number twelve. The usual way of dividing a day into twenty-four hours, an hour into sixty minutes, and a minute into sixty seconds had been inherited from the ancient Egyptians. Forward-looking reformers wanted a better, more rational system. Like many of the utopian ideas of that era, this one did not last long. But it threw into relief the question of just how "natural" time standards really were.

We usually think of clocks as instruments that help us count and divide days, roughly defined as intervals between one sunrise and the next. But from the nineteenth century onward, time as ascertained by clocks became more important than time as determined by daily routines. A new temporal awareness based on the ascendancy of clock time affected modern life and modern science (evolutionary biology, geology, astronomy, and thermodynamics). By the first decades of the twentieth century, mechanical clocks were so regular that they started rivaling astronomically determined time. "Before, it was the astronomer who surveyed the clock," explained a French astronomer. "Now, it is the clock that frequently surveys the astronomer, and rectifies his results."[2]

Clock time versus biological time. The two often did not match. What were the benefits and disadvantages of each? Which should have priority? These questions appeared in some of the most intractable scientific and philosophical discussions of the twentieth century. The very idea of "progress," and the role of technology in it, was at stake. How could a small mechanical device carried on a pocket or strapped on a wrist have such a massive effect on our understanding of the cosmos?

Clock Time

While we now believe that clocks measure universal or "real" time only imperfectly, during the first period of their development in Western Europe, clocks were symbols of a universal order maintained and set in motion by God himself. If one accidently stumbled upon a clock, so went William Paley's popular "argument from design," one would surely think of its clockmaker, and so the cosmos itself became associated with mechanistic time and clocklike regularity.[3] Paley's argument articulated often implicit eighteenth-century associations between clocks and a proportioned, lawful, and God-governed universe that had been given support by Newton's classical mechanics.

From the eighteenth century onward, the central role of clocks in modernity was frequently acknowledged. Karl Marx highlighted the clock's importance for industrial societies. "The clock was the first automatic device to be used for practical purposes, and from it the whole theory of the production of regular motion evolved," he wrote.[4] In agreement, the historian of technology Lewis Mumford stated, "The clock, not the steam engine, is the key-machine of the modern industrial age."[5] In many of these accounts, the industrialization of the modern world was described as unfurling with clocklike regularity. "It is the mechanical clock that made possible, for better or worse, a civilization attentive to the passage of time, hence to productivity and performance," explained the economist David S. Landes.[6]

A historically important change in timekeeping practices came at the end of the nineteenth century, with the development of new networks of electrically coordinated clocks. These networks expanded quickly after the Great War with the spread of radio-based wireless technologies. For example, civilian use of wireless time signals emanating from the Eiffel Tower became common in Paris after 1910. With the invention of the triode (three-electrode lamp), transmission improved so much that by the early 1920s it was possible to reach the North and South Poles and even intercept the waves in Australia. To manage this new distribution of time in the post-WWI era, the International Time Commission (ITC) was organized. The bulk of its work was carried out by the Paris-based International Time Bureau (ITB) located at the Paris Observatory.[7] The Paris facility received time signals from international observatories worldwide, "harmonized" them with their own, and sent them back to the world as "universal time."[8] Efficient time determination and distribution was considered a contribution to—and, in some instances, even a prerequisite for—world peace. Because uncertainties in the determination of time led to uncertainties in the determination of longitude, the transmission and standardization of time had clear geopolitical consequences not only for mapmaking but for the determination and maintenance of national borders. Settling border disputes was a pressing need during the century, particularly in reference to the border in Morocco between France and Spain, the Congo-Cameroon border disputed by France and Germany, and the territorial disputes in Tunisia between France and Italy. Regularizing clock time was necessary for maintaining a peaceful world order.

Wristwatches started to become common in the 1920s, when the general public adopted the fashion of soldiers who had found the old pocket models too cumbersome, and yet philosophers, for the most part, ignored the profound changes in the culture of timekeeping taking place around them. There were exceptions, as I discuss below. Phenomenology, from Husserl to Merleau-Ponty, stressed closer links to life and the living than to the mechanical and clocklike. Martin Heidegger's work started as part of a broader initiative aimed at recovering "lived time" and restoring its importance vis-à-vis "clock time." His project involved editing and publishing Husserl's *Vorlesungen zur Phänomenologie des inneren Zeitbewusstseins* (in 1928). Bergson, Husserl, and Heidegger became key references for subsequent explorations of "lived time."

During the 1960s one of the most important changes in our understanding of time took place. In 1959 scientists at the National Physical Laboratory in charge of the Standards Division of the United Kingdom explained to readers of the *General Science Journal* the revolution that was about to take place: "During the next few years there is every prospect that new definitions of the metre and the second will become adopted which will be expressed in terms of certain fundamental characteristics of the atom."[9] The change in how standards of time were defined had been brewing since the first decade of the century, but deficiencies in the current system were now too apparent to be ignored. "Existing definitions," of the second of time, explained the authors, "are now in question."[10] On October 14, 1960, at the 11th General Conference on Weights and Measures in Paris the wavelength of Krypton-86 orange light was adopted as the new international standard on which all length measurements would be based. While most historians have focused on the implications of this new definition for standards of length, it is important to remember that it was also used for standards "of frequency (or [for] its inverse, [the] time-interval)."[11] How did this new definition affect our general understanding of time? While ostensibly tucked away in national laboratories, the new definition of time set in the 1960s— one which continues to be valid today—affected not only how time was thought of, used, and experienced, but more importantly, it reflected profound transformations in the role of scientific authority and expertise, and privacy and intimacy in social and civic life.

In industrialized nations over the course of the twentieth century, then, time supplanted space as the very backbone of science, public life, community, and global sociability. Indeed, by the late twentieth century—with the sprawl of cities worldwide and the growth of a range of communication and transportation technologies—meetings among individuals could be more efficiently arranged on the basis of time rather than place (such as the coffee shop), as individuals arranged to meet or talk *then*, rather than to simply meet *there*.

Lived Time

In the last decades of the twentieth century, clock time and lived time also appeared more distant than ever. The standard account of their separation is traced back to the 1830s discovery of the "personal equation," which described the difference between time assessed by a person and time marked by a machine. When compared to automatic inscription devices, human assessments of rapid, fleeting events were found to have idiosyncratic variations and a lag time of about a tenth of a second. Experimental psychologists blamed these delays on the speed of "reaction time," part of which was due to the surprisingly slow "speed of thought."[12] While the personal equation had been a problem for scientists, by the 1930s it also became a problem for philosophers and sociologists. The sociologist Robert K. Merton cited it to show how politics or individual biases could enter into science, affecting its purported "objectivity."[13]

Clock time and lived time were often associated with other binary oppositions such as machine-human, matter-mind, objective-subjective, physical-psychological, public-private, outer-inner. The first of the binary terms was usually understood by reference to a homogeneous notion of time and causal effects, while the second term was attached to living beings, salient moments, and a sense of indeterminacy and heterogeneity. At different moments in history, these oppositions obtained specifically gendered and hierarchical valances, as they were applied to different referents.[14]

The roots of the modern conflict between "lived time" and "clock time" were deeply immersed in a longer philosophical tradition, where the works of Parmenides and Heraclitus, Aristotle and Augustine were central. In the twentieth century, this conflict resurfaced in a famous debate about time between Albert Einstein and Henri Bergson that exploded when Einstein visited Paris in 1922.[15] Bergson had the opportunity to express his views directly to the physicist. He found Einstein's tackling of time solely by reference to clocks completely aberrant. In contrast, Bergson argued for the need to deal with time—and particularly "lived time"—philosophically, stating, "The time that the astronomer uses in his formulas, the time that clocks divide in equal parts, that time, one can say, is something else."[16] Bergson controversially claimed that Einstein's theory of relativity was undergirded by a more basic sense of simultaneity that would explain why clocks were invented in the first place. That certain correspondences between events could be significant for us, while most others were not, explained why our intuition of this simultaneity remained relevant.

Likewise, when the physicist Paul Langevin declared to a room full of philosophers in 1911, "But we are ourselves clocks," he strove to defend the value of Einstein's theory of relativity.[17] The audience protested, unconvinced by Langevin's call to understand humans as clocks, a philosophical stance connected to the modern rationalism and secular materialism he defended as a scientist. We are *not* clocks, retorted the philosopher Léon Brunschvicg, we are clock *makers*.[18] The assertion that "we are ourselves clocks" seemed tightly connected to a philosophical materialism that was amply debated. Decades later, in 1949, the philosopher Gilbert Ryle would argue that if scientists adopted this materialistic stance and eliminated the concept of the "soul," they would leave behind an even stranger "dogma of the ghost in the machine."[19]

At the end of the twentieth century, the philosopher Hans Blumenberg returned to the "personal equation" as a classic case for understanding the separation between "world time" and "lived time" and how the latter appeared to have a certain "granularity" at its most basic level, related to the "tenth of a second."[20] In Blumenberg's account, the personal equation appeared as one episode of the seemingly intransigent divide between people and things, and thus was at the root of some of the most difficult questions about the nature of the world and human consciousness in it.

Indeed, in the post-WWII period a growing number of historians started focusing in this way on the history of chronology and of timekeeping (from sundials to clocks). Their insights brought new lessons that reflected back on the discipline of history. In twentieth-century political philosophy, there were, for instance, new perspectives on the history of time emerging from the Marxist tradition. Marx had not only considered clocks to be essential for the establishment of new production methods; he also fought against nonchronological explanations in historical writing.[21] Yet he could not have anticipated the extent to which the post-WWII era depended on organizing time on a standard "timeline" using fixed intervals extrapolated backward and forward in history. The "timeline" was so perfect that by the 1950s it permitted the historian Fernand Braudel to consider it *as independent from history* and disconnected from the politics of his time. He started advocating a particular way of writing history at a level "where its time does not match our traditional measures."[22] The historical method he advocated was based on a timeline that was so long that the scale of historical narratives had little to do with the scale of mortals.

The use of timelines as a way to represent a chronological sequence of events was hardly new.[23] But in the 1950s Braudel could think of a timeline on which to map historical events that were not brought about by humans, promoting instead "anonymous" *longue durée* history characterized by slow environmental and socioeconomic changes that he contrasted to the deceptive history of events (*histoire événementielle*) of identifiable, named individuals. Braudel forcefully rebelled against the method associated with Heinrich von Treitschke that "men made history," adopting instead a scale that surpassed that of the lives of men and came closer to the astronomical scales of the earth and planetary sciences. The later *Annales* historians started to consider the role of timekeeping technologies, noticing an important historical transformation that was brought about by the introduction of clocks. Jacques Le Goff saw the medieval conflict between the time of the church (with its tolling of bells) and the time of merchants (based on the clock) as one of the major events of the mental history of those centuries. Le Goff soon had to ask more complicated questions about our sense of historical time and had to find alternative ways for understanding chronology in communities that were not ruled by clocks—for example, by tracing how different epochs could be determined by how confession manuals defined and categorized sins.

Working from a tradition that increasingly wanted to recover the contributions of the working classes and other neglected historical actors, the historian E. P. Thompson similarly became interested in the history of timekeeping. In his magisterial essay "Time, Work-Discipline, and Industrial Capitalism" (1967), Thompson contrasted the "clock" time of the industrial capitalist against the "natural" rhythms of craft workers, which he associated with women and children. But "history from below," as work on the working class, women, and other groups became known, would not adopt time from below. Social history, for decades to come, adhered strictly to the chronological sequences demanded by Marx himself.

What did historians learn about history once they included clocks in their accounts? As they delved into the history of time, thinking of it in relation to particular devices, the question of the temporal structure of historical writing itself became ever more pressing. In France, the philosopher Paul Ricoeur, for example, was intrigued by how *Annales* historians in the 1970s focused on "the representation of time" of different eras. It was "amusing" to read their accounts and see how they failed to notice that the temporal structure used to organize *their* narratives was similarly dependent on a particular conception of time characteristic of *their* own historical situation.[24] Ricoeur conceived of a solution for restoring to "historical time" its "scientific ambition" while reconciling it with our sense of lived time.[25] Historical narratives gained meaning only because they "derived" from a certain sense of temporal experience, agency, and causal efficacy. These conditions were what linked the intrigue of a historical account with its historical referent, and why its structure remained different from fictional narratives. Ricoeur liked to focus on how historians frequently "shortened" episodes that "mutated" too slowly. Their job was not unlike that of a film editor, who cut episodes "by an effect of cinematographic acceleration."[26]

The German historian Reinhart Koselleck similarly analyzed the relation between historiographical traditions and particular chronometric practices.[27] Koselleck argued that "clock time" was itself a historical event: "The mechanical clock, once it had been invented, descended from the church tower to town and city halls, then moved into the living rooms of the wealthy and the bourgeoisie, and finally found its way into watch pockets."[28] The role played by timekeeping instruments in the construction of history itself, which had been kept aside in the early literature on the topic, entered with full force. Each historical epoch had a certain way of conceiving its own temporality, and this particular perspective affected the historical narratives of that time. The reign of "clock time" appeared as a short episode that fell within the longer reign of "natural time" that was "physically or astronomically processed" and that could be traced further back, to the invention of the sundial.

The discipline of history, according to Koselleck, depended on the idea that certain chronologies were nearly "absolute" and independent of history: "To make meaningful statements, we need to tie each of our relative chronologies back to a chronology that is as 'absolute' as possible and independent of history."[29] But because no chronological method could be absolutely independent of history, Koselleck chose instead to see history in connection to changes in our sense of *acceleration and surprise*. If historians could not free themselves from understanding the past in terms of their own senses of time, they could become more explicit about their biases by thinking about "historicity" in terms of the more general categories of *experience and expectation*. While some historians strove to fill out sequences of events and place them in chronological order, Koselleck insisted that these compilations could never amount to a meaningful historical account. The practice of history was notably different from that of the "Ideal Chronicler" described by the philosopher Arthur C. Danto.[30]

While philosophers and historians worked on the long-term dynamics of lived time, anthropologists were confronting its political implications. They had a long history to work upon. In the seventeenth century, for instance, Father Matteo Ricci told the story of the Jesuits bringing a mechanical clock to the Imperial Palace in Beijing, to the amusement of the Chinese emperor, who was not unfamiliar with the device. This has remained one of the exemplary narratives about the role of technology in colonial encounters. Historians were quick to notice that clock time in such instances was tightly coupled with the notion of "civilization," thus bringing questions about the role of modern machinery in industrialization to the fore. But through such examples, anthropological studies of time were making the case that clocks were "little boxes that measured not merely time, but men; and which were always wound up to do battle on foreign shores."[31] These *agonistic machines* measured the temporal distance between modernity and the dark ages only by positing an analogous distance between moderns and other people.[32] They represented conflict rather than progress, denigrating the "Other" in colonialized territories because of how modernity appeared at the head of a long, teleological, progressive narrative of events, places, and people. A technologically determined "now" is always at the head of the race, peeking out from the crowd and heading for the finish line. The philosopher Michel Serres has argued that just as the ancients used to "place the Earth at the center of everything," a merely cursory analysis reveals how moderns see themselves as living within a teleological progression heading toward the very summit of time. This high-altitude position comes with epistemological benefits, which scientists, in particular, have efficiently harvested. "It follows that we are always right, for the simple banal, and naïve reason that we are living in the present moment," he stated.[33]

Laboratories work efficiently on the past. With up-to-date measuring techniques, science changes our understanding of the deep past, yet it does so by taking for granted a progressive model of culture that lends it historical authority. Historians and anthropologists of science consider changes in our redefinition of past events as intimately tied to present advances in scientific practices and technologies. The philosopher and anthropologist of science Bruno Latour has stressed this dependency. Ramses II, he argues, "fell ill three thousand years *after his death*," since the effects of tuberculosis on the ruler technically occurred when scientists brought his mummy to Paris and determined he had been infected.[34] For Latour, the fiction of "time advancing clockwise" is only maintained thanks to an armamentarium of brick-and-mortar institutions—of networks—that protect a Kafkaesque castle of simple "inscription" techniques and devices that include those of clock time. In consequence, he exposes the fragility of so many of those universalist claims that have characterized modernity in its bureaucratic splendor. While anthropologists had previously considered the difference between mythical cultures and historical civilizations in terms of essential cognitive divisions and technological prowess, by tracing what sustains "clock time," Latour shows how these divisions arise in connection to particular forms of social and material organization. The empire of time—with its divide between moderns and primitives—has no clothes.

Yet if some anthropologists successfully put clock time back into its case and coat pocket, crushing its hopes for universality or its bestowing of laurels upon only a few "modern" societies, sociologists go further by highlighting the connection between the modern notion of time and particular technologies and practices, not all of which are favorable to civic life.[35] The role of the clock in factory organization from the Industrial Revolution to Taylorism, for example, was debated as historians and economists placed differing levels of importance on "technological" versus "social" progress. Thinkers started considering the concept of "efficiency" not only as quantitatively determinable via clock time, but also in terms of "human" costs in terms of lived time.

Clocking Lived Time, Living Clock Time

The tension between clock time and lived time motivated some of the most important contributions to twentieth-century philosophy. Martin Heidegger noticed something about time that neither Einstein nor Bergson had considered. "The very determining of time," he observed, "should claim as little time as possible."[36] For this very pragmatic reason, tied to their appeal and popularity, clocks were able to proliferate and conceal alternative conceptions of time. Heidegger had found a clue permitting him to move beyond ancient dichotomies. "As regards the title 'Being and time,' 'time' means neither the calculated time of the 'clock,' nor 'lived time' in the sense of Bergson and others," he explained, referring to his magnum opus, *Being and Time* (1927).[37] During those years Heidegger warned about the spread of many other new technologies, which were "intermediate" things—characteristically "modern"—that lay somewhere "between a tool and a machine." These things concealed themselves "in the midst of [their] very obtrusiveness" and "transformed the relation of Being to his essence."[38] Heidegger sought to resolve the impasse between "clock time" and "lived time," between the "rational" realm of science and the ground of lived experience that often was relegated to the realm of the "irrational." Clock time, considered alone, was a grossly inadequate concept for understanding time: "Once time has been defined as clock time then there is no hope of ever arriving at its original meaning again."[39]

Heidegger's investigations into the relation of clock time and lived time brought important insights to his philosophical understanding of technology more generally. In later work, he explained how one could no longer "pretend as if 'technology' and 'man' were two 'masses.'" These two elements were complexly intertwined. Their interrelation was the reason why "the much discussed question of whether technology makes man its slave or whether man will be able to be the master of technology is already a superficial question." Heidegger instead expressed the need to "ponder the 'concrete' . . . and to remove the concealment thrust upon things by mere use and consumption." Categories such as "filmic time" and "literary time" emerged as potentially mediating concepts, signaling, perhaps, the slow dissolution of one of the central dichotomies of modernity.

From Bergson to Deleuze, for example, scholars have focused on filmic time to better understand the relation between temporality and technology. Picture a gun aimed at the head of a girl. She has been kidnapped and is tied up, unable to move. The gun is linked to a clock that will go off unless it is disconnected before noontime. In this scenario, time passes differently for the victim, who focuses on the countdown; for the rescuers, who are driving the action; and for the viewers of the film. This is a scene from D. W. Griffith's *The Final Hour* (1908), which showed a dramatic trope and temporal configuration that would appear in many action movies to come. Since the early days of Georges Méliès, moviemakers used scenes such as this one and techniques such as cuts and edits to speed up and slow down action. *L'Homme à la manivelle* (Pathé 1907), for example, displayed comical effects surfacing in connection to changes in the speed of hand cranking. Variable-rate hand-cranked cameras became obsolete with the introduction of sound in film in the 1920s. Once the speed of recording and projection speeds were standardized, the contrast between the clocklike regularity of the apparatus and the fluid time on the screen became evident for a growing number of viewers. For this reason, the Russian master of montage film V. I. Pudovkin extolled the benefits of "filmic time" because of how it could be liberated from "real time." Yet "cinema pur"—based on the long take and widely employed in "realistic" documentaries—shunned filmic time, and started to be defined against it. Two kinds of film techniques—one associated with lived (and narrative) time and the other with clock (and real) time—suddenly faced off.[40]

Analyses of the temporality of film were accompanied by new investigations into the time of literature. Literature and the movies introduced a new way of thinking about clock time and lived time.[41] In a 1952 linguistic study of clock time, Thomas Storer wrote that "the description of a ticking clock as consisting of 'a tick followed by a tock followed by a tick followed by a tock followed by a tick followed by a tock' is . . . identical in structure with the ticking of a clock."[42] When spoken or written down, a "tick tock" echoes the mineral beats of the clock.[43] Yet most of language follows a temporal structure that hardly matches the enviable regularity of clocks.

* * *

At the dawn of the 1968 protests, the time of philosophers, humanists, and artists continued to be set starkly apart from that of Cold War physicists. Scholars continued to take sides, some aligning themselves with physicists' clock time and some protesting against it, pointing out how clock time is often connected to war (agonistics), disease (parasites), noise (entropic and auditory), and the creating of a subaltern "Other." We now have the possibility of thinking about time in new ways, far from the conceptual binaries of clock time and lived time.

Clock time and lived time overlap in the legal brief, the battle plan, the assassination plot, the successful invasion, the lethal infection, and the strategic rendezvous. Scholars today have sharpened their investigations of this overlap, of what may lie between clock time and lived time or how the two have been merging into new configurations alongside new technologies. Temporal awareness and chronologies emerge in cases when exact comparisons between "before" and "after" obtain relevance. In the twentieth century, "clocks" changed from mechanical and astronomical-based systems to electromagnetic and atomic-based ones, while "life" was increasingly studied in terms of its microbiological structure and cellular movement. These changes, sustaining the binary split of clock time/lived time, depended on particularly modern technologies—from radio to film.

Instead of attempting to map more and more detailed slices of human time on top of an ever-extending sequence of perfect clock-time intervals, we need to ask why this particular hierarchy gained such a potent force in modernity.[44] Cultural narratives and beliefs change drastically in connection to new timekeeping practices.[45] The phenomenological subject and the cosmos (against which it defines itself) are both shaped by timing media that play an essential role in the formation of subjectivity, objectivity, *and* their historicity. Practices and technologies (some as simple as pushbuttons and triggers, doors, and ordered lists, and others as complex as a particle accelerator or large-scale telescope) create temporal asymmetries (between the past, present, and future) and irreversibilities (that is, one-way directionality). The structure of time in modernity—and its division of clock and lived time—is sustained by the understanding and experience of causality and effective agency that cuts across science, history, literature, and film, as much as through the present, the past, and the future.

NOTES

1 "Shortening the Second," *Science News*, December 18, 1971, 408.
2 Charles Nordmann, *Notre maître le temps, Le Roman de la science* (Paris: Hachette, 1924), 136.
3 William Paley, *Natural Theology: Or, Evidence of the Existence and Attributes of the Deity, Collected from the Appearances of Nature* (London: R. Faulder, 1802).
4 Karl Marx to Fredrich Engels, "28 January 1863 [London]," in Karl Marx and Friedrich Engels, *Collected Works*, trans. Richard Dixon, vol. 41, *Correspondence 1860–64* (New York: International Publishers, 1971–2004), 448.
5 Lewis Mumford, *Technics and Civilization* (London: Routledge, 1934), 14–15.
6 David S. Landes, *Revolution in Time: Clocks and the Making of the Modern World* (Cambridge, MA: Belknap/Harvard University Press, 1983).
7 W. W. Campbell and Joel Stebbins, "Report on the Organization of the International Astronomical Union," *Proceedings of the National Academy of Sciences of the United States of America* 6, no. 6 (1920): 358.
8 Campbell and Stebbins, "Report," 358.
9 H. Barrell and L. Essen, "Atomic Standards of Length and Time," *General Science Journal* 47, no. 186 (1959): 209.
10 Barrell and Essen, "Atomic Standards," 210.
11 Barrell and Essen, "Atomic Standards," 212.
12 Jimena Canales, *A Tenth of a Second: A History* (Chicago: University of Chicago Press, 2009).
13 See Robert K. Merton, "Science and the Social Order," *Philosophy of Science* 5, no. 3 (1938).
14 This is discussed by David Couzens Hoy, *The Time of Our Lives: A Critical History of Temporality* (Cambridge, MA: MIT Press, 2009), and Hans Blumenberg, *Lebenszeit und Weltzeit* (Frankfurt am Main: Suhrkamp, 1986).
15 Jimena Canales, *The Physicist and The Philosopher: Einstein, Bergson, and the Debate That Changed Our Understanding of Time* (Princeton, NJ: Princeton University Press, 2015).

16 Henri Bergson, *Essai sur les données immédiates de la conscience*, ed. Arnaud Bouaniche, 9th ed. (Paris: Quadrige/Presses Universitaires de France, 2011), 80.
17 Paul Langevin, "Le temps, l'espace et la causalité dans la physique moderne," *Bulletin de la Société française de philosophie* 12 (1912): 42.
18 "Shortening the Second," 408.
19 Gilbert Ryle, *The Concept of Mind* (London: Hutchinson's University Library, 1949), 32, 35.
20 Blumenberg, *Lebenszeit und Weltzeit*, 253–54.
21 Karl Marx, *The Poverty of Philosophy* (New York: International Publishers, 1936). Originally published in 1847.
22 Fernand Braudel, *Écrits sur l'histoire* (Paris: Flammarion, 1969), 24. See also Paul Ricoeur, *Temps et récit*, vol. 1 (Paris: Éditions du Seuil, 1983), 187.
23 For discussion, see Daniel Rosenberg and Anthony Grafton, *Cartographies of Time* (New York: Princeton Architectural Press, 2010).
24 Ricoeur, *Temps et récit*, 195.
25 Reinhart Koselleck, "The Eighteenth Century as the Beginning of Modernity," trans. Todd Presner, in *The Practice of Conceptual History: Timing History, Spacing Concepts* (Stanford, CA: Stanford University Press, 2002), 166.
26 Ricoeur, *Temps et récit*, 196.
27 Reinhart Koselleck, "Time and History," in *The Practice of Conceptual History: Timing History, Spacing Concepts*, trans. Kerstin Behnke (Stanford, CA: Stanford University Press, 2002), 104.
28 Koselleck, "Time and History," 104.
29 Koselleck, "Time and History," 106.
30 Arthur C. Danto, *Analytical Philosophy of History* (Cambridge, UK: Cambridge University Press, 1965), 149.
31 D. Graham Burnett, "Mapping Time: Chronometry on Top of the World," *Daedalus* 132, no. 2 (2003): 19.
32 Johannes Fabian, *Time and the Other: How Anthropology Makes Its Object* (New York: Columbia University Press, 1983).
33 Michel Serres and Bruno Latour, *Conversations on Science, Culture, and Time* (Ann Arbor: University of Michigan Press, 1995). 48.
34 Bruno Latour, "On the Partial Existence of Existing and Nonexisting Objects," in *Biographies of Scientific Objects*, ed. Lorraine Daston (Chicago: Chicago University Press, 2000), 248 (image caption).
35 Judy Wajcman, "Life in the Fast Lane? Towards a Sociology of Technology and Time," *British Journal of Sociology* 59, no. 1 (2008).
36 Martin Heidegger, *Being and Time*, trans. John Macquarrie and Edward Robinson, rev. ed. (New York: Harper and Row, 1962), 499n4. Translation of *Sein und Zeit* (Tübingen: Neomarius Verlag, 1927), 7th ed.
37 Martin Heidegger, *Parmenides*, trans. André Schuwer and Richard Rojcewicz (Bloomington: Indiana University Press, 1992), 77. Translated from the German as *Parmenides* (Frankfurt am Main: Vittorio Klostermann, 1982).

38 Heidegger, *Parmenides*, 77.
39 Martin Heidegger, "The Concept of Time in the Science of History (1915)," trans. Harry S. Taylor, Hans W. Uffelmann, and John van Buren, in *Supplements: From the Earliest Essays to* Being and Time *and Beyond*, ed. John van Buren (Albany: State University of New York Press, 2002), 55.
40 Analyses of the temporality of film were accompanied by new investigations into the time of literature. See the classic text by Georges Poulet, *Études sur le temps humain* (Paris: Plon, 1952); Thomas Storer, "Linguistic Isomorphisms," *Philosophy of Science* 19, no. 1 (1952), 80; Hayden White, "The Narrativization of Real Events," *Critical Inquiry* 7, no. 4 (1981), and *The Content of the Form: Narrative Discourse and Historical Representation* (Baltimore and London: Johns Hopkins University Press, 1987), 5.
41 See Poulet, *Études sur le temps humain*.
42 Storer, "Linguistic Isomorphisms," 80.
43 According to *OED*, the echoic term "tick tock" arose during the mid-nineteenth century.
44 Canales, *A Tenth of a Second*.
45 Jack Goody, "The Time of Telling and the Telling of Time in Written and Oral Cultures," in *Chronotypes: The Construction of Time*, ed. John B. Bender and David E. Wellbery (Stanford, CA: Stanford University Press, 1991).

7

Synchronic / Anachronic

ELIZABETH FREEMAN

The terms "synchronic" and "anachronic" look symmetrical and antonymic to the eye. Both are forms of *chronos*, time. But even their prefixes suggest that they are not simple opposites. The Greek *syn-* means "together" or "with," such that synchrony, "with time," is a matter of rhythm. And the Greek *ana-* does not mean "apart" or "without." Instead it means "back" or "against," such that anachrony, "back in time," is a matter of propulsion and sequence. Thus the opposite of synchrony is not anachrony but asynchrony, "no-time" or "without-time." And the opposite of anachrony (or at least of anachronism, the misplacement of an older, often obsolete object or idea in a new time) is the little-used term "prochronism," the placement of an object or idea into a time in which it was not yet invented.[1]

Nonetheless, pairing synchrony and anachrony points us toward the paradoxical power relations engaged and made possible by these modes of time. Both are, in some ways, modes of ahistoricism recognized by particular disciplines: in linguistics since Saussure, synchronic analyses focus on the state and structure of language in the present, without reference to the reverberations of the past, such as etymology or philology, that are part of a diachronic analysis.[2] Meanwhile, the discipline of history warns against anachronism, or the understanding of the past in terms only possible to think within the present.[3] Thus both the synchronic and the anachronic can subjugate the past to the present, distorting the past in ways that mitigate against understanding how the present came to be. Additionally, synchrony subjugates the individual to a larger scheme like clock time, musical beats, and so on, flattening out individuals' particular temporalities.

Yet both the synchronic and the anachronic also offer up ways to conceptualize freedom—and the differences in their liberatory potential are somewhat counterintuitive. Both are corporeal modes of time, sites through which bodies are dominated and resist domination, and for that reason alone political. But both the synchronic and the anachronic are also bound up in specific iterations of the social, i.e., capitalist and national modernity. From an anthropological perspective, we might say that subjects become legible by inhabiting first their family's, then their larger culture's prevailing daily rhythms, and by participating in the events that organize their culture's trajectory for a meaningful life. But from a historical perspective, it is clear that the modern period—the period from industrialization in Europe during the mid-1700s to the present—saw the emergence of institutionalized techniques of biography such as the psychoanalytic case study, the parole hearing, and the petition for asylum, wherein access to protection or care is granted to subjects who can narrate and document the trajectory of their lives in particular ways. Dana Luciano has usefully termed this technique "chronobiopolitics," or the institutional arrangement of the time of life itself.[4] Finally, the contemporary era—comprising, for the purposes of this essay, the last thirty years or so—has been marked by a certain atemporal instantaneity, for which Twitter and the stock market might serve as useful figures.[5] Under the latter two conditions of chronobiopolitics and instantaneity, anachronistic irruptions of other forms of time or historical moments can certainly become sites of critique or alternative imagining, and synchrony has more power for contemporary politics than might be evident at first glance.

Synchrony

To synchronize something means to cause it to operate at the same time, rate, or pace as something else. And indeed, synchrony may be a condition of being both biological and social creatures. Our bodies come pre-synchronized, from the coordination of the heart's pumping with the lungs' need for oxygen, to the body's ability to ensure safe sleep by paralyzing the limbs of the sleeper. Other forms of synchronization are social, behavior learned in order to survive at all, and especially to live with others. Work in psychology on animal and infant communication, and in neuroscience on mirror neurons, has established that the capacity to synchronize one's movements and/or the rhythm of one's vocalizations with another person's facilitates a form of what Anna Gibbs calls "affective attunement," or feeling- and being-with another, and what Pierre Bourdieu calls *habitus*.[6] This is how individual parents and even entire cultures reproduce themselves, and it is surprising to realize that *habitus* is predominantly a matter of timing. What cultures do, then, is re-reproduce people, realigning their members' gestures and bodily stances so as to imitate—visually, somatically, and *temporally*—that culture's norms.

Over time, institutional versions of clock time have been layered over these biological and cultural forms of synchronization; for example, even ancient military exercises demanded that bodies keep time with one another in marches and drills, and medieval monastic life involved coordinating a group's activities to the timing of bells.[7] But in modernity, synchronic time arguably dominates all other forms. The philosopher Walter Benjamin describes modernity as the gradual emergence of a "homogeneous, empty time," a calendrical and chronometric time against which bodies, activities, and events can be measured and, significantly, within which they can be attuned to one another abstractly.[8] For instance, while not all cultures use the Gregorian calendar established in 1582, a great many do, and international business and politics are conducted according to it. Likewise, Greenwich Mean Time, established in 1898, is a global chronometric. But prototypes of this homogenous, empty time appear before the standardization of national or world calendars and clocks. For example, Benedict Anderson describes the rise in the seventeenth century and after of "imagined communities" of people. By reading in news journals of events happening on the same day in far-flung locations, voting on the same day no matter what state they live in, daily pledging allegiance to the national flag in schoolrooms across the country, and other recurrent public activities, such communities come to understand themselves to be part of a nation.[9] The tempos of reading, voting, pledging, and other activities understood to happen simultaneously (whether or not they actually do) project the idea of a virtual space: a synchronic polis paradoxically unbounded by geographical boundaries, whose people owe it allegiance and feel a sense of belonging to it even when abroad.

Synchronic time is also the time of industrial capitalism. According to Marx, the factory system abstracted bodies both in the name of the commodity and as, themselves, commodities in the form of their labor power and, even more crucially, their labor time.[10] Knowing that it took the average worker five minutes to hammer a sole onto a shoe, a factory manager could synchronize workers so that each was expected to work for no longer than that amount of time, in tandem with other workers doing the same thing or as part of a sequence of workers putting the shoe together. Eventually, reduction of the cost of that labor time lay not only in technological progress, but also in so tightly coordinating workers' movements that no energy was lost. This culminated in the late-nineteenth-century Taylorist system of manufacture, which calculated and then synergized human movements on the factory floor so that each worker might perform only one movement in the most efficient manner possible, over and over again like a piston in an engine.[11] Here, synchronic time enables all bodies—or at least, all of a given *part* of some bodies, like backs or hands—to be imagined as the same, moving in tandem. We might say, then, that industrial capitalism somatized synchronic time in particularly violent ways.

By the late 1920s, the Frankfurt School recognized that the aesthetic realm, particularly popular culture, revealed more about a historical moment than did its official doctrine—and that synchronic time organized early-twentieth-century aesthetics and economics alike. Siegfried Kracauer was perhaps the most astute theorist of synchrony, seeing the "mass ornament," or the aestheticized spectacles of people moving in concert that were so popular in the Weimar and Nazi eras, as symptomatic of the triumph of commodity capitalism. Writing of the American dance troupe the Tiller Girls in Europe after World War I, he declared that

> [n]ot only were they American products, at the same time they demonstrated the greatness of American production. . . . When they formed an undulating snake, they radiantly illustrated the virtues of the conveyor belt; when they tapped their feet in fast tempo, it sounded like *business, business*; when they kicked their legs high with mathematical precision, they joyously affirmed the progress of rationalization; and when they kept repeating the same movements without ever interrupting their routine, one envisioned an uninterrupted chain of autos gliding from factories into the world. . . .[12]

As Karsten Witte observes, Kracauer's most original insight among the Frankfurt School critics was that synchrony—*Gleichzeitigkeit*—underlay both work and leisure time in late modernity, and paved the way for Fascism.[13] She reminds us, too, that Benjamin, Kracauer's contemporary and interlocutor, located in the spectacle of mass simultaneity the people's self-expression at the expense of their actual rights.[14] In other words, rather than driving the proletariat to fight existing property relations, the mass ornament left the latter intact while giving the proletariat an image of their own reabsorption into and vitality for capitalism as synchronized bodies. Kracauer's mass ornament is Benjamin's empty, homogenous time *embodied*, as in a Busby Berkeley musical of the 1930s.

Thus from the tuning of bodies toward maternal, familial, and cultural rhythms, through the rise of national feeling across geographic borders, through the standardization of calendars and clocks, through the factory system of production and the aesthetics that accompanied it, synchrony *creates* the social. Synchrony is key to establishing a sense of engroupment, to implanting the affects and movements that make a person feel connected to something larger than him- or herself. Synchrony as a somatic and psychic mode of belonging is perhaps nowhere so evident in contemporary Western culture as in social media. The late twentieth and early twenty-first centuries have seen an expansion of synchrony's powers in such phenomena as networked communities, online gaming (which allows for "real time" play), and even devices like the Wii that coordinate body movements with onscreen images of oneself and/or other players. Social media is a particularly potent example of what might be called the sociosynchronic: readers' comments on a blog are written at different moments but look as if they are voiced simultaneously from a public, or news of a celebrity death received by people on Facebook at the same time, and their ability to comment immediately, makes them feel as if they have come together to mourn rather than just to hear the news. In short, social media greatly increases the simultaneity of action that makes for a sense of collectivity. This, I shall argue, is not only synchrony's problem but its promise—but not before I turn to the politics of anachronism.

Anachronism: Toward Anachrony

Anachronism, aligned with a notion of "wrong time," seems to be a more modern concept than synchrony, for the former cannot exist without a notion of time as sequential and forward moving, which is to say, without a Western, Enlightenment conceptualization of history as empty, homogenous, and linear. Certainly the planets, the seasons, and individual life spans offered pre-modern humans some notion of forward-moving time, even as natural rhythms tap into temporalities like the cyclical and the diurnal as much as they do the linear. But it is difficult to think of, say, an eclipse, or a snowstorm in June, or a teenager wearing a baby bonnet as precisely anachronistic. Neither, in the West, did ancient notions of chronology (things happening on specific dates understood to be more or less universal) or of dynasty (families succeeding one another, as in the "begats" of Genesis) bring with them a culturally pervasive idea that something could appear in the wrong time.[15] Even biblical time is simultaneous rather than sequential, with events prefiguring and repeating one another (Julia Kristeva calls this "monumental time"), and therefore events are unable to show up in an inappropriate or undesignated historical moment.[16]

In contrast to these kinds of time, according to the philosopher of history Reinhart Koselleck, history is bound to the homogeneous, empty time of nationhood: "Historical time, if the concept has a specific meaning, is bound up with social and political actions, with concretely acting and suffering human beings and their institutions and organizations."[17] Whereas there may be biological modes of synchrony, anachrony is fundamentally unnatural; indeed, the concept is most fully articulated in Vico's *New Science* (1725) as a problem of historiography. The idea of something appearing historically out of place depends on conceptualizing a large, extra-familial group's collective experience that—crucially—does not repeat itself. Koselleck dates this concept to the period between 1500 and 1800. In his view the development of historical consciousness proceeds from a pre-modern sense that past events could teach us how to act in the present precisely because human nature is consistent, to a modern sense that past events and peoples are unique and have contributed to—indeed, are a causal factor in—a larger scheme that cannot be known in advance, in which any given individual is playing only a part, and which involves unseen and unknown others.

This is modern "progressive" time, within which the possibility of anachronism appears. As Johannes Fabian has demonstrated, Enlightenment-era European explorers, following the stadialist doctrine of history, understood the human race to be developing from a savage to a civilized state.[18] On this model, which paradoxically returned history to biology, whole groups of humans could be understood as living in another era, or an earlier place along the continuum, or in humankind's childhood—anywhere but in the historical moment contemporaneous with Europe's. This "denial of coevality," as Fabian calls it, constructed people of color as living anachronisms destined to accede to, or more often to disappear as a result of, modern times—though indeed, as Srinivas Aravamudan argues, colonialism involved, precisely, the project of forcefully reconciling multiple and coexisting temporalities.[19] Siobhan Somerville and others have shown that nineteenth-century "race science" and sexology moved the locus of anachronism from the species as a whole to the type or even the individual: women, criminals, people of color, homosexuals, and the poor were figured temporally, in ways that often overlapped, and this inflected the case studies that comprised the archive and method of these "sciences."[20] The eugenic movement, for instance, warned against atavism, or the reappearance in individuals of "savage" traits more proper to humankind's beginnings than to the present; equally threatening was decadence, or the tipping point whereby a person indulged so highly in the fruits of "civilization" that he or she began to revert to so-called primitive ways.[21] Here again, time was thought in corporeal terms, as bodies were understood to express historical periods not their own.

What, then, is the political work of anachronism? On the one hand, anachronism can signal nostalgia for a time understood as simpler or more civilized—indeed, Ernst Bloch argued that Fascism relied upon connecting populations left out of new economic formations to a manufactured German folk past—and was thereby to be avoided.[22] On the other, as Bloch also recognized, even the most rose-tinted picture of the past can serve as a critique of conditions in the present, and of course there is a more self-reflexive form of anachronism that uses the past to estrange the present. A contemporary manifestation of this is steampunk, which welds the detritus of Victorian industrial culture onto the sleek technology of the 1980s and beyond as a way to critique mass production and commodity fetishism.

The same historical period that steampunk often fetishizes also saw the invention, it might be said, of anachrony as the fundamental condition of modern subjectivity. The name for it was the unconscious. If the Frankfurt School recognized synchrony as the aesthetic and economic principle of modernity, Sigmund Freud recognized anachrony as the condition from which moderns suffered and through which modern subjects emerged. His *Studies on Hysteria* (1895) and many other works theorized that pleasurable and traumatic events, usually those of childhood, were experienced twice: once when they happened, during which period the child did not have the cognitive or emotional ability to make meaning out of them, and again when they returned in the form of a symptom that in some way somatically matched the earlier experience. Freud's name for this was *Nachträglichkeit*, or afterwardness, later translated by Lacan as the *après-coup* (afterblow, deferred action). A symptom, then, was an eruption of the past into the present, an anachronism experienced on the body. Freud also suggested that some symptoms were wishes, in which case what had not yet come to be, the future, was nevertheless appearing as a somatic event for the sufferer: history appeared not *as* a body, but *in* or *on* one. These symptoms were also signs that a part of the human psyche operated outside the laws of progressive time. In Freud's view, the unconscious preserved past material, invented fantastic alternate futures, and juxtaposed events and emotions without regard to their place in the individual's chronology. It had, as Carl Jung famously put it, "no time," but this is an expression of its historical depth and heterogeneity rather than of its mythological character.[23]

Again, the Frankfurt School produced some of the best thinking on the politics of anachrony. Throughout much of his work, but especially in "Theses on the Philosophy of History," Walter Benjamin conceptualized his own era's aesthetic of juxtaposed temporalities, found in what he called "constellations" of mixed historical references such as the montages of Surrealism and the kaleidoscopic displays of the Paris arcades. "The past," he writes, "can be seized only as an image which flashes up at the instant when it can be recognized and is never seen again."[24] The sudden meeting of the past and present in cultural objects, images, rhetoric, and so on, is akin to the Freudian symptom, a sign of unrealized possibilities or undigested experiences in the past. Their political power, though, lies not in curing them but in reanimating those possibilities and experiences to transform the present.

Following Benjamin, much recent critical theory, especially postcolonial theory, has focused on the trope of haunting, with the way figures such as the ghost or the undead can index a political unconscious, in Frederic Jameson's terms, or what he also calls the history that "hurts."[25] But in contemporary culture, the ghost itself actually feels like an anachronistic way of figuring the problem of history—ghost stories are a branch of Gothic literature, and arguably contemporary phenomena like *Twilight* and *Interview with a Vampire* are less self-consciously engaged with the problem of history than were older Gothic works. Instead, a recent spate of millennial films such as *Groundhog Day* (1993), *The Matrix* (1999), *The Bourne Identity* (2002), *50 First Dates* (2004), *Eternal Sunshine of the Spotless Mind* (2004),[26] and others have thematized implanted or lost memory. Several successful novels and films have also run time backward, including the novels *How the García Girls Lost Their Accents* (1991) and *Time's Arrow* (1991), and the films *The Sweet Hereafter* (1997) and *Memento* (2000).[27] The popularity of these works suggests two things. First, denizens of the contemporary sometimes accept the idea of *time* as nonlinear and heterogeneous; though the works that merely run time backward do accede to linear time, in works where memories are lost and found mixed time is not always a crisis. Second, and conversely, in these pieces *historical thinking* still remains a problem: progressive time clearly won't do, even when earlier moments on the timeline reappear to trouble the present, but how do we encounter history other than through a timeline? If time is mixed, how do we know what is historical?

The Politics of Anachrony, the Politics of Synchrony

The previous discussion suggests that anachrony seems more conducive to critical thought, political action, and the work of freedom: anachrony figures the possibility of time escaping the rationalization and calculability that modernity has imposed upon it. It opens up the possibility of a heterogeneous coevality, as when, for example, Dipesh Chakrabarty theorizes a "History 2" that runs alongside but is not coterminous with capitalism's empty, homogenous time, organized instead around affects and affinities.[28] Anachrony promises ways of using the past differently: neither as an example that the present must follow, nor as a golden time to return to, but as a means of politically reframing the present and what seem like its imperatives.

But it is also true, as Freud recognized, that one can be bound to the past in ways that paralyze thought and action. Consider, for example, the character Rogue from the Marvel Comics series *The X-Men* and its spinoffs.[29] A mutant, Rogue has the power to absorb the memories and the superpowers of anyone she touches. Her first kiss, with a fellow teenager named Cody, leaves him in a lifelong coma and her flooded with his feelings and experiences. From that point on, Rogue wears gloves, long sleeves, and long pants. She does not allow anyone to touch her. While her superpowers remain intact, and are often augmented when she accidentally or purposefully touches someone, she remains perpetually haunted, perpetually bombarded by other people's memories, and unable to connect with her fellow mutants or with humans. Rogue is a good reminder that, as Friedrich Nietzsche wrote, amnesia is sometimes a necessary way out of a historical impasse.[30]

And if synchrony appears in the previous discussion to be only a mode of domination, *X-Men* superhero Synch offers a good counterexample for thinking it otherwise. Synch can align his aura with those of other mutants, synergizing with their powers and thereby absorbing them. Synch, who is African American, also identifies with the working class. (When his love interest Monet St. Croix tries to buy him new clothes, he says, "Perhaps people who aren't *born* rich like you might actually want to *work* for what they have?")[31] In light of Synch's distinctly human, racialized, and class-marked identities, his power to "synch" offers a way to think about synchrony beyond its tendencies to turn humans into a mass ornament. I have said that synchrony creates the social as a somatic and political form of temporality in capitalist and national modernity. More than that, though, synchrony creates a nonbiological ability to physiologically reproduce. Indeed, when Synch synchs with Monet, also of African descent, combining his aura with hers in order to absorb her invulnerability, she is twinned.[32]

Synchrony is, to put it simply, a different form of breeding: *The X-Men* can only imagine this as intraracial, but synchrony is actually a way *out* of eugenic thinking about embodied continuity over time. This process may be crucial for imagining engroupment and descent among populations who do not biologically reproduce themselves, such as queers, or who are stigmatized by the idea that they overreproduce, such as poor people and people of color. If the limp wrist or lilting speech of "the homosexual," the crotch-grabbing swagger or the diva stance of "the black person" are stereotypes, they are also codes signaling insider status, and hinting at ways that bodies can synch up so as to oppose mainstream modes of being. Indeed, William Condon describes "interactional synchrony," in which the bodies of people listening in conversation operate "in *organizations of change* which reflect the microstructure of what is being said."[33] This occurs, of course, in large-scale as well as small-scale interactions. And Condon's phrase *organizations of change* is striking, for it suggests a politics of synchrony.

Bodies moving in concert may suggest mindless submission, but bodies do not have to be entrained or synchronized only in order to present people with an image of themselves that belies their structural place in a class system. They can move in concert *against* that system while simultaneously producing other social forms. We see hints of the power of synchrony in late-twentieth-century and twenty-first-century forms such as lip-synching, where a performer channels the voice and body movements of another performer, thereby drafting the earlier performer into the later one's world view, and transforming the later performer's bodily repertoire. A more recent and even more powerful example of a politics of synchrony is the flashmob: using highly synched technology such as mass texting, people show up en masse in a single locale and do something together—dancing, marching, performing. ACT UP's Die-Ins were an early example; organizing by phone tree, large numbers of people converged on the streets of New York and other major cities. Often in front of buildings that housed institutions and corporations that ACT UP had charged with contributing to the AIDS epidemic through neglect or obstructionism, the gathered people all fell in concert onto the street, blocking traffic.[34] Of course the flashmob technique is ripe for cooptation. There are even marketing firms that will coordinate a flashmob for companies, for example, and the often touching performances of the "dancing inmates" of the Cebu Provincial Detention and Rehabilitation Center are coordinated from the top by prison officials to showcase the power of exercise and music in rehabilitation.[35] Notwithstanding synchrony's uses in projects of domination, however, it has the power to join bodies in the somatic attunements that are the bases of intimate belonging, and thereby to subvert the dominance of both "family" and "nation" in conceptualizing affinity and descent, or even dissent. Under the twenty-first-century regime of global financialization and late capitalism, several scholars have argued, affect itself is often distributed more or less synchronously, providing those living precariously with new modes of becoming collective, or attaining a kind of historical/political consciousness, however different from the traditional Marxist model this might look—consider, for instance, Lauren Berlant's "affect worlds," Sianne Ngai's exploration of the public feelings arising from suspended agency, and John Protevi's "political affect."[36]

Perhaps the most profound recent example of this form of synchrony as becoming-collective is the Occupy Wall Street movement that began in 2011, particularly the encampments of that year. The encampments' many simultaneous and concomitant activities created a new culture whose politics lay both in the explicit messages it promulgated and in the way it modeled what the United States could be.[37] Occupy's "human microphone" makes this mode of extra-familial, extra-national socio-synchronicity especially clear: as Occupiers repeated a speaker's words in unison, they expanded the resonance of these words *in other bodies* across both the time it took the chant to reach the outer ring of the audience, and the space delineated by the encampment itself. Seen through the human microphone, Occupy is not just a vision of a better *polis*, it is, for a time, that polis. Occupiers and the 99% they stood for were not just an abstract idea of "the people"; they also were a new form of peoplehood.[38]

* * *

Synchrony and anachrony do not, in the end, mirror one another. Rather, they asymmetrically hinge upon one another. Without synchrony, there can be no engroupment of the sort that is a prerequisite for the historical thinking that makes anachronism, and eventually anachrony, possible. Without anachrony, it is difficult to imagine ways to rupture the homogeneous, empty time that synchrony produces and depends upon. Each of these forms of time creates modes of temporal being that both critique and reorganize the times in which and according to which we live. But surprisingly, "now" may be a time when synchrony produces effects that we are only just enacting and apprehending.

NOTES

1. My thanks to Colin Milburn for the loan of his *X-Men* comics collection, to Matt Franks for assistance with preparing this manuscript, and to Amy Elias and Joel Burges for astute editorial help.
2. On the synchronic and diachronic methods, see Ferdinand de Saussure, *Course in General Linguistics* (Chicago: Open Court Publishing Company, 1998), 89–93. Originally published as *Cours de linguistique générale* (Lausanne and Paris: Librairie Payot & Cie, 1916). On "ahistoricism," see Valery Rohy, "Ahistorical," *GLQ* 12 (2006): 61–83.
3. A good example of scholarship warning against anachronism is David Halperin, *How to Do the History of Homosexuality* (Chicago: University of Chicago Press, 2004). This form of anachronism might more properly be termed "prochronism," when something appears in a historical context in which it has not yet been invented.
4. See Dana Luciano, *Arranging Grief: Sacred Time and the Body in Nineteenth-Century America* (New York: New York University Press, 2007).
5. On instantaneity, see Jonathan Crary, *24/7: Late Capitalism and the Ends of Sleep* (New York: Verso, 2014).
6. Anna Gibbs, "After Affect: Sympathy, Synchrony, and Mimetic Communication," in *The Affect Theory Reader*, ed. Melissa Gregg and Gregory J. Seigworth (Durham, NC: Duke University Press, 2010), 186–205. On *habitus*, see Pierre Bourdieu, *Outline of a Theory of Practice* (Cambridge, UK: Cambridge University Press, 1977). For a perspective from the hard sciences, see Steven H. Strogatz, *Sync: How Order Emerges from Chaos in the Universe, Nature, and Daily Life* (New York: Hachette Books, 2004).
7. On military drills, see William H. McNeill, *Keeping Together in Time: Dance and Drill in Human History* (Cambridge, MA: Harvard University Press, 1995). On monastic time, see Eviatar Zerubavel, *Hidden Rhythms: Schedules and Calendars in Social Life* (Chicago, IL: University of Chicago Press, 1981).
8. Walter Benjamin, "Theses on the Philosophy of History," in *Illuminations*, ed. Hannah Arendt, trans. Harry Zohn (New York: Schocken Books, 1969), 253–64.
9. Benedict Anderson, *Imagined Communities: Reflections on the Origin and Spread of Nationalism*, rev. ed. (Brooklyn, NY: Verso, 1991).
10. See, e.g., Marx, *Capital, Vol. 1: A Critique of Political Economy* (1867; repr. New York: Penguin, 1992).
11. Edward Winslow Taylor, *The Principles of Scientific Management* (1911; repr. New York: Harper, 1929).
12. Siegfried Kracauer, "Girls und Krise" (*Frankfurter Zeitung*, 27 May 1931), quoted in Karsten Witte, "Introduction to Siegfried Kracauer's 'The Mass Ornament,'" trans. Barbara Correll and Jack Zipes, *New German Critique* 5 (1975): 63.
13. Witte, "Introduction," 64.

14 Witte, "Introduction," 62. She quotes and refers to Benjamin's "Epilogue" to "The Work of Art in the Age of Mechanical Reproduction," *Illuminations*, 241.
15 Cultures that understand the dead to coexist with the living are outside the scope of this essay, but it's imaginable according to their logic that a spirit could return to the wrong moment.
16 See Julia Kristeva, "Women's Time," trans. Alice Jardine and Harry Blake, *Signs* 7, no. 2 (1981): 13–35.
17 Reinhart Koselleck, *Futures Past: On the Semantics of Historical Time*, trans. Keith Tribe (Cambridge, MA: MIT Press, 1985), xxii.
18 Johannes Fabian, *Time and the Other: How Anthropology Makes Its Object* (New York: Columbia University Press, 2002).
19 Srinivas Aravamudan, "The Return of Anachronism," *MLQ* 62, no. 4 (2001): 331–53.
20 Siobhan Somerville, *Queering the Color Line: Race and the Invention of Homosexuality in American Culture* (Durham, NC: Duke University Press, 2000). See also Sander Gilman, *Difference and Pathology: Stereotypes of Sexuality, Race, and Madness* (Ithaca, NY: Cornell University Press, 1985).
21 On atavism and decadence, see Dana Seitler, *Atavistic Tendencies: The Cultural Science of American Modernity* (Minneapolis: University of Minnesota Press, 2008).
22 See Ernst Bloch, *Heritage of Our Times* (Cambridge, UK: Polity Press, 2009).
23 Carl Jung, *The Symbolic Life: The Collected Works of Carl Jung*, vol. 18 (Princeton, NJ: Princeton University Press, 2014), 287.
24 Benjamin, "Theses," 255.
25 Fredric Jameson, *The Political Unconscious: Narrative as a Socially Symbolic Act* (Ithaca, NY: Cornell University Press, 1982), 102. On the ghost as social critique, see, e.g., Jacques Derrida, *Spectres of Marx: The State of the Debt, the Work of Mourning, and the New International* (New York: Routledge, 2008); Avery Gordon, *Ghostly Matters: Haunting and the Sociological Imagination*, 2d ed. (Minneapolis: University of Minnesota Press, 2008); Molly McGarry, *Ghosts of Futures Past: Spiritualism and the Cultural Politics of Nineteenth Century America* (Berkeley: University of California Press, 2008).
26 *Groundhog Day*, directed by Harold Ramis (1993; Culver City, CA: Sony Pictures Home Entertainment, 2001), DVD; *The Matrix*, directed by Andy Wachowski and Lana Wachowski (1999; Burbank, CA: Warner Home Video, 2001), DVD; *The Bourne Identity*, directed by Doug Liman (2002; Universal City, CA: Universal Studios Home Video, 2004), DVD; *50 First Dates*, directed by Peter Segal (2004; Culver City, CA: Sony Pictures Home Entertainment, 2004), DVD; *Eternal Sunshine of the Spotless Mind*, directed by Michel Gondry (2004; Universal City, CA: Universal Studios Home Video, 2004), DVD; *Source Code*, directed by Duncan Jones (2011; Universal City, CA: Summit Entertainment, 2011), DVD.
27 Julia Alvarez, *How the García Girls Lost Their Accents* (Chapel Hill, NC: Algonquin Books of Chapel Hill, 1991); Martin Amis, *Time's Arrow, or, The Nature of the*

Offense (New York: Harmony Books, 1991). See also *The Sweet Hereafter,* directed by Atom Egoyan (1997; Los Angeles: New Line Home Video, 1998), DVD; and *Memento,* directed by Christopher Nolan (2000; Culver City, CA: Columbia TriStar Home Entertainment, 2001), DVD.

28 Dipesh Chakrabarty, *Provincializing Europe: Postcolonial Thought and Historical Difference* (Princeton, NJ: Princeton University Press, 2000).
29 Rogue first appeared in *Avengers Annual* 10 (New York: Marvel Comics, November 1981), and joined the X-Men in *Uncanny X-Men* 1, no. 171 (New York: Marvel Comics, July 1983). She has remained a consistent character in various X-Men titles, and appeared in her own spin-off comic, *Rogue,* beginning in 1995.
30 Friedrich Nietzsche, "The Uses and Disadvantages of History for Life," in *Untimely Meditations*, ed. Daniel Breazeale, trans. R. J. Hollingdale (Cambridge, UK: Cambridge University Press, 1997).
31 *Uncanny X-Men* 1, no. 318 (New York: Marvel Comics, 1994), 10.
32 *Generation X* 1, no. 31 (New York: Marvel Comics, 1994), n.p.
33 William S. Condon, "Communication and Empathy," in *Empathy* 2, ed. Joseph Lichtenberg, Melvin Bornstein, and Donald Silver (Hilldale, NJ: Analytic Press, 1984), 36–37. Quoted in Gibbs, "After Affect," 197.
34 For documentation of Die-Ins, see the ACT UP Oral History Project, coordinated Jim Hubbard and Sarah Schulman, http://www.actuporalhistory.org/index1.html.
35 See the original upload of CPDRC inmates' "Thriller," YouTube video, 4:26, posted by byronfgarcia, July 17, 2007, http://www.youtube.com/watch?v=hMnk7lh9M3o. For an early, albeit uncritical, discussion of the Cebu "dancing inmates," see Ian McKinnon, "Jailhouse Rocks: Philippine Inmates' Thriller Routine an Internet Hit," *Guardian*, July 27 2007, at http://www.theguardian.com/world/2007/jul/27/ianmackinnon.
36 Lauren Berlant, *Cruel Optimism* (Durham, NC: Duke University Press, 2011); Sianne Ngai, *Ugly Feelings* (Cambridge, MA: Harvard University Press, 2007); John Protevi, *Political Affect: Connecting the Social and the Somatic* (Minneapolis: University of Minnesota Press, 2009).
37 Judith Butler, "Bodies in Alliance and the Politics of the Street," lecture held in Venice, September 7, 2011, in the framework of the series "The State of Things," organized by the Office for Contemporary Art Norway (OCA), at http://www.eipcp.net/transversal/1011/butler/en. For an astute analysis of this lecture, see Michael O'Rourke, "Time's Tangles," *Social Text*, "Social Text: Periscope" dossier on Elizabeth Freeman's *Time Binds, July 10, 2014, at http://socialtextjournal.org/periscope_article/times-tangles/
38 For a sage analysis of the human microphone, see John Protevi, "Assembly, Political Space and the Human Microphone," *New APPS: Arts, Politics, Philosophy, Science*, October 24, 2011, at http://www.newappsblog.com/2011/10/assembly-political-space-and-the-human-microphone.html. My thanks to Michael O'Rourke for the reference.

8

Human / Planetary

HEATHER HOUSER

After exploring the landscapes and histories of the Lake Superior region, poet Lorine Niedecker announces in a letter to Cid Corman, "I'm going into a kind of retreat so far as time (going to be geologic time from now on!) is concerned."[1] Writing in 1966, at the dawn of so-called modern environmentalism,[2] she wishes to break through the dam separating the human and the planetary. Conceptual silos enclose these domains despite the obvious fact that (for now) being human requires being of planet Earth. Niedecker's project of geologic timekeeping names time as one of the categories of thought that has opposed both human and nonhuman time. Human biological and planetary rhythms partially harmonize around seasonality, but technological interventions—from the factory clock to the electric light bulb—have increasingly moved humanity out of temporal alignment with the nonhuman. And yet human time and "nonhuman" technological time are not tidy synonyms either. Rather, a third term—the inhuman time of instrumentation, computation, and mathematicization—mediates the entrenched binary of human and planetary time. Inhuman time widens the cognitive and affective gap between humans and the geophysical and ecological processes to which we would be fully subject without machine aids.

Time-inflected understandings of humanity and planetarity are not new to the contemporary, nor are the processes of modernity that cleave them. What is emergent are climate realities that show the human-planetary opposition to have been an incomplete one. Conceptualizing climate change requires thinking about the fusions of human and nonhuman systems in ways that chime with Niedecker's exclamation. The rallying cry "geologic time from now on!" concisely captures this fusion. "Geologic" typically evokes a past whose residues await discovery in the present; it thus captures the then, the "now," and what carries "on" into the future. We all (and this is a planetary "all") must get on geologic time if we're to understand and address climatic disturbance, the social and political meanings of which have gained urgency since climate scientists such as James Hansen placed global warming on the U.S. political agenda in the late 1980s. In this same period, the technologies and media by which nonspecialists access information about climate change have diversified and played a greater role in public discourse. If the first concern of this essay is with the (incomplete) bifurcation of human and planetary time, the second concern is with how the representational tools of inhuman time give access at once to the rift between these domains and to their integration.

Human

As the number and variety of keywords in this volume attest, the category of the human is shot through with time in myriad ways. Opening the aperture widely, we can say that humans are accustomed to operating on timescales organized around the individual, family, community, fiscal unit, or legislative entity. The categories of time might be biological (menstrual cycle, lifetime), chronometric (second, hour), calendrical (day, decade), historical (election cycles, technological development), or economic (business cycles), but they are all to some extent phenomenological. To take work as an example, we calibrate experience and expectations to the pulses of the lunch break, the financial quarter, and the career. We are aware of and manipulate these tempos, and this awareness often indexes human exceptionalism, as do capacities of retrospection and prospection. Indeed, philosopher Charles Taylor hangs the possibility for self-identity and agency on faculties of futural thought.[3] Crucial to this process of identity-formation are representing and giving shape to time. We are said to work on it and against it in ways that even sophisticated creatures that plan for the future (for example, birds building nests for their young) do not.

One of the unique characteristics of the present, however, is that the range of time concepts keyed to human phenomenological experience will not suffice for apprehending environmental crises. To bend historian Fernand Braudel's words for contemporary climate disturbance, "A day, a year once seemed useful gauges. Time, after all, was made up of an accumulation of days," but living in our weirding world "demand[s] much wider terms of reference."[4]

Enter the Anthropocene.

The latest buzzword in environmental discourse, "Anthropocene" accentuates humanity and planet as interdigitating concepts of time.[5] It designates an epoch in which human activities, such as releasing greenhouse gases (GHGs) into the atmosphere, substantially alter geophysical systems. The concept makes of humans the modifiers of the planet itself—ratios of atmospheric compounds, surface air temperature, mass species migration—and not only of its land and seas. If in Braudel's vivid formulation, "[e]ach social reality secretes its own peculiar time, or time scale, like common snails," then current climate reality secretes a many-layered shell.[6] The time concept of the Anthropocene is both popular and controversial because it engenders complex time problems of the sort Niedecker pondered: how to think humanity as geologic and vice versa. This confrontation is present in the Holocene, which began 11,500 years ago. But while Holocene humans had effects "likely significant for the ecology of . . . continents over large areas, there is no evidence that they had any appreciable impact on the functioning of the Earth system as a whole."[7]

Chemist Paul Crutzen and biologist Eugene Stoermer proposed the neologism "Anthropocene" in the *Global Change Newsletter* in 2000, and Crutzen elaborated on the concept for *Nature* in 2002. But it has only recently—and with gusto—entered the lexicon of environmental researchers and spread widely.[8] In 2015, the International Commission on Stratigraphy was in a prolonged debate over whether to add the epoch to the official Geologic Time Scale (GTS) after the Holocene. Of course, the fossil record is rich with evidence of human and nonhuman species terraforming Earth; after all, it's the intensity, not the fact, of anthropogenic transformations—specifically agriculture and civilization building—that differentiates the Holocene from the Pleistocene. The Anthropocene sets itself off from the preceding epoch for the number of human-altered processes it entails: the climate, certainly, but also "element cycles, such as nitrogen, phosphorus and sulphur . . . the terrestrial water cycle . . . and . . . the sixth major extinction event in Earth history."[9] Historians and geologists volley proposals for a start date to the Anthropocene but commonly place it in the early nineteenth century, when the intensive burning of coal for industrial production began elevating atmospheric carbon dioxide (CO_2) concentrations.[10] The beginning of the Cold War is an alternative origin point, or at least a pivot within the epoch designated "the Great Acceleration." As Will Steffen et al. remark, "every indicator of human activity underwent a sharp increase in rate around 1950": economic and population growth, urban migration, fossil fuel burning, agricultural industrialization, air and automotive mobility, and technological and communications innovation.[11] With the Anthropocene, the GTS will formalize the planetary impacts of these human processes. And it will inscribe them as irreversible; after all, thus far the GTS travels in only one direction.

Scholars within the humanities have studied conditions contributing to the Anthropocene for decades, but Dipesh Chakrabarty's 2009 article "The Climate of History: Four Theses" sent the concept echoing through academic halls.[12] This is the challenge of the proposed epoch: how can humans retain their "historical sensibility," especially with respect to political economy and anthropological difference, while also understanding ourselves as agents who alter the planet *as a species*?[13] How can historians and cultural producers "put global histories of capital in conversation with the species history of humans"?[14] As Chakrabarty elaborates in a 2014 talk on the topic, to adapt to the interpenetration of geophysical processes and human history, historiography must develop an analytics of capital that incorporates a phenomenon that exceeds the history of capital.[15] His questions about whether it is phenomenologically and historiographically possible for humanity to conceive of itself as a species are squarely anthropocentric.[16] For some environmentalists, though, promoting the human through the Anthropocene brings concern that, instead of eliciting humility and responsibility for planetary harms, the concept will encourage "species narcissism."[17]

Inhuman

How might climate change media conceptualize humanity in terms of the planetary or block that correspondence? I argue that a third temporal form, that encoded in inhuman devices, mediates the dance between human and planet. Inhuman time variously jams this ontological distinction or enforces human exceptionalism. Inhabiting and representing the odd pulses of climate crisis will increasingly be key experiences of twenty-first-century existence, and will be aided by advanced supercomputers and graphics technologies. Carbon footprint calculators and climate model visualizations exhibit these pulses and cross the domains of scientific research, commerce, and culture. Their aesthetic strategies for figuring time activate cognitive procedures that alternately set in motion or stall the human-planetary dialectic. In all cases, climate change visualizations establish that the human-planetary relation depends as much on inhuman temporal framing as on familiar, if contested, differentiations between types of species and matter.

Carbon calculators, for example, are overtly commercial and put climate threat directly into the wallets of individuals, businesses, and municipalities in their roles as consumers and emitters. Calculators body forth the species *homo economicus*. This living being drains fossil fuel reserves and threatens human and nonhuman species survival, but the calculators primarily emphasize users' contributions to GHGs as a drain on fiscal reserves. Doing so, they rescale and familiarize climate change, defying environmental critics' widespread evaluation that "[a]ny phenomenology of climate change . . . inheres in an eerie disconnection from sensate experience."[18] Might carbon calculators make climate threat sense-able by quantifying it in the dollars and cents that fuel households, businesses, and governments rather than the datasets that fuel climatological research and modeling?

Global Footprint Network (GFN) is a nonprofit consultancy that, in this way, calculates groups' contributions to climate disruption and, for a fee, advises how to reduce it. Its goal is to make the metric of carbon emissions as meaningful to policy and business planning as gross domestic product currently is. All carbon calculators are overtly prescriptive; they posit tidy correlations between inputs and outputs and thus show climate disturbance to be as manageable as a checking account. GFN's calculator "measures how much nature we have, how much we use, and who uses what."[19] The open online version computes how many "earths" are necessary to support the user's lifestyle for a year. The result for the average American will shatter all optimism for steering humanity out of the Anthropocene. Yet calculators such as GFN's eschew pessimism and instead suggest lifestyle modifications that anyone can institute to trim off carbon excess while fattening one's wallet.

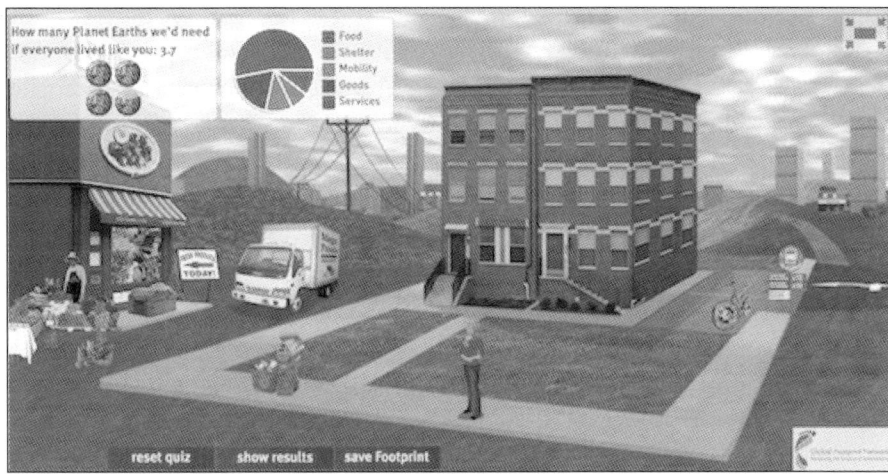

Figure 8.1. "Footprint Calculator," *Footprintnetwork.org*, Global Footprint Network, last modified September 8, 2011, http://www.footprintnetwork.org/en/index.php/GFN/page/calculators/

Turning calculation into a game, GFN alleviates the negative feelings that arise when visualizing one's responsibility for eating up the earth. To the accompaniment of ambient music, the user creates an avatar that walks through a graphical background onto which her or his resource-intensive environment grows (see figure 8.1). GFN is one of the few calculators that marshal metaphor to envision the effects of burning hydrocarbons. The visualization tool converts a figure for GHG emissions (e.g., ten thousand pounds of CO_2 emitted per year) into consumption of the planet itself. In most calculators, the number stands on its own; it has meaning insomuch as it reflects well or poorly on the user—akin to weight, cholesterol, or debt amount. Measuring ongoing patterns of carbon use and then adjusting those patterns in the future are behaviors of the same type as frequenting a gym, attending Weight Watchers, or reviewing a household budget. These tools share a family resemblance with genres of self-improvement and self-accounting, such as diaries and checkbooks, self-help manuals and lifestyle magazines. Calculators punctuate actions in the now that, in effect, make one a measurably "better" person.

Not surprisingly, they are thus pegged to a personal continuum of human time. GFN nods to the planetary scope of climate change by conceptualizing CO2 emissions in terms of the regenerative powers of the Earth. Without adopting the term, it seems to endorse an Anthropocene perspective. Ultimately, however, the calculator paints a picture of a lived present extending out only to the next electric bill or quarterly financial statement and retains faith in the powers of self-determination and -actualization. Elevated GHG concentrations in the atmosphere have immediate solutions, and the *oikos* holds the tools to make those repairs. The calculators are inherently optimistic. While past emissions are fixed, these cultural objects of quantification goad users to project an improved personal scenario. They underscore the traffic between economy and ecology evident in the words' shared etymon and hail an economic subject who can alleviate the planetary harms of contemporary life.

Even as the GFN calculator describes GHG emissions as a release of Earth matter, it does not acknowledge the *protracted* nature of their effects. Specifically, calculators gesture toward but do not address a knotty time problem within climate ethics: we won't reap the rewards of present GHG asceticism. And this is where the shoe pinches. Present CO2 reductions will do little for those currently living on the planet because the carbon cycle moves to tempos different from human lifecycles. As geoscientist David Archer instructs, "After several centuries when the oceans have inhaled their fill, a significant fraction of the fossil fuel CO2 will remain in the atmosphere, affecting the climate for millennia into the future."[20] Calculators are oriented toward an extended present and leave this lag out of the equation. In the final analysis, their temporal orientation correlates to an isolated and unitary conception of *homo economicus*.

Planetary / Nonhuman

As carbon calculators encode it, the Anthropocene bespeaks human self-regard and perhaps even hubris: we made this epoch and, by God, we'll unmake it! But the concept can also position humanity in multidimensional planetarity. Indeed, the promise of the Anthropocene is that it could point out how—or at least that—humanity embeds itself in several layered and seeping temporal dimensions without suggesting that human agency unidirectionally and uniformly determines them.

The tight opposition between human and nonhuman that modernization has soldered might then unfix.

Sociocultural theory has a vocabulary—in fact, many vocabularies—for expressing the temporal dynamism of humanness. By contrast, planetary time too easily reduces to a form of stasis-in-flux. That is, it pulses with the cyclical cadences of diurnality, seasonality, jet streams, and even ice ages, but these cadences become imperceptible under the weight of the planet's imagined solidity and constancy.[21] Particularly in its iconography, the planet tends toward fixity rather than dynamism. In NASA's famous "Blue Marble" photograph from 1972, Earth hovers in faint relief against a black background. Cloud formations intimate air flows and meteorological flux, but the predominant sensation is of a material object unchanging in space and time. If we were to drill into this object, we would collide with the geological record that counters this perception: mineral strata, ice cores, and hydrocarbon reserves are repositories of time. This record inscribes human history but does not reduce to it. In fact, the concept of the "planetary" captures forces, processes, and artifacts such as thermodynamics or Earth rotation that always exceed human making but that cannot always resist it. Above all, the planetary is subject to alteration but is rarely at risk of utter annihilation.

After humans have extracted and released geologic reserves from the earth for centuries, planetary time becomes irrefutably visible. GHG emissions might be additive from an atmospheric perspective, but they are subtractive from a geospheric one. The most mundane acts that one performs—taking a flight or eating vegetables—require devouring the prehistoric as materialized in hydrocarbon deposits and aquifers. To put it in literary scholar Michael Ziser's more numinous terms, "Our Western bodies, homes, communities, governments, arts, and ideologies are—in a sense that is still mysterious and infrequently spoken of—expressions of the mystical surplus of energy that is fossil fuel."[22] Merely existing in late-capitalist modernity places one in a historical present of extracting, processing, and consuming hydrocarbons. The Anthropocene then concatenates the enduring geologic past and long present and, rhetorically, orients toward a degraded future to be averted or an alternative one to be realized. As we visualize releasing fossilized time into the atmosphere through human activity and formalize the consequences in the Anthropocene, how might the technologies of inhuman time overcome a trenchant conceptual cleavage of human and planet and achieve their ontological integration?

When one poses this question, it becomes apparent that one of the ironies of visualizing the Anthropocene is that it relies on the technologies of the very Great Acceleration that has led to the epochal break. I speak here specifically of climate system models. Broadly considered, climate models use principles of physics and geoscientific datasets to simulate climate processes and perturbations. They developed out of regional weather forecasting when computer processing power swelled in the 1960s and have become heuristics for understanding the whole Earth's climate system.[23] Models are technological marvels and morasses: marvels for the numerous domains of data and expertise they incorporate and the stochastic and even chaotic climate mechanisms for which they account; morasses because they are contested reference points in climate policy debates. Today's most discussed models account for the sensitivity of climate to changing anthropic forcings—that is, those agents that affect the system's energy balance and raise or lower temperatures. Designed in principle for specialist audiences of climate scientists and policy wonks, models in practice circulate widely and detach from their explanatory contexts. Consequently, the representational conventions that model visualizations employ bear an even greater burden of signification.

Cultural as well as quantitative objects, models simulate not only the climate system and its sensitivity to anthropic forcings; their visualizations also simulate the temporalities of the Anthropocene. How and whether planetary and human time interlock in these artifacts depends on how inhuman time cuts across them. Fundamentally, climate modeling advances only when computer processing speeds accelerate, and two ways in which they advance is by incorporating more historical data and projecting further into the future. The inhuman temporalities of modeling are intricate, however, and progress is not linear. As models get more complex by covering longer timespans and running calculations at smaller time intervals, processing time slows down accordingly. Multiple temporal seams thus run through models, and the geologic and computational temporal strata that inscribe human time become particularly poignant in the model visualizations. If we think of "modeling as world building" in line with historian Paul Edwards, then models are artifacts of the imagination as much as they are symbiotic with data.[24] As they mediate Anthropocene data and experience through strategies of temporal representation, they do the conceptual and imaginative work of engendering human-planetary ontologies.

The style of model visualizations ranges from graphically simple but efficient line graphs to stunning multimodal animations. As one of the world's leading modeling organizations, the Geophysical Fluid Dynamics Laboratory (GFDL) of the U.S. National Oceanic and Atmospheric Association (NOAA) generates some of the most sophisticated visualizations. "Extinction Optical Depth by Aerosol in 2012," by atmospheric scientist Paul Ginoux and his team, employs the "Blue Marble" photograph as its base layer (see figure 8.2).[25] With this iconic image, the visualizers eschew a diagrammatic style geared only to specialists. The whole earth image notwithstanding, the visualization seems to depict a fantastical land. In the foreground, an animation shows the absorption and scattering of aerosols such as dust (red), sulfate from fossil fuel combustion (gray), sea salt (blue), and burning carbonaceous matter (green). These are all compounds that act as naturally occurring or anthropic forcing agents on climate. The idea of a "forcing" takes physical shape in the green cloud forms that resemble a disquieting miasma blanketing regions of Earth. The swirling action of the aerosols is hypnotic, fixing attention on whether they are streaming across the screen or throbbing in place. Following the aerosols' motion immerses the viewer in the rhythm of the visualization. Along with the constant throbs and flows, flashes of bright color—red over Baja and North Africa, gray over East Asia—indicate when a new compound peaks and atmospheric composition consequently mutates.

The asynchronous movements and ethereal palette take viewers out of the idiom of scientific representation and also obscure data elements lying at the edge of the image. Inhuman timepieces enframe the temporalities of aerosol flows; they both reintroduce the idiom of technoscience and place geophysical activity on a human timescale of seconds, hours, days, and months. In the lower border, a timer counts out the model period: every day of 2012 divided into six-hour segments. Below that—or superimposed over that, depending on the media player and browser one uses—a time slider ticks off the seconds of the animation. As these inhuman temporal devices position the viewer in varied human

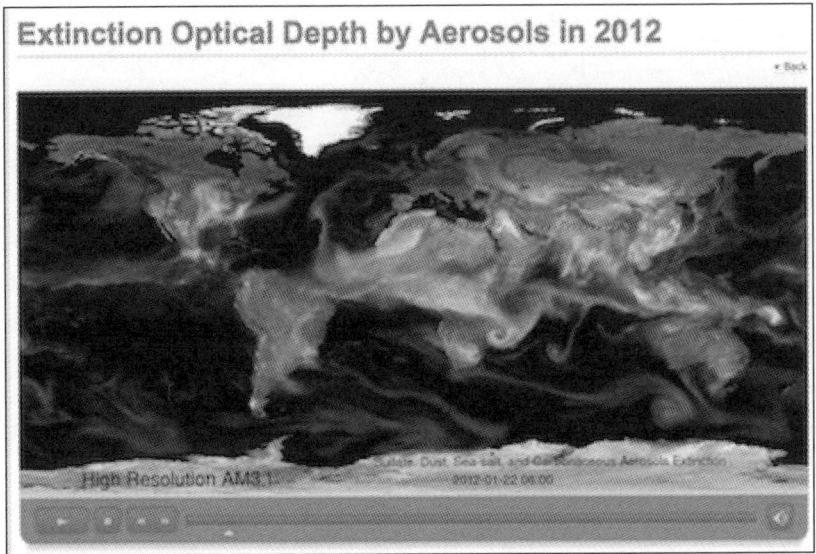

Figure 8.2. Paul Ginoux, "Extinction Optical Depth by Aerosols in 2012," NOAA/GFDL, created December 2012, http://www.gfdl.noaa.gov/visualizations-aerosols-and-clouds

timescales, they also dislocate and even frustrate time sense. Holding all components of the data visualization in view at once is impossible. The user must pause the animation and convert it into a fixed frame, collocate the two time stamps, and align them with the geographical location of the colors and the colors' meaning.

Pausing, collocating, and correlating, toggling between stillness and movement: this viewing procedure is also an epistemological one. Without it, the piece cannot cohere as a knowledge object rather than just an alluring picture. This epistemological procedure is common to other climate visualizations, notably GFDL's "Surface Air Temperature Anomalies" (2005).[26] In addition to toggling between several data elements, the viewer also places the present in a continuum spanning back to 1971 and projecting into a future just outside her or his own lifespan, 2100. As it traverses 140 years in one-year increments, the upper-left-hand animation ranges through a familiar color scheme for representing temperature: from cerulean blue to orange-red. Each color corresponds to anomalies relative to the mean established from 1971 to 2000. With few exceptions, each region of Earth tears toward the red, a color whose mundane association with heat is in this case outstripped by its association with threat and passionate emotion. The scenario that unfolds confirms the observation by climate scientists Noah Diffenbaugh and Christopher Field that, "despite important uncertainties about the magnitude of future global warming, several sources of inertia make some future climate change a virtual certainty."[27] As the animated globe reddens, the viewer juggles an even greater number of components than the aerosol piece contains: a line graph depicting temperature anomalies, a legend for the color scale, a more schematic graph of the anomalies by latitude, and finally, the timer ticking out the model year.

In these pieces, as in a preponderance of model visualizations, what seems a smooth, sophisticated image is striated with several disjunctions, or "sutures," to borrow from art historian Barbara Stafford. These sutures "force homogeneous data to exhibit its heterogeneity."[28] And, in the case of models, they force an apparently seamless image to exhibit the heterogeneities of its data. The visualizations incorporate time-based components that do not readily fuse. The sutures that surface through the temporal aesthetics do not simply disturb knowledge; they also enable a productive form of epistemological friction that reorients human self-understanding. Sutures weaken any sense that the data situation—and that of the imperiled planet—has been mastered, a sense that the computational sophistication of the technology encourages. They underscore the limitations of human capacities to quantify, display, and control climate perturbations. Out of the hermeneutic disturbance that arises from crossing model temporalities, the human acquires a curious ontological status. If, as argued above, calculators body forth a *homo economicus*, these models body forth a being that is both human and not, planetary and not.

The very processes of modeling and visualizing exhibit their creators' desires to control the datascape of the climate system. Yet mastery is elusive because models are also always records of uncertainties, errors, and incompletenesses that counterbalance epistemological mastery. For example, the mind-blowing number of code lines that comprise a model—hundreds of thousands to millions—signal computational sophistication, but the potential for error is proportionally vast. A tension between mastery and humility is also apparent in the aerial perspective most visualizations employ. The view from above implies epistemologi-

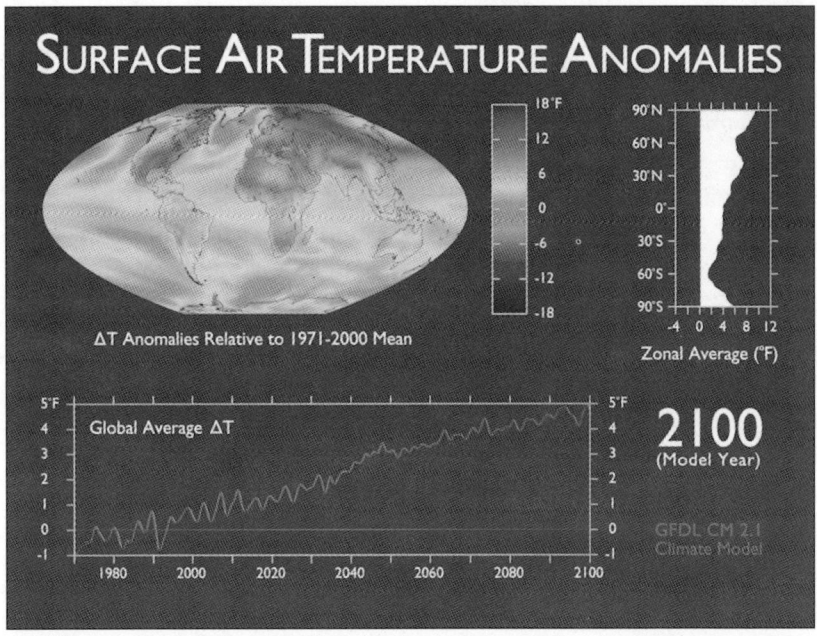

Figure 8.3. Keith Dixon, "Surface Air Temperature Anomalies," NOAA/GFDL, created October 2005, http://www.gfdl.noaa.gov/visualizations-climate-prediction

cal privilege.[29] Flattened for scopic consumption, the Earth appears to be a quickly knowable quantity. However, the sutures in the recording data and the hermeneutic cobbling necessary for grasping the image's meaning make the assembly more jagged than smooth. Though "[c]limate simulations are based on the assumption that nature can be quantified," as Myanna Lahsen writes,[30] the inhuman mediation of human and planetary time ultimately underscores limitations of human capacities to mathematize and display. Unlike carbon calculators, which soothe by making climate disruption manageable, model visualizations agitate by producing ontological and epistemological uncertainty *at the same time as* they aim to manage climate information.

Just as sutures in the image suspend the viewer between epistemological mastery and humility, they also suspend her or him between ontological positions. Humans hold a peculiar place: we are the only entities that simultaneously create and manipulate the data and are implicated in it. On the one hand, we are merely one of many cogs in the geophysical machinery. Models render humans nonhuman by grouping us with other planetary forcing agents. On the other hand, the reason we need models at all is humanity's outsized influence on Earth. In essence, the excesses of the human, our large-scale, Earth-altering industry, have transformed us into something other than ourselves. In this respect, we inhabit dualities.

Models provoke Slavoj Žižek's question: "To what extent can we say that, in confronting the Otherness of Nature, humanity is confronting its own essence?"[31] In climate visualizations, this confrontation occurs through frictions of planetary and human temporalities that inhuman time sets in motion. These climate media put the varied rhythms of environmental change into relief and show those rhythms to be at once antagonistic and intertwined as well as partially synchronous. In the friction between data components, the "human" splits, but not into mutually exclusive components. Rather, as viewers travel between temporal domains, they hold in tension human exceptionality and planetary inclusion, epistemological control and failure.

As Rob Nixon reminds us, there is no tipping point from quantification to social change.[32] Nor can the artifacts that mediate data and the phenomenal world simply reverse the course of the Anthropocene. Though by no means a panacea for climate action and the inconvenient changes that will mitigate GHG emissions, digital climate media pulse to time signatures that animate human-planetary dynamics. These artifacts and their inhuman temporalities either fix the horizon of prosperity to the next bank statement, or they imply that this horizon has been set too close and must expand out to the next epoch. The human becomes other to itself through the fractures in temporality that model visualizations encode, or it coheres around the timescale of personal economy in calculators. These positions are inhabited as dualities just as climate consciousness is shot through, to revive Niedecker's terms, with the geologic, the now, and the from-now-on.

NOTES

1 Lorine Niedecker, *Lake Superior: Lorine Niedecker's Poem and Journal along with Other Sources, Documents, and Readings* (Seattle: Wave Books, 2013), 53.
2 Scholars typically date this era of environmentalism in the U.S. to the publication of Rachel Carson's *Silent Spring* (Boston: Houghton Mifflin, 1962) and U.S. legislation such as the Clean Air (1963), Wilderness (1964), and Water (1977) Acts.

3   Charles Taylor, *Sources of the Self: The Making of Modern Identity* (Cambridge, MA: Harvard University Press, 1989).
4   Fernand Braudel, "The *Longue Durée*," in *Historical Methods in the Social Sciences*, ed. John A. Hall and Joseph M. Bryant, vol. 2 (London: Sage, 2005), 248.
5   For the purposes of this essay, I use "human" as a unified biological designation rather than a strongly socially inflected one. It's important to recall, however, that the "species story" that the Anthropocene tells does not easily account for the "divergence story" of how greater disparities stratify populations, particularly rich and poor, North and South. See Rob Nixon, "This Brief Multitude: The Anthropocene and Our Age of Disparity," paper presented at ASLE Tenth Biennial Conference, Lawrence, KS, May 29, 2013.
6   Braudel, "The *Longue Durée*," 269.
7   Will Steffen, Jacques Grinevald, Paul J. Crutzen, and John McNeill, "The Anthropocene: Conceptual and Historical Perspectives," *Philosophical Transactions of the Royal Society A* 369, no. 1938 (2011): 847.
8   Paul J. Crutzen and Eugene F. Stoermer, "The 'Anthropocene,'" *Global Change Newsletter* 41 (2000): 17–18; Paul J. Crutzen, "Geology of Mankind," *Nature* 415, no. 6867 (2002): 23. There are antecedent concepts. Journalist Andrew Revkin coined the misnomer "Anthrocene" in 1992 in *Global Warming: Understanding the Forecast* (New York: Abbeville Press, 1992), 55. Further back, George Perkins Marsh conceived of humans as planetary agents; see *Man and Nature* (1864), ed. David Lowenthal (Seattle: University of Washington Press, 2003).
9   Steffen et al., "The Anthropocene," 843.
10 Steffen et al., "The Anthropocene," 849.
11 Steffen et al., "The Anthropocene," 849.
12 Dipesh Chakrabarty, "The Climate of History: Four Theses," *Critical Inquiry* 35, no. 2 (2009): 197–222.
13 Chakrabarty, "The Climate of History," 197. See also Chakrabarty, "Postcolonial Studies and the Challenge of Climate Change," *New Literary History* 43, no. 1 (2012): 1–18. Other historians and theorists, notably Richard White and Slavoj Žižek, have challenged the newness, or the nowness, of Chakrabarty's theses. See White, "Does Global Climate Change Change History?" (response to paper presented by Dipesh Chakrabarty at the Environmental Humanities Project, Stanford University, Stanford, CA, April 20, 2010); and Slavoj Žižek, *Living in the End Times* (London: Verso, 2010).
14 Chakrabarty, "The Climate of History," 212. Two decades before Chakrabarty posed his theses, Michel Serres observed the conditions that inspire them, writing, "Global history enters nature; global nature enters history: this is something utterly new in philosophy," *Natural Contract*, trans. Elizabeth MacArthur and William Paulson (Ann Arbor: University of Michigan Press, 1990), 4.
15 Dipesh Chakrabarty, "Beyond Capital: The Climate Crisis as a Challenge to Social Thought" (paper presented at Mahindra Humanities Center, Harvard University, Cambridge, MA, February 7, 2014).

16 Chakrabarty's lecture addresses the "rift" between anthropocentrism and ecocentrism and comes to the point that we need an "enlightened" anthropocentrism, one that recognizes that a more ecocentric orientation is in our best interests (Chakrabarty, "Beyond Capital," n.p.).
17 Nixon, "This Brief Multitude," n.p.
18 Jesse Oak Taylor, "The Novel as Climate Model: Realism and the Greenhouse Effect in *Bleak House*," *Novel* 46, no. 1 (2013): 1.
19 "At a Glance," *Footprintnetwork.org*, Global Footprint Network, last modified December 16, 2010, http://www.footprintnetwork.org/en/index.php/GFN/page/at_a_glance/.
20 David Archer, *The Long Thaw: How Humans Are Changing the Next 100,000 Years of Earth's Climate* (Princeton, NJ: Princeton University Press, 2009), 113.
21 What Mark McGurl calls "the new cultural geology" is a notable exception to a view of the planet that correlates to human exceptionalism. "Cultural geology" is "a range of theoretical and other initiatives that position culture in a time-frame large enough to crack open the carapace of human self-concern, exposing it to the idea . . . of its external ontological preconditions" (Mark McGurl, "The New Cultural Geology," *Twentieth-Century Literature* 57, no. 3–4 [2011]: 380). Wai Chee Dimock is a motive force of this method; see Dimock, *Through Other Continents: American Literature across Deep Time* (Princeton, NJ: Princeton University Press, 2006).
22 Michael Ziser, "Home Again: Peak Oil, Climate Change, and the Aesthetics of Transition," in *Environmental Criticism for the Twenty-First Century*, ed. Stephanie LeMenager, Teresa Shewry, and Ken Hiltner (New York: Routledge, 2011), 183.
23 I cannot detail here the mathematics and climate science involved in climate modeling. For more on the history, politics, and technologies of climate modeling, see Amy Dahan Dalmedico, "Models and Simulations in Climate Change: Historical, Epistemological, Anthropological, and Political Aspects," in *Science without Laws: Model Systems, Cases, Exemplary Narratives*, ed. Angela Creager, Elizabeth Lunbeck, and M. Norton Wise (Durham, NC: Duke University Press, 2007), 125–56; Andrew Dessler and Edward A. Parson, *The Science and Politics of Global Climate Change: A Guide to the Debate*, 2d ed. (New York: Cambridge University Press, 2010); Paul N. Edwards, *A Vast Machine: Computer Models, Climate Data, and the Politics of Global Warming* (Cambridge, MA: MIT Press, 2010); and Kirsten Hastrup and Martin Skrydstrup, eds., *The Social Life of Climate Change Models: Anticipating Nature* (New York: Routledge, 2013).
24 Paul N. Edwards, "Representing the Global Atmosphere: Computer Models, Data, and Knowledge about Climate Change," in *Changing the Atmosphere: Expert Knowledge and Environmental Governance*, ed. Clark A. Miller and Paul N. Edwards (Cambridge, MA: MIT Press, 2001), 62.
25 "Extinction Optical Depth by Aerosol in 2012" is a high-resolution atmospheric model (HiRAM). Atmospheric models are built with equations "governing the flow of the atmosphere," as outlined by Peter Lynch, "The Origins of Computer

Weather Prediction and Climate Modeling," *Journal of Computational Physics* 227 (2008): 3432. For a breakdown of climate model types, see Edwards, *Vast Machine*, xiv–xvi.

26 "Surface Air Temperature Anomalies" is a high-resolution coupled climate model.
27 Noah S. Diffenbaugh and Christopher B. Field, "Changes in Ecologically Critical Terrestrial Climate Conditions," *Science* 341 (2013): 489.
28 Barbara Maria Stafford, *Good Looking: Essays on the Virtue of Images* (Cambridge, MA: MIT Press, 1996), 78, 77.
29 On the epistemologies and politics of the aerial perspective, see Denis Cosgrove, *Apollo's Eye: A Cartographic Genealogy of the Earth in the Western Imagination* (Baltimore: Johns Hopkins University Press, 2001); Heather Houser, "The Aesthetics of Environmental Visualizations: More than Information Ecstasy?" *Public Culture* 26, no. 2 (2014): 321–39; and Sheila Jasanoff and Marybeth Long Martello, "Heaven and Earth: The Politics of Environmental Images," in *Earthly Politics: Local and Global in Environmental Governance*, ed. Sheila Jasanoff (Cambridge, MA: MIT Press, 2004), 31–52.
30 Myanna Lahsen, "Seductive Simulations? Uncertainty Distribution around Climate Models," *Social Studies of Science* 35, no. 6 (2005): 899.
31 Žižek, *Living in the End Times*, 336, n22.
32 Nixon, "This Brief Multitude," n.p.

9

Serial / Simultaneous

JARED GARDNER

1905 is famously Albert Einstein's *annus mirabilis*, the year in which the patent office worker wrote a series of papers that would change forever the way in which physicists understood the universe, especially how time and its navigation would be reimagined in the twentieth century. After centuries of Newtonian physics, time was no longer absolute, nor was it any longer separate from the three dimensions that defined space. Einstein's special theory of relativity highlighted the paradoxical relationship between two seemingly contradictory models of time: *seriality*, as the model that corresponds with how we experience time; and *simultaneity*, as the model of time that emerges from Einstein's insights into relativity (and which would be subsequently reinforced—and complicated—by the quantum mechanics that followed).

It would take years for physicists fully to come to terms with the consequences of Einstein's insights, and longer still for these insights to make their way into popular culture. And more than a century later, our lived *experience* of time remains unable to match the insights of 1905. However, even before Einstein's discoveries, new narrative media were already engaged in experiments that would begin the slow process of bringing these two seemingly irreconcilable models of time into productive dialogue. Comics and film had been born a few decades earlier, themselves the product of a century of experimentation with optics and new insights into the nature of human vision. In 1905, the first nickelodeons opened, featuring as their first "hit" Edwin S. Porter's *The Great Train Robbery* and its pioneering innovations in parallel editing—what Tom Gunning describes as "a dialectical leap in the portrayal of space and time."[1] In the other new narrative medium of the twentieth century, comics, the first experiments with long-form storytelling began that same year, resulting in a new narrative form: the open-ended serial.[2] And this was to be only the beginning of the temporal experiments of the comics form. A century later we can now recognize that comics would be the form to make available storytelling and reading practices that finally fully embrace the insights into time Einstein first described in 1905.

Serial

The serial version of time we experience every day is often termed "tensed time." Our very grammar is adapted to naturalize it. This is the time of the moving present, that node at which a fixed but always-already lost past and an as-yet unreal future converge. Tensed time—what McTaggart called "A-series" time—imagines events as occupying variable positions "from the far past through the near past to the present, and then from the present to the near future and the far future."[3]

Despite being so deeply felt, this model of time has troubled philosophers and scientists for millennia. Relying on a model of movement—away from a past and toward a future—the question necessarily arises: at what rate does time move? The natural answer—time moves at one second per second—is necessarily unsatisfactory. As the earliest classical philosophers recognized, time as we experience it does not hold up to analysis, giving rise to the very different accounts of time offered by Parmenides on one hand and Heraclitus on the other. While their accounts of time could not have been more different—for Parmenides time is an illusion and only the *now* is real, while for Heraclitus, time is change and only *change* is real—both were premised on a similar revelation: time as we experience it simply does not make logical sense, therefore it must be either an illusion or a window into the divine (that is, the Truth).[4]

More than two millennia of scientific and philosophical inquiry have resulted in remarkable transformations in our understanding and perception of the world, but the conundrum of time confronted by Heraclitus and Parmenides remains. Despite numerous experiments and theories arguing for the incoherence in our "natural" sense of time, we continue to experience time in "tensed" terms, as a *moving* Now ceaselessly carried toward a Future and away from a Past, or, alternately, as a *fixed* Now toward which a moving Future relentlessly rolls while the Past recedes behind us. Time moves at one second per second, because, science and philosophy notwithstanding, this tautology continues to describe our *ex-*

perience of time (even if our experience is *not* in fact scientifically "true"), allowing us to coordinate an increasingly complex, networked society.

It is useful here to recall, as historians of time have demonstrated, that while time has troubled philosophy in changing ways, our *experience* of time has not been static. For example, Jacques Le Goff influentially described the shift in the late Middle Ages from "Church time" to "merchant's time."[5] Later, Niklas Luhmann argued that our modern conception of the future emerges only with the rise of bourgeois society in the seventeenth century and the newly found model of historical progress.[6] And with the *series rerum* as a model for temporal progress, we see the rise of racial pseudo-science, the modern nation state, and global capitalism. Print facilitated the imagination of a new experience of serial time, culminating in the rise of the novel and the emergence of its industrial—and serialized—forms (Dickens, the story paper, the dime novel) in the nineteenth century.[7]

Thus since the "discovery" of the future, storytelling has devoted itself to reinforcing a causal-temporal model of serial progress. As will be discussed below, film, one of the two new narrative media of the twentieth century, would ultimately prove a most powerful ally of serial time. Although born of experiments that might have led the medium in radically different directions, film in the early years of the twentieth century began exploring its ability to re-create time as we experience it—that is, as a series of events witnessed from the perspective of a moving now.

Simultaneous

If seriality describes our experience of time, the model of time premised on simultaneity is one that is seemingly beyond our experiential capacity. This model of time had circulated in philosophy and at the margins of theoretical physics long before Einstein, but it was his *annus mirabilis* that would provide the foundation for the first time for experiments that would prove this model's accuracy.

Einstein's thought experiment famously imagines two lightning strikes and two observers, one in a moving train and one on a platform.[8] For the observer on the platform, standing directly between the two strikes as they hit, the strikes are perceived as simultaneous, as the light moves toward the observer at the speed of light. For the observer in the train, moving forward in space toward the strike on the front of the car, the strikes are perceived as *serial*—lightning first striking the front of the train followed by the rear of the train. Science requires that we know which of the observers is right; the scientific answer, Einstein assures us, is that they both are true.

Einstein's special theory of relativity was conceived while he was working in a Swiss patent office, looking at countless new inventions for ever-more accurate clocks and related technologies designed to coordinate time across increasingly complex networks of trains, cities, and commerce.[9] What emerged was a new conception of time. Contrary to Newton's classical mechanics, and in violation of our faith in time as a universal measure, Einstein described how a body's movement through space impacts its movement through time. Ironically, it would be the atomic clock—the most accurate timepiece to date—that proved his theory. In 1971, two of the devices were flown around the world; when finally compared to devices safely on the ground, the clocks recorded different times.[10] No longer was time a universal constant: it was relative.

Volumes have been written about the philosophical and scientific work that laid the groundwork for Einstein's breakthrough,[11] but less recognized is the impact of an earlier but related shift growing out of optics, another field where Newton had long had the last word. Jonathan Crary has described the phenomenological adjustment that resulted from the displacement of the long-held "camera obscura" model of vision with one that acknowledged the idiosyncrasies of individual eyes and perspective.[12] It is from these insights that film and sequential comics were born at the end of the nineteenth century; moreover, this changing vision of vision has proved foundational for new representations of time in the twentieth and twenty-first centuries. As Arthur J. Miller has argued, Picasso's cubism and Einstein's relativity both were generated from a newfound need "to confront the concept of simultaneity."[13] But such experiments were not the province of science and high art alone, as related investigations in the popular arts would come to shape the arts of the present still more profoundly over the course of the next century.

Here we might begin, for example, with Muybridge's famous 1878 experiment with a series of sequential photographs capturing the running of a horse, designed to settle a bet, or so the legend has it, that all four of the beast's legs do indeed leave the ground simultaneously while in full stride:

Figure 9.1. Muybridge's sequential photographs of a horse in stride

As with Einstein's theory a generation later, this work served not to fulfill the Enlightenment promise that objective observation of reality was possible, but instead demonstrated the gap between our perception and the world itself. The demonstration, moreover, turned on a serial representation in search of a potentially simultaneous act in nature.

Narrative film as we know it—and, at least as we experience it within the traditional cinematic apparatus, with its powerful reification of tensed time—would famously emerge from this moment over the course of a generation, but in fact Muybridge's experiments initially made a more persuasive case for a very different model, as the sculptor Auguste Rodin recognized in denouncing Muybridge for attempting to make time "stand still."[14] Indeed, it is perhaps ironic that Muybridge's legacy would be the development of a cinema that prosthetically reinforced conventional time, when his experiments resulted in a representation of time that in many ways supported those—from Zeno to Einstein—who argued that our experience of time was not in fact the thing itself.

However, a different and new narrative form actually followed through on the latter possibilities. In comics, the individual moments of time are *not* projected at twenty-four frames per second to create the illusion of movement, but instead, as Scott Bukatman puts it, "time in comics is represented as territory in space."[15] From the start, comics laid bare the fault lines in the way we experience time—in our faith in a meaningful distinction between past, present, and future. Much as Hoffmann and Poe described the unconscious well before Freud mapped it, comics had been exploring this version of time before Einstein had a theory (or Picasso an aesthetic) to describe it.

Within its dominant traditions, we experience film not as narrative—the telling of a past event—but as an event unfolding in the present. The traditional cinematic apparatus is designed to reinforce the illusion of being strapped to the theater seat in the dark, moving inexorably forward in time. Film and comics might share the genetic code of sequential images epitomized by Muybridge's horse, but film after 1905 worked to make that code invisible to the viewer. Comics was the first narrative medium dedicated to imaginatively exploring a model of time that allowed for the past, present, and future to exist simultaneously. The reasons for this were primarily formal. For example, the space between panels is known in comics as the "gutter"; in film it is not known at all, because it is not visible once the film is projected. Thus in classical Hollywood cinema the frame effectively disappears from the viewer's focus, while it remains visible in comics, often penetrated or fractured in ways that serve only to call further attention to its material boundaries. And then of course there is the fact that sequential comics largely involve multiple panels laid out on a single page, or on a double-spread of pages in a longer narrative, such that we speak of *mise-en-page* as opposed to film's *mise-en-scène*.

Perhaps no distinction between film and comics better underscores the fundamental differences of their approach to storytelling than does the question of time. The time between film frames is fixed—conventionally twenty-four frames per second. But what of the time between panels on the sequential comics page? In the gutter between each panel we must measure time relative to the information presented in the two juxtaposed slices of time (each panel itself, as Scott McCloud has demonstrated, often requiring further determinations as to how narrative time passes within each panel).[16]

To read comics is to engage, at least on the level of the page, with tenseless time. After all, to read comics is always to see past, present, and future in a glance—whether via the initial navigation of the page layout as a whole or via the more focused reading of a single panel, which always brings with it visual information from the panels preceding and succeeding the panel we are reading—the "present" that here never can entirely command our complete attention. We always inhabit multiple temporalities when reading comics—not just imaginatively, as when a film or novel encourages us to imagine what will happen or to recall an earlier event. We actually see past, present, and future laid out before us in space-time with every page.

It is no coincidence that special relativity—and indeed all attempts to explain time as *simultaneous*—so often requires a turn to the comics form for its representation. Here, for example is how Einstein's thought experiment is conventionally represented:

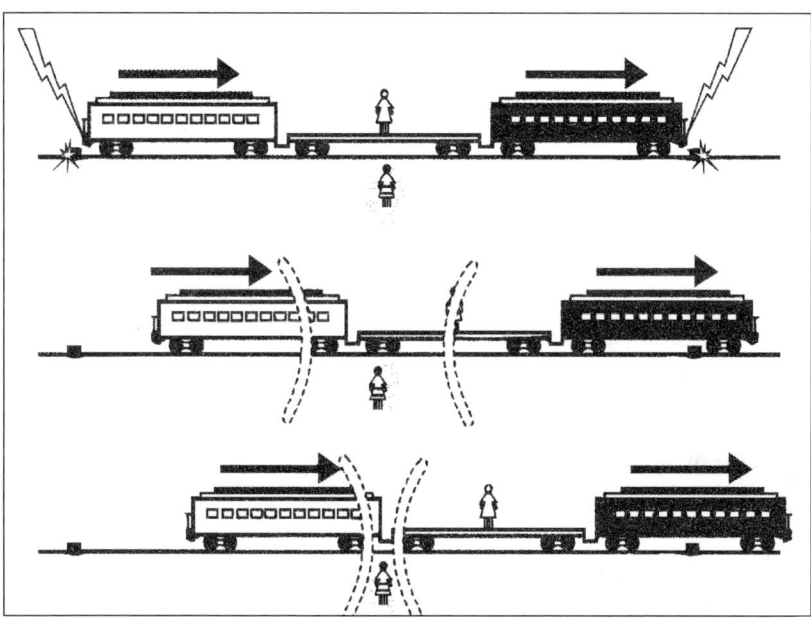

Figure 9.2. Graphic representation of Einstein's special theory of relativity

In three sequential images we can see how the observer on the train experiences the lightning flashes (serially) and how the observer on the platform experiences them (simultaneously). In film, the same information requires a minimum of two retellings of the event, one in which we are presented the flashes from the point of view of the observer on the platform and another in which we see from the perspective of the observer on the train. In comics, past, present, and future can coexist and two different points of view can be mapped simultaneously.

In tenseless time, past, present, and future are simultaneously real, and time is mappable, like space. That we cannot visit the past or the future should in this model be no more troubling than the fact that we cannot at this moment visit Mars. Of course, a century after special relativity, our inability to travel to the past or future remains deeply troubling to our belief in this model of time in a way that the inaccessibility of Mars never is to our sense of space. But however glacial our progress toward an ability to conceive of time as simultaneous might be, there is evidence that we have been moving slowly in that direction over the course of the last century, guided first and foremost by the seemingly marginal cultural form of comics.

However, just as Einstein's thought experiment ultimately illustrated that the observers' experiences of the lightning strikes as serial and simultaneous were *both* true, so is it important to recognize that special relativity does not make obsolete serial, "conventional" time. The three panels above *do* allow the reader to see the past, present, and future of the event at once, and as we cannot in our conventional time, we are here able to move backward and forward in time as we make sense of the information being presented. However, serial time remains necessary to meaning: the story being told ultimately would not make sense were the panels to be reordered. Although aligned with scientific leaps forward in physics, advancing to an understanding of time as simultaneous does not superannuate serial time. Comics is the first medium, therefore, to attempt to represent time as both simultaneous *and* serial. If film reinforces our tensed model of serial time, comics from early on explored how both models of time can be deployed at once.

Serial-Simultaneous

As David Wittenberg suggests, time-travel narratives emerge almost simultaneously with Einstein's theories and intensify in complexity over the course of the twentieth century. Wittenberg compellingly reads time-travel stories as laboratory experiments with narrative time itself, as a way of thinking about how *all* narrative involves a kind of time travel.[17] However, as the century progresses, the creation of and audience for increasingly complex (and time-consuming) imaginative exercises in time travel outstrips such explanations. Time travel may indeed be fundamental to all narrative, but the vast majority of narratives—like the vast majority of commercial narrative film—have not asked us to move beyond our conventional temporal models. Increasingly, and especially beginning in the second half of the twentieth century, we find narratives that take decades to produce *and* consume and that tell stories that cross millennia and numerous parallel timelines.

Here I am thinking, for example, of the extended and profoundly intricate experiments with the multiverse that begin to take concrete shape in the pages of superhero comics in the early 1960s—eventually culminating in a storyverse that covers numerous "earths" at different times and with different outcomes imagined from identical events, across which characters travel seemingly effortlessly in narratives that cross or overlap in byzantine ways. Or, beginning at almost the same time, we have the elaborate time games of *Doctor Who*, a story told over the course of a half-century in a range of media, about a time traveler who brings his companions and audiences from the dawn of time to the heat death of the universe and everywhere in between. Both of these examples, and the intensely committed fan cultures that have grown up around them, underscore the cognitive and collaborative pleasures in working through not only narrative threads and plotlines that often require intense untangling, but also what Doctor Who refers to as the "wibbly-wobbly, timey-wimey stuff" of nonlinear, tenseless time.[18]

The twentieth-century exploration of "timey-wimey stuff" started with comics, and a significant reason for the growing visibility of comics in recent years after a century on the far side of cultural respectability is the growing desire for and sense of familiarity with the kind of temporal navigations the form engages. As Bukatman puts it in *The Poetics of Slumberland*, "Modern culture from the late nineteenth century forward oscillated between the sense of time as unbound, mutable, and multiple, and time as rigid, deterministic, and most insistently bound to linear coherence," and it was comics that first followed the radical possibilities of modeling time as irrevocably bound to space.[19] However, comics was not dedicated exclusively to a mappable, simultaneous time model—what Bukatman describes as the new "mutable and multiple" model of time and what Einstein calls "space-time." What comics explored were not arguments for choosing one model over the other but strategies for navigating storyworlds using both in concert. From the early experiments with open-ended seriality—beginning with the weekly serial *Little Nemo* in 1905 and taking off fully with the daily serial comic strip a couple of years later—the comic strip began some of its most radical experiments in negotiating time complexly across multiple frames. The daily serial strip from the start established a syncopated temporal structure: on the one hand, the daily rhythm of the newspaper in which the day's panels would appear; and on the other hand, the stagger-step rhythm of the panels themselves, which each day needed to establish connections to the previous day's events (past), move the story forward (present), and bait the narrative hook for the next day's events (future)—all times simultaneously (if not equally) present and accessible on the page.

Of course, the newspaper comic strip was just one of several forms comics has explored over the course of a century, each with its own temporal affordances. The double-time of the daily comic strip itself would give rise to the comic book (birthplace of the multiverse), in no small measure because of the scrapbooking habits the open-ended serial strips inspired (and often explicitly encouraged), as readers sought out more complex and long-ranging temporal maps than the daily installment provided. The comic book itself inspired new kinds of reading practices. As Fredric Wertham, the form's most infamous early critic, declared in *Seduction of the Innocent*, comic book readers would spend "an inordinate amount of time" with their comic books, engaging in a practice of compulsive rereading that Wertham found at least as disturbing as the often sensational subject matter of the comic books themselves.[20]

With the rise in the last generation of the so-called "graphic novel," it might at first appear that comics has moved to emulate the narrative conventions and temporal disciplines of the traditional novel. But comics' obsession with modeling and navigating seemingly incommensurate models of time is as pronounced in today's graphic narrative as it was a century ago. A quick survey of those texts that have been canonized in the syllabi of the emerging field of comics studies illustrates ongoing and even intensifying interests in such navigations of time as *both* tensed and tenseless, serial and simultaneous: Alan Moore and Dave Gibbons's *Watchmen*, Art Spiegelman's *Maus*, Chris Ware's *Jimmy Corrigan*, Alison Bechdel's *Fun Home*—all of these require of the reader (as they surely did of their creators) the ability to read through both tensed and tenseless time in order to navigate and participate in the necessarily collaborative work of meaning making.²¹

Richard McGuire's six-page story "Here" (1989) perhaps best represents the coalescing of the experiments with time in the comic strips and comic books of the previous generations and the new temporal experiments of the comics of the present.²² In this story, over the course of thirty-six panels, one corner of a room is represented across millennia of time, with multiple moments of time frequently layering one on top of another within the crowded panel:

Figure 9.3. Illustration from Richard McGuire, "Here," *RAW* 2, no. 1 (1989): 70

The story represents time as *both* serial and simultaneous, as the panels are laid out in a conventionally arranged six-panel grid, beginning with an undated panel in the empty corner of the room and ending with a panel that represents "here" simultaneously at 500,957,406,073

BC—when the earth is a swirling mass of molten metals and gasses—and 1945 AD—when a sailor home from the war sits reading a paper in his living room.

Of course, it is not only alternative comics that have been doing this work. So-called mainstream superhero comics have continued their own explorations, as repeated attempts to collapse the multiverse back to a singular narrative timeline give way to a seemingly inevitable return of the multiverse, one that has today survived repeated crises on infinite earths. In addition to superheroes, the long-running zombie comic book *The Walking Dead*, created by Robert Kirkman, has spawned a parallel storyworld in the AMC TV series where similar characters make different decisions and experience different outcomes. The effect is to allow audiences consuming the transmedial story to engage in complex acts of travel across parallel worlds not so very different from the DC superheroes crossing from Earth-1 to Earth-2 and back again.

In the digital age, the temporal powers and challenges comics first opened up a century ago are now a regular feature of popular storytelling, requiring the user to embrace competing models of time simultaneously, often across different media. For example, narrative video games almost always require replayings, as the player navigates the storyworld in different ways until achieving the desired outcome—and with it, the movement to the next level or stage of the game. Today we might overhear conversations that would be unimaginable a generation ago: "I died thirty times yesterday before I finally beat the boss" or "I needed to go back to the previous day's save in order to finally progress to the next level."

The stories we tell about video games involve the mapping of seemingly contradictory modes of temporal narration. There are "speedrun" videos, for example, designed to show off the fastest way through a video game; "let's play" (LP) videos designed to share one user's subjective experience of a videogame's gameplay (often with audio commentary from the player); videos that warn of glitches that trap users in an "eternal present" from which the only way to "progress" into the "future" is to return to the "past" (in the form of a save from an earlier point in the gameplay). And there is "glitch art," which treats these ruptures in narrative time as an end in itself. More recently we find a growing library of video games—from *Prince of Persia: Sands of Time* (2003) to *To the Moon* (2011)—in which navigating serial and simultaneous time is integral not only to the gameplay and the narratives the play inspires but also to the narrative of the game itself.[23]

Similarly, we also see in the twenty-first century the rise of so-called puzzle films, movies that invite—even require—the viewer to watch in tensed time and then again, remote in hand, mining pockets of time available simultaneously in the digital age for clues.[24] With *Memento* (2000) and *Mulholland Drive* (2001), we see the emergence of a cycle of films that not only explicitly invite users to deploy the active reading affordances of the new DVD technology, but that explicitly make nonlinear time sampling central to the story being told.[25] Increasingly as the cycle progresses, time theory moves to the center of the films themselves, as in *The Butterfly Effect* (2004), *Primer* (2004), *Los Cronocrímenes* (*Timecrimes*) (2007), and *Looper* (2012).[26] In all of these examples, the audience is essentially left with a film that remains incomplete or "unsolved" if experienced only serially via the conventional moving present that the cinematic apparatus had reinforced for a century. Instead, like comics, these films now require rereading using an alternative approach to time in which past, present, and future are simultaneously accessible (and remixable).

Perhaps nowhere today do we see the growing market for opportunities to straddle these two models of time than in long-form serial TV. In the digital age, viewing practices have changed dramatically, from being bound to the synchronized time of national syndication schedules to twenty-first-century practices that include a wide range of consumption rituals. Even while watching "live TV," many DVRs offer the opportunity to "rewind," to recover a "past" that had been historically lost in live broadcast television. And most visible in our current moment we see the shift from traditionally chronometric consumption to "binging"—the practice of consuming serial TV as quickly as possible.

Finally, we must look to the language of contemporary fandom, and particularly the practice of fanfiction. For example, we can examine alternate timeline fanfiction that eschews "canonical" storytime in favor of "what-if elseworlds." If certain events or pairings happen with sufficient frequency, they can and often will enter the "fanon," an alternate timeline agreed upon by fans that exists in parallel with the officially authorized "canon." Recognized genres of fanfiction include "backstory" (devoted to filling in past history not developed in the authorized storyworld), "vignette" (focused on a particular slice in time), and most famously "ships" (in which characters not in sexual or romantic relationships in the authorized storyworld are paired up). The lexicon used to categorize these engagements with "canonical" popular culture texts demonstrates that fans are increasingly able to navigate and create within multiple timelines, but it also reveals that opportunities for such creative engagements are a key source of what organizes these fan communities. All describe the desire to have (and preserve) *both* serial "continuity" and the demand for simultaneous access to past, present, and future as navigable spaces always open to reimagining. At the heart of the practices and communities forged initially out of comics and increasingly found in a diverse range of media in the twenty-first century is the power of storytellers and audiences to travel from present to future to past as easily as traveling from New York to Baltimore.

Key to these fantasies are the pleasures and rewards of imaginatively inhabiting a tenseless time without surrendering the conventional models of time upon which our social and psychological fabrics depend. Instead of serving as the *prosthesis* that film sought to be in the wake of discoveries of the fallibilities of human vision and our experience of time, comics and the storytelling practices to which it has given birth in the arts of the present have worked to make accessible and available Einstein's impossible insights into time as *supplement*. Today we increasingly seek out narrative that offers us both ways at once—tensed, serial time *and* tenseless, simultaneous time. We spent the last century preparing ourselves for this moment, in the highly complex laboratories of quantum physics and the relatively marginalized laboratories of narrative comics. But the arts of the present suggest that we are eagerly seeking out opportunities to explore tenseless time serially, and serial time simultaneously. And if the past is any promise of the future, the arts of the present will lead us to discoveries and insights we haven't even begun to imagine yet—even if, as we surely will, we find out we have been drawing, playing, and remixing them for years without knowing it.

NOTES

1 Tom Gunning, *D. W. Griffith and the Origins of American Narrative Film: The Early Years at Biograph* (Urbana-Champaign: University of Illinois Press, 1994), 77.

2 For a discussion of the emergence of open-ended seriality, see Jared Gardner, *Projections: Comics and the History of 21st-Century Storytelling* (Palo Alto, CA: Stanford University Press, 2012), chapter 2.
3 J. E. McTaggart, "The Unreality of Time," *Mind* 17, no. 68 (October 1908), 458.
4 See Ronald C. Hoy, "Heraclitus and Parmenides," in *A Companion to the Philosophy of Time*, ed. Heather Dyke and Adrian Bardon (West Sussex: Wiley-Blackwell, 2013), 9–29.
5 See Jacques Le Goff, "Merchant's Time and Church's Time in the Middle Ages," in *Time, Work & Culture in the Middle Ages* (Chicago: University of Chicago Press, 1980), 29–42.
6 Niklas Luhmann, "The Future Cannot Begin: Temporal Structures in Modern Society," *Social Research* 43 (Spring 1976): 130–52.
7 While Benedict Anderson has popularized the association of the rise of the realist novel with the rise of the modern nation through their shared investment in the simultaneity of "meanwhile time," of course it is *seriality* that was the necessary precondition for both nation and realist novel. From the *American Magazine* in 1741 through the Federalist Papers of 1787, it was in periodical, *serial* print that the modern nation came first to imagine and define itself. And of course the first Anglophone realist novels were originally conceived and consumed as serial texts in serial forms.
8 From early on, this thought experiment has been rendered in sequential images, showing the natural affinities between comics and theoretical physics, especially in relationship to nonlinear representations of time.
9 See Walter Isaacson, *Einstein: His Life and Universe* (New York: Simon & Schuster, 2007). Bern was unparalleled at the turn of the century for its obsessions with time and its synchronization, possessing arguably the most sophisticated urban time networks of the age. As Isaacson points out, "Einstein's chief duty at the patent office ... was evaluating electromechanical devices. This included a flood of applications for ways to synchronize clocks by using electric signals" (126).
10 See Paul Davies, *About Time: Einstein's Unfinished Revolution* (New York: Simon & Schuster, 1995), 57.
11 Especially relevant for this volume, see Peter Galison, *Einstein's Clocks and Poincaré's Maps: Empires of Time* (New York: W. W. Norton, 2003).
12 See Jonathan Crary, *Techniques of the Observer: On Vision and Modernity in the 19th Century* (Cambridge, MA: MIT Press, 1992).
13 Arthur I. Miller, *Einstein, Picasso: Space, Time, and the Beauty That Causes Havoc* (New York: Basic Books, 2001), 239.
14 Rebecca Solnit, *River of Shadows: Eadweard Muybridge and the Technological Wild West* (New York: Penguin, 2004), 196.
15 Scott Bukatman, *The Poetics of Slumberland: Animated Spirits and the Animating Spirit* (Berkeley: University of California Press, 2012), 31.
16 See chapter 4, "Time Frames," in Scott McCloud, *Understanding Comics: The Invisible Art* (New York: Harper Perennial, 1994).

17 David Wittenberg, *Time Travel: The Popular Philosophy of Narrative* (New York: Fordham University Press, 2013).
18 The quote is from the 2007 episode "Blink" in which the Doctor attempts to explain how he sees time: "People assume that time is a strict progression of cause to effect, but *actually* from a nonlinear, nonsubjective viewpoint—it's more like a big ball of wibbly-wobbly, timey-wimey . . . stuff."
19 Bukatman, *The Poetics of Slumberland*, 31.
20 Fredric Wertham, *Seduction of the Innocent* (New York: Rhinehart & Co., 1954), 11.
21 Alan Moore and Dave Gibbons, *Watchmen* (New York: DC Comics, 1987); Art Spiegelman, *Maus* (New York: Pantheon, 1991); Chris Ware, *Jimmy Corrigan, the Smartest Kid on Earth* (New York, Pantheon, 2000); Alison Bechdel, *Fun Home: A Family Tragicomic* (New York: Houghton Mifflin, 2006).
22 Richard McGuire, "Here," *RAW* 2, no. 1 (1989): 69–74. In 2015, McGuire published with Pantheon a full-color, book-length reimagining of *Here*.
23 Jordan Mechner, et al., *Prince of Persia: The Sands of Time* (Montreal: Ubisoft, 2003); Kan Gao, et al., *To the Moon* (Toronto: Freebird Games, 2011).
24 See Warren Buckland, ed., *Puzzle Films: Complex Storytelling in Contemporary Cinema* (West Sussex: Wiley-Blackwell, 2009), and Graeme Harper, "DVD and the New Cinema of Complexity," in *New Punk Cinema*, ed. Nicholas Rombes (Edinburgh: Edinburgh University Press, 2005), 89–101.
25 *Memento*, directed by Christopher Nolan (Los Angeles, CA: Newmarket Films, 2000); *Mulholland Drive*, directed by David Lynch (Universal City, CA: Universal, 2001).
26 *The Butterfly Effect*, directed by Eric Bress and J. Mackye Gruber (Los Angeles, CA: New Line, 2004), *Primer*, directed by Shane Carruth (New York: ThinkFilm, 2004); *Los Cronocrímenes,* directed by Nacho Vigalondo (Bilbao, Spain: Karbo Vantas Entertainment, 2007); and *Looper*, directed by Rian Johnson (Culver City, CA: TriStar, 2012).

10

Emergency / Everyday

BEN ANDERSON

How should we draw the temporal line between the emergency and the everyday? Does emergency designate a rare and exceptional interruption that threatens to transform the comforting routines that make up the everyday? Is the everyday something finalized and predictable, or is it a realm of potentiality and possibility from which emergencies arise? In what ways are the relations today between the everyday and emergency unique, as people learn to live with perpetual uncertainty in the midst of crises that have become normal?

Drawing a line between everyday and emergency as different but related temporal registers is central to a twenty-first-century politics of emergency. Common-sense narratives teach us that emergency is a rarity that arises unexpectedly. Emergencies are supposedly exceptions to the everyday and, as such, require some form of exceptional response. The everyday is that to which one returns once emergencies are brought to an end. But upon reflection, one realizes that it is also from within the everyday that emergencies supposedly arise, and this raises a question of to what extent the everyday *is* (or contains within itself) an emergency. Today, in fact, the time of the everyday and the time of emergency are frequently conjoined in claims that in the midst of various ecological and financial crises, the contemporary condition of human life is life lived in uncertainty. Recalling the roots of the term "emergency" in *emergere* ("arise, bring to light"), the lines that surround and demarcate emergencies become blurred.

How to draw a line, then, between "emergency" and "everyday" in temporal terms?

Emergency

Emergencies disrupt the everyday. They surprise. Their temporality is of the not-yet. They break with, interrupt, overturn, or problematize a state of affairs that had appeared to be settled. They demand some form of immediate response. Arising suddenly, happening unexpectedly, the characteristic affect of emergencies is of a bad surprise. Emergencies may be ordinary occurrences such as a traffic accident or a fire, or they may intensify in the midst of roiling crises, but what marks them out is that something about them "overtakes" the here and now.[1] Exceeding attempts to anticipate and prepare for events—to seize possession of events before they happen—surprise is an affective register of an emergency's evental quality. Life changes, if only momentarily, as the effects of an emergency unpredictably arise. For example, the typical use of the term "emergency" by the emergency services of Europe and North America (such as the UK fire and rescue services) signifies a punctual event happening at a single site that can be bounded within a "scene." Yet at that scene, to paraphrase Slavoj Žižek, ordinary life is shattered.[2]

In the early twenty-first century, the term "emergency" is used in relation to multiple events or situations across different domains of life and across different functional sectors. It has become as ubiquitous as terms such as "disaster," "catastrophe," and "crisis." Although a genealogy of the term remains to be written, what is assumed to be common across the events or situations named as emergencies today is a particular *quality*. It is a quality of unpredictable, rapid change and the time of a turning point. An event or situation is named as an emergency if urgent, time-limited action is deemed necessary to forestall, stop, or otherwise affect some kind of undesired future. Central to uses of the term "emergency" is, then, a sense that something valued (life, health, security) is at risk and, importantly, a sense that there is a limited time within which to curtail irreparable harm or damage to that value. The temporality of emergency is, then, irreducible to linear or cyclical time. Characterized by rupture, emergencies interrupt the smooth, continuous progression from past to present to future. They also place in question any return to normality after the emergency has been brought to an end, although so often the promise of return is part of how emergencies are managed by governments. Emergencies may be governed by being reinscribed within a linear-causal chain of response and recovery, but the return to normality is never guaranteed.

The first characteristic temporality of emergency is, then, of the exception: emergencies arise unexpectedly as exceptions to some kind of pre-existing order. An emergency constitutes an "exception" that breaks, overturns, or disrupts normal life. The most significant treatment of emergency as exception is by Carl Schmitt in the context of his infamous definition of sovereignty: "Sovereign is he who decides on the exception."[3] Normally we might think that this means that a sovereign—a king, a ruler, a government—decides when an emergency or exceptional event has occurred and then decides on the exceptional measures that will be used (by services, military troops, legislatures, etc.) to handle this emergency. Yet in the background to some of Schmitt's comments on constitutional liberalism in *Political Theology* is the idea that the event itself *provides* the exception (an exception that both pre-exists the sovereign decision and is intensified and transformed by the decision). Schmitt stresses that the exception "is more interesting than the rule. The rule proves nothing; the exception proves everything. It confirms not only the rule but also its existence, which derives only from the exception. In the exception the power of real life breaks through the crust of a mechanism that has become torpid by repetition."[4] For Schmitt, the exception erupts into and transforms the "torpid" liberal, constitutional order. As a "power of real life" that "breaks through," an emergency thus corresponds to the miracle in Christian theology: an interruption that exceeds and transforms a state of affairs.

In other words, the everyday is made by the emergency. Tied to a confirmation and intensification of sovereignty, normal life is suspended in the moment of emergency, but it is also confirmed, felt, made visible at that same moment. The moment of break makes perceptible, or even defines, what constituted continuity before it. The category of emergency does not, however, name only an exception that interrupts "normal life" and issues in a time out of time. "Emergency" is a term that is inseparable from a series of other temporalities. To designate an event or situation as an emergency is to demand an urgent response: the claim is that action is necessary immediately in order to meet the exception.

This involves two other temporalities in addition to exceptionality. First, it involves the presence of (or construction of a sense of) an onrushing future that severs the present from the past and compresses the time for decision and action. The first time, then, is the time of an omnipresent present: there is no time except the time of *now* that requires some form of urgent action. Actions proper to an emergency, such as the dispensing of emergency medicine or shelter, share a capacity to act in accord with exigency. The urgency of the temporary event necessitates and calls forth similarly urgent action. In short, the time to act is compressed, and pauses in action—as introduced by procedures of deliberation or dissensus—become luxuries that threaten delay. Delay is a risk. There is not time to wait. Elaine Scarry has illustrated this by showing how "claims of emergency" function through an affect of urgency that forestalls processes of deliberation and dissensus.[5] Democratic procedures and habits become impediments to timely action, since "the unspoken presumption is that either one can think or one can act, and given that it is absolutely mandatory that an action be performed, thinking must fall away."[6] The characteristic temporality of emergency is thus a suspension of time's unfolding. Emergency is characterized by a stretching of the present and a temporary suspension of the transition to a future, even as a threatening future looms over the here and now.

The second concept of time embodied in "emergency" and acting alongside its quality of urgent action is the *interval*, the time during which action can still make a difference. If action is decisive and happens at the correct time, then the emergency can be brought to an end without loss, harm or damage. Like the state of exception that is the emergency, the interval is time out of time: it defines a state of flux, the moments between the disruption of the present and the solidification of a new future state of social being. Emergency is inseparable from demands that action must happen in this "interval" and a faith that action can make a difference, that the future is not pre-determined. An emergency is brought to an end and becomes something else only when action can no longer make a difference. Emergency becomes a heroic memory in the future if action is effective and things turn out well; emergency becomes a disaster or catastrophe when time has run out and harm, damage, or loss materializes—when, in short, time materializes imminently and threateningly.

The quality of urgency that is inseparable from emergency, and the attendant opening up of an interval of and for action, distinguishes it from other terms with which it is often used interchangeably. For example, similarly applied across multiple domains of life, "crisis" differs from emergency in the demand that each generates for action. While clearly connected to a point of decision, crisis can be a spatially and temporally dispersed condition. Crises are atmospheric in that sense: distributed, they surround and envelope. For example, one can be in the midst of a financial crisis and anxiously feel it slowly unfold, before it intensifies into a topic of extreme concern. By comparison, emergencies involve a demand for immediate, urgent action without delay. In an emergency, there is no time, except the time of now, a time that is running out. Emergencies are, in this sense, activating: they are events or situations where action can still make a difference and demands to act are imperative.

We could thus say that inseparable from the category of emergency is a species of hope: though the outcome of an event or situation is uncertain, correct action may make a difference, and that which is threatened might be averted. In a situation of emergency, the future is alterable, even as it looms over a suspended present. A world of emergencies is far away from a world of pre-ordained fate in which the future is already given. This takes us back to Schmitt's comment that "real life breaks through." For the final characteristic of the temporality of emergency is *emergence*. In an emergency, some kind of harm, damage, or loss to a personal or social value is in the midst of emerging, as is a new spatial and temporal configuration of the present that will form post-emergency. Threat comes to loom over the here and now and blur with the everyday, but at the same time an emergency roots bodies in the facticity of a changing present. This means that what is also emerging in an emergency, or at least is demanded in situations where a responsibility to protect and an imperative to act remains, is action taken to stop, halt, or otherwise affect the emergency. Emergency and the response to an emergency emerge together, both interrupting the everyday.

Everyday

In common thought, "the everyday" is associated with the mundane, trivial, and inconsequential. The temporality of "the everyday" is the temporality of the quotidian, that of ordinary, familiar activities. "The everyday" is what happens when emergencies do not. It consists of a series of repeated activities that constitute a taken-for-granted world enveloped by something like an atmosphere of repetition and recognition. "The everyday" is prosaic and ordinary, a dimension of lived experience supposedly made through innumerable mundane repetitions. We might think here of boredom, with its stilling and slowing of time, as the characteristic affect of the uneventful, recurrent time of the everyday. To study the everyday is to study how nothing happens, or what Blanchot calls banality: "what lags and falls back, the residual life with which our trash cans and cemeteries are filled: scrap and refuse."[7]

In the twenty-first-century expansion of the category of emergency to more and more contexts, emergencies arise within the everyday, but emergencies are not of the everyday. Rather, the everyday has tended to be used to name that which an emergency interrupts. Emergency statements apparently break the torpor of the everyday and are designed to galvanize action dulled by the routine of everyday habits. We might think here of how political statements around climate change so often resort to a tone of emergency to cut through public indifference (a strategy that sometimes backfires). If emergencies are temporary occurrences of an unknown but limited duration, then "the everyday" is the temporal rule that emergencies reveal. In their rarity and transformational qualities, emergencies reveal the taken-for-granted of the everyday: "nothing happens, this is the everyday."[8]

We might say, then, that the exceptional, emergent time of "emergency" has emerged alongside but in contrast to the normal, deadened time of "the everyday." For the time of "the everyday" might appear to be everything that "emergency" is not, and vice versa. The everyday is what goes unnoticed, whereas an emergency demands attention. An emergency comes and goes, whereas "the everyday" is what persists and remains. "The everyday" has a quality of diffuse vagueness and indeterminacy, while an emergency is a punctual disruption.[9] To place action "in" everyday life is, seemingly, to drain it of the qualities of dynamic, timely urgency that have been associated with the time of emergency. While he does not simply equate everyday times with the dullness of repetition, for Henri Lefebvre, attending to the "everyday life" of, for example, the army (among other institutions) is a way of downplaying the heroic, dramatic moment of decisive action in favor of attending to the unheralded, the unforeseen, and the forgotten: "The army prepares itself for war; that is its aim and its purpose. And yet moments of combat and opportunities to be heroic are thin on the ground. The army has its everyday life: life in barracks and more precisely life among the troops. . . . There is a saying that army life is made up of a lot of boredom and a couple of dangerous moments."[10]

The category of "the everyday" is valued by Lefebvre, and a diverse tradition in his wake, precisely because it does not lead to a focus exclusively on the exceptional.[11] But this can lead to a crude distinction between emergency times and everyday times. This is most intense in claims that "the everyday" has been subjected to rationalizing forces that homogenize life and render everyday existence banal (think today of critiques of mass culture). On this account, "the everyday" as a term emerges as part of a conceptual-political vocabulary for critiquing the effects of modernization after 1945 (particularly around the emergence of a consumer society) and how they play out in daily existence. In a critique that resonates with Schmitt's lament of "a mechanism that has become torpid by repetition," Lefebvre, for example, argues that the incorporation of everyday life into production-consumption-production relations makes the everyday into "an object of consideration" and "the province of organization."[12] The repeated, noneventual qualities of the everyday are denaturalized and made a consequence of processes of functionalization and homogenization associated with capitalist forms of production, including cultural production. Everyday life is made into an "object" shorn of creativity and spontaneity. Nothing is left "unforeseen," as everyday time becomes an effect of "systems that aim at systematizing thought and structuralizing action."[13] Daily life becomes routinized and formulaic, in part under the homogenizing spell of mass culture and the commodity form, and in part because everyday life is caught up in systems of bureaucracy. Despite the fact that the everyday harbors the possibilities of transformation and is the ground from which "higher" activities emerge, Lefebvre writes as if it must be riven by an exceptional event—revolution—that would "put an end to the everyday by shattering all constraints, and by investing the everyday immediately or gradually, with the values of prodigality and waste."[14]

But the everyday is not and has never been a simple cognate of ordinariness.[15] As a critical and artistic category with multiple genealogies, it is inseparable from a range of practices that, when taken together, finds hope in the indeterminacy, or elusiveness, of the common. We might think, for example, of artistic projects that attempt to "apprehend," in Sheringham's terms, the "ordinary miracles" that persist within the insignificance of the everyday, without lapsing into an emphasis on the spectacular and eventful.[16] Similarly, John Roberts explicitly rejects the equation between everyday and ordinary (or the popular, in some UK and North American cultural studies). Roberts stresses the long connection between the category of "the everyday" and revolutionary praxis, including what he describes as "the promise of total dehierarchization and dealienation of capitalist production and social relations" and the connected but not equivalent promise of "socialist democratization through cultural change."[17] The everyday becomes a realm for interventions that might both reveal its hidden possibilities and effect some form of transformation.

The everyday is here made into the realm of the always unfinished. It becomes infused with moments in which something different, perhaps something better, may emerge, even if they may not be recognized and may not live on. These temporal moments are not other to an emphasis on the repetitive nature of everyday life as it is lived. Rather, repetitions provide the very ground for traces of something different to arise. Blanchot expresses the temporal indeterminacy of the everyday—an indeterminacy that we might contrast with the clear demand to respond that can issue forth during an emergency: "Always the two sides meet: the daily with its tedious side, painful and sordid (the amorphous, the stagnant), and the inexhaustible, irrecusable, always unfinished daily that always escapes forms or structures."[18]

We might conclude, then, that the time of the "everyday" is ambiguous. Everyday time is the time of repetition felt in the stilling and slowing of boredom and depreciated as inconsequential or trivial. And yet, everyday time is also a time of emergence, the time of the "unfinished daily" that always "escapes" attempts to fix and reduce it.

Everyday Emergency

Emergency is what is not everyday; everyday is what is not emergency. Or so we presume if emergency is understood as a temporary exception to the everyday that, unlike daily activities, it is impossible to overlook or ignore. Blanchot's emphasis on the unfinished, indeterminate qualities of the everyday, however, might give us pause, given the connection between emergency and emergence. Perhaps the lines between emergency times and everyday times have always been more osmotic that we have assumed and, perhaps, those lines have been breaking down.

One example of where lines have long blurred is in the seemingly paradoxical phenomenon of a "permanent state of emergency" in which states of war and peace blur with one another. We might think of countries where a permanent state of emergency obtains as an occasion where emergency measures become part of the indeterminacy of the everyday.[19] Among numerous examples of the permanency of emergency powers across different types of political regime, Mark Neocleous gives the example of Paraguay under Alfredo Stroessner, where a "state of siege" (a variant of "state of emergency" legislation) was renewed every ninety days between 1954 and 1987 on the basis of the looming, continual threat of communism (but lifted for one day every four years when elections took place).[20] Similarly, central to diagnoses of the contemporary condition after 9/11 has been the claim that a temporary measure or paradigm—the state of emergency—has become a normal part of contemporary liberal-democratic states.[21] While a formal state of emergency has not been in force for the entire period, the "war on terror" in the U.S. has involved the production of something like an atmosphere in which life is lived always on the verge of emergency.

Perhaps a permanent "state of emergency" is but one example of how the "everyday" and "emergency" blur when the everyday is lived as a series of emergencies that unpredictably emerge in the midst of a background of perpetual instability and change. We might think here of how social security has morphed into a matter of emergency relief, with the consequence that a stable everyday life is a rare, precarious achievement crafted in the midst of emergency healthcare, emergency shelter, and emergency food. Perhaps the everyday may no longer be described (if it ever could be) in terms of the seeming security of repetition formed through routine. For example, the naming of "precarity" as something close to a contemporary structure of feeling simultaneously heralds a changed everyday constituted through insecurity, while also revealing that, for many peoples, a stable everyday has, at best, been present only as a future promise.[22]

In other words, one thing that the perpetual sense of "everyday emergency" (which seems to hold in the U.S. and elsewhere) may prove is that a stable everyday was perhaps a mid-twentieth-century, middle-class reality predicated on specific, and perhaps historically anomalous, facts of wealth distribution, resource accumulation, and postwar politics. The distinction between everyday and emergency outlined above has only ever been available to some and is produced at the cost of making life into a perpetual emergency for others. For rather than the future being present through the ominous heralds of catastrophe or the dream of progress, precarity makes present an unstable here-and-now vaguely menaced by the force of an uncertain future. If Richard Sennett is right that "instability is meant to be normal" in precarity, then the diagnosis of precarity for those who once had the luxury of a secure everyday life coexists with the recognition that, for some, everyday times and emergency times have never been separate from one another.[23]

For example, Kathleen Miller uses the term "everyday emergencies," after Ben Penglase, to describe the "disrupted life" of workers who collect and sell recyclables on Rio de Janeiro's garbage dump, and how "disruption and insecurity not only suspend but also constitute 'normality' in Rio's favelas" as the regularity of "everyday emergencies" interrupts participation in formal labor relations.[24] In this example, and doubtless countless others, a stable, secured normality is not the rule that is taken-for-granted until it is revealed in and by the exceptional times of a rare, punctual emergency. The precariousness of the everyday may be the lesson being learned by the wealthiest nations today as they enter a perpetual atmosphere of economic emergency, real or fabricated, felt or claimed. A category like "everyday emergency" is a temporary solution to the breaking down of lines between "emergency" and "everyday" and a sign that, in the midst of diagnoses of insecure times, the terms might be reaching their end as useful means for measuring distinct temporalities.

But of course states are constantly drawing the line between emergencies and a normal state of affairs in ways that reaffirm the equation between emergencies and exceptional time. For example, for the UK emergency services, emergencies are a normal, anticipated feature of working life. For them, emergencies are everyday, and their time is one of regularity and predictability. The lines between the everyday and emergencies blur for the emergency services, with the result that new categorizations of emergency are needed to demarcate emergencies from one another. For example, the operational term "major incident" is a special type of emergency that is defined as an exception to the ordinary capacities of the emergency services. The term (and its use) reinscribes the line that separates emergency and everyday, but also acknowledges that certain types of emergencies are everyday—in the sense that they happen routinely and, while potentially devastating for those who go through them, can be handled in the normal times of work.

Governing emergencies may involve drawing lines between the time of emergency and the time of the everyday. However (and perhaps more important for our purposes), perpetual preparation for emergency causes a reversal of normal "everyday/emergency" temporality. Operationalizing definitions of emergencies is a matter of making preparation for emergencies an accepted part of the everyday. For example, preparedness as a governmental apparatus involves undertaking various types of preparatory action in the present.[25] Action for an emergency— say the provision of emergency shelter—is planned and rehearsed in the present, bringing the future emergency into the here and now, but it also is designed to ensure that action in the "interval" of emergency has the qualities that mix emergency and everyday. Action should be drained of the drama of emergency: panic should be absent. But action should also be shorn of the dullness of the everyday: response in emergency must have qualities of urgency and purposefulness to meet the event. The paradigmatic technique here is the *exercise*: live rehearsals of response that stage and perform future emergencies in order to ensure that emergencies are prepared for and therefore do not surprise.[26] Exercises work by making exceptional response a routine matter of correct protocols and procedures, and emergency in general becomes something close to an atmosphere within which life is lived. Kathleen Woodward puts this well when she talks about how in a time of risk, "the normal seems virtually sure to turn catastrophically into its opposite at any moment."[27]

Woodward is only half right in identifying something like an atmosphere of risk that acts as the affective correlate to efforts to render emergencies everyday. The ubiquity of preparations for emergency breaks down the line between emergency and everyday and ushers in a series of other times, as emergencies are staged and performed within the everyday. Consider, for example, a UK exercise undertaken in preparation for a global avian flu pandemic.²⁸ The event exercised protocols for the prioritization of scarce resources (principally hospital beds) in a pandemic. Based on a scenario and involving live role-play, the exercise staged and performed the escalation of the event before the emergency ran out of control. At various points, updates to a scenario were used to stage life-or-death decisions: who gets treated? The exercise was designed, in part, to enable familiarization with the event and response, to ensure there were no bad surprises. Thusly the constant anticipation of future emergencies renders those emergencies everyday: they arise within the everyday as expected occurrences that can be handled within existing techniques. In the avian flu exercise, responders would be familiar with the exceptional response, including protocols for the prioritization of scarce beds in a situation of excessive and escalating illness. More than anything, though, responders had to be familiar with the demands that might be placed on them in an emergency. What was rehearsed, what was made everyday, was how to respond with exigency in an emergency.

Richard Grusin describes this temporal effect in slightly different terms than do Woodward or Cooper when he identifies the double presence of a "low level anxiety" *and* a "sense of continuity" that results from the "everyday-ification" of techniques for anticipating emergencies (of which exercises are but one). He is writing on media techniques, but the identification of "premediation" as a style of relating to the future (a "logic" in Grusin's terms) holds for other techniques for governing emergencies that may give publics and those who govern emergencies "the feeling of assurance that there will not be another catastrophic surprise."²⁹ Put simply, the effort is to ensure that emergencies are not emergencies. Premediation does not work to predict futures. Rather, premediation involves an effort to "preclude the possibility of an unmediated future, or unmediated real" by proliferating multiple, sometimes contradictory, versions of what might potentially happen.³⁰ Any future emergency, say a trans-species pandemic, will already have been premediated in that a range of future effects and impacts will have been staged in the present. We could say that what is produced by the government of emergencies are "everyday emergencies" in three ways: emergencies happen to and within the everyday, future emergencies exist as possibilities within the everyday, and attempts are made to drain emergencies of their evental qualities.

We see the shifting distinctions and equivalences between everyday times and emergency times that is characteristic of premediation if we turn to the end of the avian flu exercise. After responding to a series of life-and-death scenarios, players in the exercise had applied a number of familiar protocols based on the logistics of mass casualty care. Future death and illness had become a regular occurrence. The event no longer surprised. What was staged at the end of the exercise, in a final update to the fictional scenario, was an escalation of the pandemic, an escalation designed by the exercise planners to simulate unpredictability and intended to render existing everyday protocols inoperable. It did, and the exercise ended with the failure of existing protocols in the face of the emergency, alongside a familiarization with how to respond in a future emergency. The future emergency (and response to it) was staged and rehearsed in the everyday, or in Grusin's terms was premediated, without the emergency being reducible to an everyday occurrence. It became an "everyday emergency" in the paradoxical sense of mixing what were once thought of as "everyday" and "emergency" times without rendering the two equivalent.

* * *

Emergency and everyday are understandable only in relation, a relation that takes many, qualitatively distinct, forms. The terms name different temporal qualities but share a series of hidden affinities: both are ways of apprehending and articulating the unfinished and open dimension of life. The complexities of their relation today are best exemplified by a now ubiquitous device: the green or red emergency exit sign. Emergency exit signs suggest, rather than prescribe, proper movement in an emergency. The sign exemplifies today's (in)distinction of emergency and everyday: an emergency may arise from within the everyday and may be an ever-present possibility, but it nevertheless threatens to transform the everyday. Emergencies are not only rare, punctual exceptions to normality that reveal the noneventual character of an everyday drained of liveliness by repetition and routine. The times of emergency and the everyday in fact merge and blur, as future emergencies are made almost but not quite everyday, while, for many, the everyday remains or becomes an emergency.

NOTES

1. Craig Calhoun, "Contemporary States of Exception," in *Contemporary States of Emergency: The Politics of Military and Humanitarian Interventions*, ed. Didier Fassin and Mariella Pandolfi (New York: Zone Books, 2010), 29–58.
2. This is a paraphrase of the thesis of Slavoj Žižek, *Event: Philosophy in Transit* (London: Penguin Books, 2014).
3. Carl Schmitt, *Political Theology: Four Chapters on the Concept of Sovereignty*, trans. George Schwab (Cambridge, MA: MIT Press, 1986; repr. Chicago: University of Chicago Press, 2006).
4. Schmitt, *Political Theology*, 15.
5. See Elaine Scarry, *Thinking in an Emergency* (New York and London: W. W. Norton, 2011).
6. Scarry, *Thinking in an Emergency*, 7.
7. Maurice Blanchot, "Everyday Speech," trans. Susan Hanson, *Yale French Studies* 73 (1987): 13. Reprinted from Maurice Blanchot, "La Parole quotidienne," in *L'Entretien infini* (Paris: Gallimard, 1959): 355–66.
8. Blanchot, "Everyday Speech," 15.
9. See Michael Sheringham, *Everyday Life: Theories and Practices from Surrealism to the Present* (Oxford: Oxford University Press, 2006).
10. Henri Lefebvre, *Critique of Everyday Life, Vol. 2: Foundations for a Sociology of the Everyday*, trans. John Moore (London and New York: Verso, 2002), 44.
11. See the discussions in Sheringham, *Everyday Life*.
12. Henri Lefebvre, *Everyday Life in the Modern World* (New Brunswick, NJ: Transaction Publishers, 1984), 72.
13. Lefebvre, *Everyday Life in the Modern World*, 72.
14. Lefebvre, *Everyday Life in the Modern World*, 73.
15. John Roberts, *Philosophizing the Everyday: Revolutionary Praxis and the Fate of Cultural Theory* (London: Pluto Press, 2006).
16. Sheringham, *Everyday Life*.
17. Roberts, *Philosophizing the Everyday*, 120, emphasis in original.
18. Maurice Blanchot, "Everyday Speech," 19.
19. For a discussion of this idea, see Giorgio Agamben, *State of Exception*, trans. Kevin Attell (Chicago: University of Chicago Press, 2005).
20. Mark Neocleous, *Critique of Security* (Edinburgh: Edinburgh University Press, 2008).
21. For a summary of these arguments, see John Armitage, "State of Emergency: Introduction," *Theory, Culture and Society* 19, no. 4 (2002): 27–34.
22. On the "promise of normativity," see Lauren Berlant, *Cruel Optimism* (Durham, NC: Duke University Press, 2010).
23. Richard Sennett, *The Corrosion of Character: The Personal Consequences of Work in the New Capitalism* (New York: W. W. Norton, 1998), 31.

24 Kathleen Miller, "The Precarious Present: Wageless Labor and Disrupted Life in Rio de Janeiro, Brazil," *Cultural Anthropology* 29, no. 1 (2014): 34.
25 Stephen J. Collier and Andrew Lakoff, "Distributed Preparedness: The Spatial Logic of Domestic Security in the United States," *Environment and Planning D: Society and Space* 26, no. 1 (2008): 7–28.
26 Ben Anderson and Peter Adey, "Governing Events and Life: 'Emergency' in UK Civil Contingencies," *Political Geography* 31, no. 1 (2012): 24–33.
27 Kathleen Woodward, *Statistical Panic: Cultural Politics and Poetics of Emergency* (Durham, NC: Duke University Press, 2009), 216.
28 The exercise was observed as part of research funded by the Economic and Social Research Council on the role of exercises in UK preparedness. See Anderson and Adey, "Governing Events and Life."
29 Richard Grusin, *Premediation: Affect and Mediality after 9/11* (London: Palgrave Macmillan, 2010), 48.
30 Grusin, *Premediation*, 45.

11

Labor / Leisure

AUBREY ANABLE

In 1930, the economist John Maynard Keynes predicted that by 2030 the wealth created by new technologies would bring about an era of universal leisure.[1] We can safely say that Keynes's prediction was way off the mark.[2] In the West, the postwar transformations of work—from the computerization of factories and offices to the outsourcing of manufacturing jobs to the global south—have changed the labor landscape dramatically since 1930. Still, labor, not leisure, structures the vast majority of people's time. Labor—a word that names both the kind of remunerated work that we do and also how that work situates us within particular historical and social matrices—is the central facet of everyday life for most people. As such, it has been the primary site of Marxist interventions into capitalism and it has long been the anchor of the "labor/leisure" dyad. Leisure, the nonremunerated time we spend doing activities that are pleasurable or otherwise different from our paid labor, has always been the ideological lightweight of the pair. Whereas leisure, too, positions us in certain ways according to economic class, rarely is this seen as more than a side effect of the centrality of labor's power to subjectivize.[3]

Since the Industrial Revolution, the boundaries between labor time and leisure time became meaningfully blurred yet still remained relatively distinct. After 1960, however, with the rise of computational systems and shifts in managerial practices, industrial production, and global economic networks, and with related shifts in the communications infrastructure on which mass and popular cultural forms depend, these two terms have become, if not indistinguishable, at least vastly more temporally and experientially confused. By the turn of the millennium, these changes engendered new considerations of the relationship between leisure time and "immaterial" labor under neoliberal modes of capitalism. While "free" time—time spent off the employer's clock—was never really free,[4] the degree to which our sociality, our bodies, our creativity, and our time are currently harnessed to digital networks that turn leisure activities into quantifiable and valuable data marks a different discursive and material relationship to the experience of labor time and leisure time. In short, leisure time has shifted from being a counterpart to labor time to being another modality of productive work available for the temporal calculus that capital loves.

Mobile phones (and their transformation from telephony devices to "smart" platforms for a variety of work and entertainment applications) are rich sites for understanding the contemporary interrelationship of labor and leisure. The global market for mobile devices has surpassed that for personal computers, and there are technological, ideological, and temporal changes that accompany this shift. Looking at casual games—the most popular type of mobile software application downloaded globally—we can see how work and play on mobile devices are qualitatively and quantitatively fused. The habitus of social class, as Pierre Bourdieu argued, is reproduced in everyday leisure activities.[5] Similarly, our daily interactions with mobile games harness our bodies to the habitual rhythms of digital capital. These ludic interludes produce data in the service of capital and also produce us as flexible, connected, and data-producing subjects. Furthermore, our engagement with mobile phones is not simply an interface between human time and computer time, but rather a different temporal experience altogether: mobile time.

Labor Time

Labor time generally refers to the time spent engaged in activities during which one's labor power is exchanged for pay to produce value for an employer.[6] In Karl Marx's double formulation of labor as concrete and abstract, this conventional notion of labor time is aligned with concrete labor in that it describes the time spent performing the actual physical labor that goes into producing a commodity as opposed to the abstraction of labor time that sets the exchange value of commodities.[7] With the post-1960 foregrounding of immaterial labor and the contemporary blurring of labor and leisure, however, we can register an experiential shift in labor time as an increasingly abstract temporality during which a wide variety of concrete and abstract activities, outside of paid labor, generate value.

The organization of work into distinct periods, set apart from the rest of life, arose during the Industrial Revolution. By the beginning of the early nineteenth century, however, social reformers in England and America began calling for limits on labor time in response to the long hours and harsh working conditions in factories. Many proposed the eight-hour workday as a healthy limit that would allow for adequate rest and leisure.[8] Just as time spent at the factory was beginning to decrease in the nineteenth century, the scientific management of labor emphasized standardized work rhythms and the close supervision of workers by managers to increase productivity. Frederick Winslow Taylor initiated the scientific management movement to track, measure, and rationalize worker productivity. Harry Braverman notes that Taylorism is often treated as the "science of work," though it was actually the science of managing others' work through the fragmentation and de-skilling of labor time.[9]

Standardized work schedules and the highly rationalized management of labor time began to fall out of fashion in some sectors of employment with the popularization of concepts like "flextime" and "telecommuting" in the 1970s. These were part of an emerging constellation of labor processes and contexts, often called "immaterial labor," that arose in the 1960s as a result of global post-Fordist manufacturing shifts, computerization of the workplace, and countercultural rejections of the "scientific management" of labor in the West.[10] Some characteristics of the immaterial mode include the reassertion of the subject and his/her communicative and affective qualities into the ideal worker; the rejoining of cognitive and physical labor (rather than their separation under Taylorization); self-management; increased precarity; and the universalization of affective labor. Maurizio Lazzarato writes, "What modern management techniques are looking for is for 'the worker's soul to become part of the factory.'"[11] The space and time of labor becomes socially diffuse in this mode, reconstituting labor as an immaterial relationship that takes place in "the society at large" through constantly innovating "forms and conditions of communication."[12]

Immaterial labor does not describe something brand new, but rather a mode of production that has come to the foreground that has always operated in older labor models. Capitalism has always depended on immaterial labor, for example, in the form of women's unpaid affective labor.[13] Additionally, the effects of immaterial modes of labor are distributed unevenly. In the ICT industries, for example, the manufacturing of the very technologies that are so central to Lazzarato's formulation depends on modes of labor, usually located in the global south, that have more in common with nineteenth-century labor conditions than with the "flexible" and "diffuse" modes he describes.

Another way to think about the bifurcation of immaterial and material labor is through Moishe Postone's reinterpretation of the relationship between abstract and concrete labor. The value of a commodity is not determined by the actual concrete *labor* that produced it, but rather by the average *time* required to produce it. Here both the time spent laboring to make something and the production process are abstract—linked only to the actual conditions of production in the abstracted form of a temporal average. In this way, Postone understands the temporality of capitalism to become more abstract as new technologies decrease the average time of the production of goods, thus producing an increasingly abstract relationship between value and labor.[14] Immaterial labor, then, in both its concrete and abstract forms, is a social relation that is symptomatic of workers' increasingly abstract relationship to time and its value.

Immaterial modes of labor are not so much a rejection of Taylorization as a personalization and digitalization of its logic. Through mobile media we can see how the measurement of time and productivity is outsourced onto the self. Melissa Gregg describes the "tyranny of the mobile phone" in her account of the ways mobile digital devices raise expectations for perpetual productivity, regardless of where one is or what time it is.[15] Mobile phones are direct technological and ideological facilitators of the diffuse and flexible workforce and significantly related to the endless workday and 24/7 productive time.[16] Making ourselves visible and making our labor visible in the network become the same things. For example, we are marketed productivity software applications (apps) for our smart phones that track and measure the time we spend doing various tasks. Similarly, there are services that measure our social media clout via algorithms that analyze our activities in online social networks and provide scores to potential employers interested in the value of our networked communication.[17] Apps for social media platforms such as Twitter, Facebook, and Instagram encourage us to update our status and post content on the go, and to participate in what Sarah Banet-Weiser critiques as the logic of self-branding.[18] Our digital labor also is made visible and valuable as part of the large data mines to which we inevitably contribute when we interact with mobile media. Mark Andrejevic writes, "Media and cultural studies, long engaged in the study of media audiences, have tended to focus on new manifestations of audience productivity rather than how these audiences are themselves *put to work* by these proliferating forms of audience monitoring."[19] These examples emphasize how the digital activities we do outside of the specific context of paid labor time, what once would have been called leisure time, are actually directly bound up with value-producing work.

Leisure Time

Leisure time refers to the time spent engaged in activities that are for pleasure and are unrelated to either paid or unpaid work, chores, or other personal obligations. Leisure time names the historically changing conditions of time spent away from work.[20] With shifts in labor time, what constitutes leisure and how it is experienced also shifts. Immaterial labor time relies on quantifiable and computable, and quite material, leisure time. More than simply being another modality of labor, it is in leisure time that we learn to produce the rhythms of digital labor, but also where we seek relief from it.

The nineteenth-century rationalization of leisure time and the related boom of amusement parks, cinemas, and other mass-cultural sites of leisure activities are directly related to the contemporary rationalization of labor time.[21] Time spent away from work had the double benefit of pleasing social reformers and scientific managers who promoted leisure as part of a productive system of work, rest, and play. In the early part of the twentieth century, the Frankfurt School's critique of the forms of leisure offered by "the culture industry" was a meaningful contribution to Marxist perspectives on ideology, casting new light on entertainment as a site of capitalist domination in that it bears the imprint of the production process. The formulation of mass culture as an "industry" firmly linked certain forms of leisure to labor as the primary structuring relation of capitalism.²² Siegfried Kracauer, for example, compared the synchronized movements of the Tiller Girls' legs to the rhythms of labor and the machinery of industry. Such spectacles of leisure time, he argued, were "the aesthetic reflex of the rationality to which the prevailing economic system aspires."²³ In the latter half of the twentieth century, scholars working under the rubric of cultural studies revised the Frankfurt School's view by reframing "popular culture," especially literature and screen cultures, as potential sites of ideological negotiation and political contestation.²⁴ By the late twentieth century, as new types of work schedules emerged, aided by networked computers and mobile phones, distinct periods and forms of leisure became more difficult to identify and rationalize.

Contemporary leisure time is predicated on a more recent technological shift with temporal implications: the shift from computer time to tablet/smart phone time. Mobile time accrues in the numerous glances at our devices in the midst of other tasks and also in our more extended and engaged interactions. It is a temporality inhabited wherever we happen to use our device, whether at work, at home, or somewhere in between. Mobile time is temporally and spatially liminal in that mobile phones are interfaces between the time and space our bodies inhabit and the multiple times and spaces to which digital devices connect us. Often, our longest engagements with mobile phones happen in between times and spaces more clearly demarcated for other activities—say, commuting between work and home, at the airport waiting to board a flight, in line for a movie, etc.²⁵ Mobile phones and the networks (both virtual and real) to which they connect us are about the constant measurement and visualization of time and its digital currencies: data and connectivity.

In a very literal sense, of course, mobile phones tell the time. Smart phones are elaborate timepieces that have made the wristwatch redundant. Mobile phones with capacitive touchscreen interfaces, accelerometers, gyroscopes, global positioning systems, and wireless Internet access are also devices that constantly and obsessively measure, calculate, record, and share our movements through time and space. The companies that provide the mobile network infrastructure measure and monitor our usage: from minutes spent talking, to length and number of text messages, to the amount of data downloaded from the web. From using our phones as alarms to wake us in the morning and calendar notifications that remind us of daily tasks to the time, date, and GPS coordinates that tag our photos, Tweets, and Facebook updates, mobile time is very much about the conscious and unconscious, visible and invisible, concrete and abstract ways mobile phones structure and measure our activities. Productivity apps such as Shift Worker and Procraster harness this constant measurement of data and connectivity for the literal measurement of labor time.[26]

The history of the mobile phone, like the longer history of telephony, is marked by a gendered transformation of the technology from a business tool to one associated with domesticity, "gossip," and leisure.[27] The same devices that have played such a large role in restructuring the space and time of labor are now also often the first devices we consult for leisure activities. The shift to quantifiable and computable leisure time can be seen, most explicitly, in the popularity of mobile gaming apps. Games for mobile devices have transformed the global video game industry and changed how we interact with our phones. In 2013, a watershed year for the industry, the game *Candy Crush Saga* had over six million active users and generated nearly two billion dollars in revenue. This phenomenal success instigated a boom in casual game production and consumption, and also a cultural shift away from casual games being explicitly coded as feminine.[28] In the U.S., the most popular mobile games of 2013, measured by number of downloads, included *Candy Crush Saga*, *Angry Birds: Star Wars*, and *Plants vs. Zombies 2*.[29] In China, where the mobile gaming market is a twenty-billion-dollar industry, these same games are also among the most downloaded, alongside domestically produced games that share similar themes and structures.[30] For example, *Tiantian Ai Xiaochu* (*Tencent*) shares *Candy Crush Saga*'s match-three structure and user interface, but swaps out candy for anime-inspired animal faces.[31]

The blurring of work and play on mobile devices seems to speak to contradictory uses and desires—productivity *and* escape from work. Yet, beyond their seemingly different uses and affects, productivity apps and game apps have similar algorithmic structures, interfaces, and quantification metrics. These similarities across apps and our alternation between them structure the temporality of mobile time. That is, we are not so much multitasking in mobile time as we are *metatasking*. In our play, we produce data about ourselves while simultaneously producing ourselves as subjects connected to the network and, thus, free from (the fixed time of) work and free to play. Elsewhere, I have argued that casual games are intimately bound up with labor time. Their particular combination of flow and interruptability speak to the way work on computers and other digital devices is often done. Additionally, their mechanics, narratives, and reward structures speak to a desire for time-bound tasks with identifiable outcomes—things often perceived as missing from contemporary digital work.[32] These qualities of casual games are related to the institutional and technological origins of all video games. Video games were always an interface between work and play.

Yet there is a tendency in analyses of video games and labor to artificially separate the act of playing from the labor that produces the game and the device on which it is played. Play is figured as immaterial and an illusion that shields us from the knowledge of the material labor that went into producing the game itself. In this formulation, video game play is seen as a form of leisure that gets co-opted by neoliberalism.[33] This version of video game history suppresses just how central computer games were to the digital transformation of work from the 1960s forward. For example, both *Tennis for Two* (created at Brookhaven National Laboratory in 1958) and *Spacewar* (created at MIT in 1961) are widely considered the first video games and were created, not as distractions from work, but as playful means to visualize and make accessible the work of computers. *Tennis for Two* and *Spacewar* were designed as demonstrations for lab visitors, as ways for nonspecialists to *see* the processing of an otherwise boring and inert machine. Despite the amount of energy that has gone into characterizing the creation of early video games as countercultural hacking,[34] both games were created as part of the broader labor of graduate students and research scientists affiliated with labs partially or wholly funded by national defense contracts. We can see both games as ways to make not just the labor of computers visible, friendly, and accessible, but also—at least in retrospect—as ways of demonstrating and justifying the work of federally funded researchers to a skeptical public.

At their origins, video games are part of the transformation of work that occurred in the postwar U.S. context that was heralded by the importance of computers for the military-industrial complex. The hacker/video game player as creative worker is an ideal that Fred Turner points out emerged from, not against, the labor contexts of the 1960s and 1970s.[35] What is missing from the immaterial labor critique of video games as well as from critiques of gamification, then, is the crucial understanding of how video games, from the very beginning, were not a rejection of work, but rather a platform for the reconceptualization of work emerging just as the context of labor in the West was also shifting from manufacturing to service and information industries.[36]

Through analysis of the mobile games *Candy Crush Saga* and *Plants vs. Zombies 2* we can see how mobile games use methods of measurement and quantification that reflect but also refract mobile time's larger quantification of productivity. *Candy Crush Saga*, *Plants vs. Zombies 2*, and most other free-to-play mobile game apps are generally classified as "casual" games. Casual games are designed to be played in short bursts of five to ten minutes, while taking a break from work or while riding public transit, and then set aside.[37] Their simple interfaces, gentle learning curves, and short levels are designed so that anyone may pick them up easily, play for a few minutes, and then stop, but be hooked just enough to want to pick the game up again whenever they have a "free" moment.

What is notable about these types of games in relation to labor/leisure is the way they construct our play—through their narratives, graphics, mechanics, and algorithms—so similarly to the ways other types of mobile apps and interfaces ask us to manage and measure our time and productivity. Unlike most apps, game apps, once opened, obscure the phone's measurement data and replace them with visualizations of measurement that are specific to gameplay. The various timers, meters, gauges, and score displays in mobile games mimic, compositionally and graphically, the symbols on our phones that measure and display various modes of connectivity to the network, time, date, and battery power. When we play games on our mobile phones, we ostensibly leave the realm of self-measurement and management of productive labor in order to play with and amongst a zany version of these very categories. For example, in *Candy Crush Saga* (CCS), a tile-matching puzzle game set in a Candyland-inspired world, the phone's conventional symbols of connectivity are replaced with gauges measuring the player's score, remaining moves, and boosters. In *Plants vs. Zombies 2* (PvZ2), a tower defense game that pits the player's strategic gardening skills against zombies, the interface measures daylight, plants available, coins

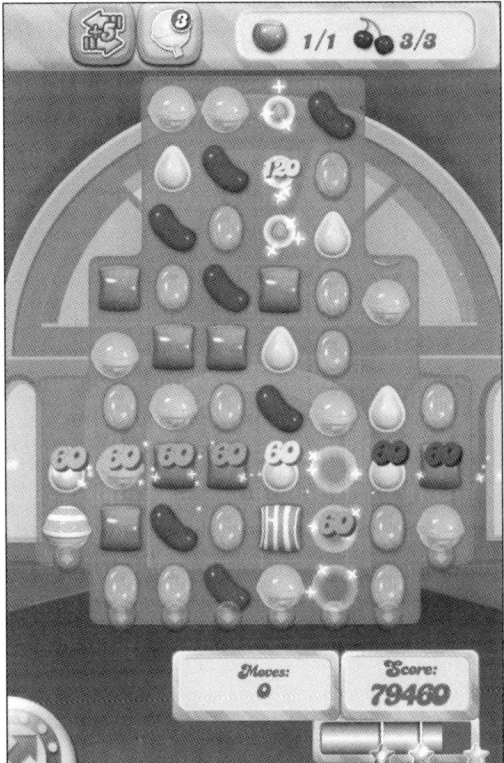

Figure 11.1. In *Candy Crush Saga* the various icons of gameplay measurement simultaneously reference and obscure the mobile phone's nongame measurement icons. (Source: author's screenshot)

earned, and zombies killed. In both games, these visualizations of game productivity are also accompanied by constant prompts to publicize one's accomplishments in the game through social media.

Thus in mobile time, our leisure activities take on the rhythms and representations of productivity. In many mobile games, for example, grids organize the action on the screen. In PvZ2, the playing grid is a checkerboard of alternating light and dark colors, referencing the game of chess. Likewise, in CCS, the candy is displayed in neat rows, and each piece has its own square. The grid, of course, is also a familiar and ubiquitous interface for the organization and measurement of data in

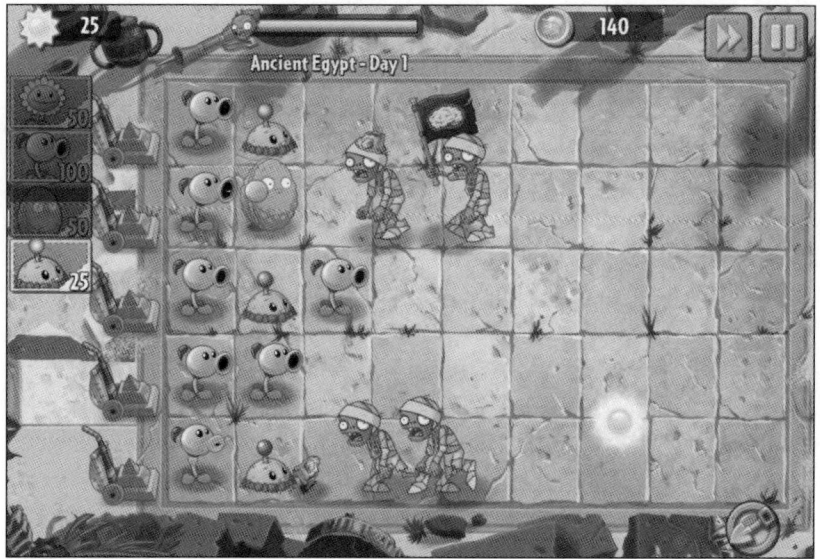

Figure 11.2. As in CCS, various icons of measurement surround a grid in *Plants vs. Zombies 2*. (Source: author's screenshot)

many productivity apps. For example, users of Shift Worker arrange colorful icons on the basic grid of the calendar to keep track of their changing work schedules. Other productivity apps, such as Procraster and Life Graphy, use grids, graphs, gauges, and colorful icons to track and display statistics like number of "productive minutes."[38] Even the names of many productivity apps point to their organizational and self-management rhetoric: OmniFocus, 24me, Grid, and so on.[39]

The grids on screen are visual links to the rhythms and mechanics of the games' algorithmic structures. The repetition and alternation between order and disorder is a common element across many mobile games. For example, the match-three structure of CCS asks the player to sort candy by type to score points and progress through the levels. Like all tile-matching games, CCS offers pleasure from the simple mechanic of creating an orderly set of objects that, once ordered, then disappear, creating a new disorderly scenario as the setting for the next match. A field of disorder is presented to the player. The player creates temporary order. A new mess is then presented for ordering. In PvZ2, the logic of disorder/order is reversed. The player begins with a clean,

Figure 11.3. The productivity app Shift Worker uses colorful gamelike icons to organize work schedules. (Source: app publisher's screenshot)

empty grid that she fills with strategically positioned plants in neat rows, but the zombies wreak havoc on the order created.

The contemporary state of the free-to-play mobile game industry illustrates both how the concrete labor of game production is profoundly alienated from its value and our consumption, and also, simultaneously, how the games that are produced reflect and refract these conditions in quite material ways. A game that costs nothing to download certainly obscures the people that mined the columbite-tantalite, the workers that assembled the phone, and even the programmers and designers that made the game. Yet, labor, like the return of the repressed, shows up in the narratives, interfaces, mechanics, and algorithms. Through game apps we are invited to engage with systems of measurement and evaluation that produce us not as concrete workers, but as subjects of the mobile and shifting interface between labor and leisure.

While mobile games are a continuation of mobile time's metrics of labor, they also skew these systems of quantification through their sped-up, exaggerated, and preposterous tasks, creating what Sianne Ngai describes as a "zany" relation to labor and productivity. Ngai writes, "[Zaniness] is really an aesthetic about work—and about a precariousness created specifically by the capitalist organization of work. More specifically . . . zaniness speaks to a politically ambiguous erosion of the distinction between playing and working."[40] Like Lucy and Ethel flailing on the candy factory assembly line in *I Love Lucy*, mobile games like CCS and PvZ2 derive their humor and enjoyment, but also their pathos, from the predicament of having one's sense of control and mastery tested by more difficult tasks, increased targets, and more limiting constraints.

In PvZ2, the limiting constraint is time. In fact, the game's subtitle is "It's About Time"—a reference to both the perception of the wait between the first and the second installments and the sequel's time-travel narrative structure. "It's About Time" is also an accurate description of what the game's graphics, algorithms, and mechanics measure, visualize, and make manifestly felt by the player: time and its passing. The game's humor resides in the premise of a high-stakes battle between two entities not known for their speed: zombies and plants. The tension resides in the slowness of the zombies' trajectory toward the player's house and how long it takes seeds to regenerate before the player can plant them again. As the player progresses through the levels, the waves of zombies and their strength increase, making her job more difficult. For added pressure, the player can tap the fast forward button to speed up the entire game. Just as Lucy and Ethel's behavior becomes zanier as the speed of the conveyor belt picks up, these games and the player's activities within them become zanier as they progress. The humor of zaniness, Ngai argues, is linked to notions of the laboring subject who becomes preposterously flexible and adaptable under the demands of capitalism. Along these lines, mobile games—through their interfaces, ordering mechanics, zany actions, and relationship to the labor of mobile devices—would seem to be purely about creating similarly flexible and adaptable subjects.

Yet zaniness's seeming hyperproductivity is deceptive. The zany performance can seem at first glance like utter commitment to productivity; however, as Ngai points out, zaniness—in all its desperate and frenzied action—is also fundamentally destructive. While Lucy and Ethel face the prospect of being fired from their jobs if they let a single piece of candy by without being wrapped, we laugh at them stuffing the candy in their mouths and down their aprons in an attempt to hide their failure. The destruction of the commodity for the sake of preserving the social relations of labor is another way of describing the zaniness of abstract labor. While our play in mobile games is always connected to the means and modes of productivity, it would be foolish not to recognize that our play is always also about taking some time away, even if very briefly, from concrete labor time and using our machines toward different ends.

* * *

The mobile phone is an important device through which to understand the convergence of labor time and leisure time in a networked society. Mobile time—the time we spend engaged with devices that seek to produce us as flexible, creative, and communicative subjects—is, paradoxically, a temporality obsessed with monitoring, measuring, and quantifying. Under shifts to self-management in a post-Taylorist immaterial labor context, the mobile phone becomes the central device for self-measurement and evaluation. Through mobile game apps we can see how the blurred distinction between labor time and leisure time is both concretely and abstractly productive. Globally popular mobile game apps like *Candy Crush Saga* and *Plants vs. Zombies 2* remediate the measurement aesthetics and mechanics of self-managed labor. At the same time, and in the name of time, their zany rhythms and aesthetics leave room for play.

NOTES

1 John Maynard Keynes, *Essays in Persuasion* (New York: W. W. Norton, 1963), 358–73.
2 See Ian Bogost, "Hyperemployment, or the Exhausting Work of the Technology User," *Atlantic* (November 8, 2013), http://www.theatlantic.com/technology/archive/2013/11/hyperemployment-or-the-exhausting-work-of-the-technology-

user/281149/, and David Graeber, "On the Phenomenon of Bullshit Jobs," *STRIKE!* (Summer 2013), http://www.strikemag.org/bullshit-jobs/.
3   Thorstein Veblen, *The Theory of the Leisure Class* (New York: Dover, 1994).
4   See discussion of this point in Theodor Adorno, "Free Time," in *The Culture Industry: Selected Essays on Mass Culture*, ed. J. M. Bernstein, 2d ed. (London and New York: Routledge, 2001), 187–96.
5   Pierre Bourdieu, *Distinction: A Social Critique of the Judgement of Taste*, trans. by Richard Nice (Cambridge, MA: Harvard University Press, 1984).
6   For a deeper discussion of this exchange, labor, and labor time, see Karl Marx, "Grundrisse," in *Karl Marx: Selected Writings*, ed. David McLellan (Oxford: Oxford University Press, 2000), 388–89, 415–17.
7   On the relationship between concrete and abstract labor see Karl Marx, "Capital," in *Karl Marx: Selected Writings*, 465–69.
8   In the U.S., the eight-hour workday did not actually gain legislative traction until the labor union movements of the early twentieth century made it a central issue. For this history, see David R. Roediger and Philip S. Foner, *Our Own Time: A History of American Labor and the Working Day* (New York: Verso, 1989), 81–99.
9   Harry Braverman, *Labor and Monopoly Capital: The Degradation of Work in the Twentieth Century* (New York: Monthly Review Press, 1975), 90.
10 For discussion of this, see Maurizio Lazzarato, "Immaterial Labor," in *Radical Thought in Italy: A Potential Politics*, ed. Paolo Virno and Michael Hardt (Minneapolis: University of Minnesota Press, 1996), 133–47.
11 Lazzarato, "Immaterial Labor," 133.
12 Lazzarato, "Immaterial Labor," 136–37.
13 Scholars discussing this in ways pertinent to the present discussion include C. Wright Mills, *White Collar: The American Middle Class* (New York: Oxford University Press, 1951); Arlie Russell Hochschild, *The Managed Heart: Commercialization of Human Feeling* (Berkeley: University of California Press, 1983); and Leopoldina Fortunati, *Arcane of Reproduction: Housework, Prostitution, Labor and Capital* (New York: Autonomedia, 1996).
14 Moishe Postone, *Time, Labor, and Social Domination: A Reinterpretation of Marx's Critical Theory* (Cambridge, UK: Cambridge University Press, 1993), 123–225.
15 Melissa Gregg, *Work's Intimacy* (Cambridge, UK: Polity, 2011), 18.
16 This kind of time is discussed by Gregg and also by Jonathan Crary, *24/7: Late Capitalism and the Ends of Sleep* (London: Verso, 2013).
17 For discussion of this, see Alison Hearn, "Structuring Feeling: Web 2.0, Online Ranking and Rating, and the Digital 'Reputation' Economy," *Ephemera* 10, no. 3/4 (2010): 421–38.
18 Sarah Banet-Weiser, "Branding the Post-Feminist Self: Girls' Video Production and YouTube," in *Mediated Girlhoods*, ed. Mary Celeste Kearney (New York: Peter Lang, 2011), 277–93.
19 Mark Andrejevic, "Privacy, Exploitation, and the Digital Enclosure," *Amsterdam Law Forum* 1, no. 4 (2009): 47–62, http://www.amsterdamlawforum.org.

20 For histories of leisure since the Industrial Revolution, see Peter Borsay, *A History of Leisure: The British Experience Since 1500* (London: Palgrave Macmillan, 2007), and Rudy Koshar, ed., *Histories of Leisure* (London: Berg, 2002).
21 See, for example, Bill Brown, *The Material Unconscious: American Amusement, Stephen Crane, and the Economics of Play* (Cambridge, MA: Harvard University Press, 1997).
22 Max Horkheimer and Theodor Adorno, "The Culture Industry: Enlightenment as Mass Deception," in *Media and Cultural Studies: Keyworks*, ed. Meenakshi Gigi Durham and Douglas M. Kellner, 2d ed. (Hoboken, NJ: Wiley-Blackwell, 2012), 53–74.
23 Siegfried Kracauer, "The Mass Ornament," in *The Mass Ornament: Weimar Essays*, trans. Thomas Y. Levin (Cambridge: Harvard University Press, 1995), 79.
24 See, for example, Raymond Williams, "Culture Is Ordinary," in *The Raymond Williams Reader*, ed. John Higgins (Malden, MA: Blackwell, 2001); Stuart Hall, "Encoding and Decoding in Television Discourse" (University of Birmingham Centre for Contemporary Cultural Studies, Stenciled Paper no. 7, 1973); Janice A. Radway, *Reading the Romance: Women, Patriarchy, and Popular Literature* (Chapel Hill: University of North Carolina Press, 1984).
25 For research on how mobile games affect notions of "presence" and "place" in mobile time, see Larissa Hjorth and Ingrid Richardson, "The Waiting Game: Complicating Notions of (Tele)presence and Gendered Distraction in Casual Mobile Gaming," *Australian Journal of Communication* 36, no. 1 (2009): 23–35.
26 Shift Worker (Production Shed Pty Ltd, 2013), iOS; Procraster (Simon Sorboe Solbakken, 2013), iOS.
27 For an account of the gendered history of telephony, see Michele Martin, "The Culture of the Telephone," in *Sex/Machine: Readings in Culture, Gender, and Technology*, ed. Patrick D. Hopkins (Bloomington: Indiana University Press, 1991), 50–74.
28 In 2013, mobile game apps constituted 63% of iOS App Store global revenue and 92% of the total revenue for Google Play. See Christel Schoger, "Publication: 2013 Year in Review," *Distimo* (blog), December 17, 2013, http://www.distimo.com/blog/2013_12_publication-2013-year-in-review/.
29 *Candy Crush Saga* (King Digital, 2012), iOS; *Angry Birds: Star Wars* (Rovio Entertainment in conjunction with LucasArts, 2012), iOS; *Plants vs. Zombies 2* (Electronic Arts/PopCap Games, 2013), iOS.
30 See Shuli Ren, "China's Mobile Games Reach $20 Billion Market Cap in 2014?," *Barron's* (blog), December 19, 2013, http://blogs.barrons.com/emergingmarketsdaily/2013/12/19/chinas-mobile-games-reach-20b-market-cap-in-2014-how-to-gain-exposure/, and Bien Perez, "Tencent Poised to Become China's Mobile Gaming Juggernaut," *South China Morning Post*, December 4, 2013, updated December 5, 2013, http://www.scmp.com/business/china-business/article/1372690/tencent-poised-become-chinas-mobile-gaming-juggernaut.
31 As a sign of how significant the Chinese mobile gaming market has become, PopCap Games released *Plants vs. Zombies 2* in China before it was released

in the United States. See Tony Leamer, "Plants vs. Zombies 2 Is Available on Android in China," *PopCap Blog Ride*, September 12, 2013, http://blog.popcap.com/2013/09/12/plants-vs-zombies-2-is-available-on-android-in-china/.

32 Aubrey Anable, "Casual Games, Time Management, and the Work of Affect," *Ada: A Journal of Gender, New Media, and Technology* 2 (June 2013), http://adanewmedia.org/2013/06/issue2-anable/.

33 For an example of this argument, see Nick Dyer-Witheford and Greig de Peuter, *Games of Empire: Global Capitalism and Video Games* (Minneapolis: University of Minnesota Press, 2009), 27–28.

34 Stewart Brand, "SPACEWAR: Fanatic Life and Symbolic Death Among the Computer Bums," *Rolling Stone* (December 7, 1972), 58. Available online at http://archive.rollingstone.com.

35 Fred Turner, *From Counterculture to Cyberculture: Stewart Brand, the Whole Earth Network, and the Rise of Digital Utopianism* (Chicago: University of Chicago Press, 2006), 24, 116.

36 For an example of the gamification argument, see Jane McGonigal, *Reality Is Broken: Why Games Make Us Better and How They Can Change the World* (New York: Penguin Books, 2011).

37 Time commitment is one of the primary distinctions between casual and hardcore games, but the distinction is also gendered. Casual games are generally associated with women players, though the popularity of mobile gaming is changing their demographics.

38 Life Graphy (ByungWook Kang, 2014), iOS.

39 OmniFocus (OmniGroup, 2013), iOS; 24me (24me Ltd, 2013), iOS; Grid (Binary Thumb, 2013) iOS.

40 Sianne Ngai, *Our Aesthetic Categories: Zany, Cute, Interesting* (Cambridge, MA: Harvard University Press, 2012), 188.

12

Real / Quality

MARK MCGURL

Among the handful of time concepts that can be said to be original to the postwar period, two are prominent enough to lay claim to decisive significance for our understanding of the historical specificity of the present. The terms "real time" and "quality time" bob near the surface of contemporary life, too compact to count as clichés but too inflated with surplus meaning to be considered simply as keywords. As distinct modifications of a more basic (and yet notoriously mysterious) noun, each emerges from a relatively specialized postwar discourse and begins to circulate promiscuously in the realm of everyday American usage, attracting further shades of meaning as it goes. While the first, marked masculine, remains associated with the technical operations of data management, war fighting, media, and markets, and the second, marked feminine, continues to be linked to the warmth of interpersonal relations, they nonetheless converge in present usage as complementary signs of the general sociopolitical and existential condition of *hurry* that Paul Virilio memorably called the "State of Emergency."[1] What's interesting about the terms "real time" and "quality time," when examined in this context, is how they are found to be not in perfect, but rather in broken or partial, opposition, as though they are the mutually reflecting shards of some prior, and presumably deeper, dialectical collision in the late capitalist life-world. Determining what that collision has consisted of—finding terms for it—will be part of the work of the pages to follow.

Real Time

To acquire information in real time is, most simply, to do so without the usual delay attendant to processes of mediation, whether technological or social. Real-time data closes the gap between the occurrence of an event and its subjective or machinic apprehension as information. In it, reality and representation crowd together in the urgent space of a perpetually self-renewing *now*. The term "real time" dates from the early 1950s, when it emerged in the bourgeoning professional literature of computer science and systems engineering. This was an important moment in the computer's evolution from an information-processing tool to a real-time communication device—the difference between these two things being, simply, the speed of the computer's cycle of inputs and outputs and semantic richness of those outputs. Any computer has at least a minimally communicative function in the display of its results, but real-time processing radicalizes that function to varying degrees and at varying paces. As a relative notion, so-called "real-time" processes run a wide spectrum of tolerance for lag, from a matter of minutes to seconds to parts of seconds. In computer science after the midcentury invention of the semiconductor, we are often talking about meaningful delays measured in nanoseconds, or billionths of a second, which can matter greatly in the context of, say, particle physics.

One can find early approaches to the concept of real-time communication stretching back through the nineteenth century, when the possibilities for the efficient coordination of industrial processes, industries, and markets first started to be realized. As is well known, real-time geographical synchronization was born in the establishment of Greenwich Mean Time at the Royal Observatory in the mid-nineteenth century, first proving helpful to ship navigation, and then becoming crucial to the development of an integrated rail system. In this historical moment, processes of transportation and communication, the movement of bodies and messages, began to converge. In 1912 the *Titanic* was the largest and fastest cruise ship ever built, but what has struck cultural historians such as Stephen Kern is how, as it sunk over the course of 160 minutes in the North Atlantic, this event was being followed "live" via wireless communication in London and New York.[2] Grounding the effect of cultural synchronicity that, as Benedict Anderson famously stated, underlies the "imagined community" of the modern nation, the regime of real time continued to spread, finally knitting the world together under the unquestioned rule of Coordinated Universal Time.[3] As Manuel Castells has made clear, the force of real time is not only intensive but extensive, facilitating the constant self-monitoring of a given network sub-system but also the continuous synchronization of all of a network's far-flung nodes.[4] In a heady moment in 1969, Marshall McLuhan expressed a similar sentiment, asserting in an interview with *Playboy* that, linking all human beings across space in real time, the "computer thus holds the promise of a technologically engendered state of universal understanding and unity, a state of absorption in the logos that could knit mankind into one family and create a perpetuity of collective harmony and peace."[5]

But this belies the fact that real-time computing and communication were fired in the crucible of war—and soon became a part of the U.S. Cold War strategy. For instance, a full-page advertisement in the January 1954 issue of *Michigan Technic*, the "oldest engineering college magazine in America," invited its readers to apply for jobs at Hughes Aircraft in Culver City upon graduation.[6] Like the full-page ads for Westinghouse, IBM, Bell Labs, RCA, Boeing, GE, and their like that precede and follow it, this one is designed to impress a group of educated young men with a great many job opportunities, for at this point both arms of the Cold War military-industrial complex were pumping at full capacity. "MORE THAN 100 TIMES FASTER THAN THE HUMAN BRAIN," the ad announces, showing what appears to be a jet cockpit seen from overhead, connecting it by jagged line to a boxy depiction of a small computer and back again. In early 1954, even to this knowledgeable group of readers, some basics needed explanation, and the "real" of real time still needed quotation marks:

> The special character of the application of an electronic digital computer for airborne automatic controls is reflected principally in the input-output units. The physical quantities defining the state of the system ... are measured by instruments whose outputs are usually in the form of mechanical displacements or voltages. These analog quantities are converted into digital numbers that are processed by the computer; it performs in "real" time the computations corresponding to the mathematical representation of the control problem.... These output numbers are converted into the analog-type signals used in the control operations.[7]

This is not scintillating ad copy, but it is a fairly good account, very near its historical origin, of the theory and practice of real-time information processing. If the computer works fast enough translating analogue reality into digital information, and back again, it can become a crucial prosthetic enhancement—and in some cases, replacement—of the laggard perceptual apparatus of the human pilot.

It is not difficult to detect the hot and cold wars hovering in the background of the brisk optimism of a magazine such as *Michigan Technic*, circa early 1954, and a familiar story about real time could be told simply by following their trace, via a detour through mass news media, through the reduced (if still significant) time lag of reporting on the Vietnam conflict, toward the eerily instant infotainment emanating, via satellite, from the first Gulf War and after. This version of real time would correspond most closely to the one originally theorized by Virilio in the 1970s, for whom its advent signifies above all the violent "reduction" or "negation" or "defeat" of space by supersonic weapons technology. The new geopolitics of mutually assured destruction occasions the installation of a "hotline," as it was newly called, for the real-time communication of the U.S. and Soviet heads of state, but the more saliently scary fact for Virilio is that real-time technologies "reduce to little or nothing the time for human decision to intervene in the system."[8] This view was essentially a perfect reversal of the utopian vision of McLuhan. Filtered through the acid bath of French theory, that is, the synchronization of the globe in real time looks not like Utopia but instead like a massive project of Western capitalist domination and deterritorialization, in which indigenous temporalities are mercilessly uprooted and coordinated with distant sources of authority, and any spot on the globe can instantly be secured as a target.

But even for those not marked for death, the military-industrial complex has gradually restructured the economy in a way it is not far-fetched to call violent, aggressively converting "rigid" work forces—which is to say, unionized, stable, and secure—into "flexible" work forces engaged in "just in time" production. The management theory associated with these terms arose later in the century, in the 1980s, with the mission of collapsing the interval between the customer's order and the company's production and delivery of it, thus reducing the drag of unexploited human labor and inventory on the firm's balance sheets. Real time is also, in turn, associated with the broadly overlapping, but still at least minimally distinct, system of real-time mass media, in which information is itself the product being sold. The Amazon.com customer interface presents an interesting switch-point between the domains of real-time commerce and real-time mass media. It is a system designed to shorten the interval between the perception of a need and the purchase that will satisfy it, but it is also an intensely graphic experience in its own right, the virtualization of window-shopping. How interesting, then, that its aspiration to the status of universal store was founded on the sale first of books, then of other media. These are now streamed as often as they are physically shipped, offering another opportunity for the collapsing of an interval, this time between purchase and consumption. Amazon's announcement, in 2013, that it already sunk large funds into developing a drone package delivery service to be called Amazon Prime Air, is almost too perfect. No wonder if, under real-time rule, the respective domains of the real and the represented start to blur in that familiar "postmodern" way: like a weird version of the film-watching phenomenon of persistence of vision, in real time these things flicker together so quickly as to seem continuous.

Quality Time

Inhabiting the incessantly hectic world where—whether for the competitive advantages they confer or simply for the excitement of following a live feed—such real-time proximities are deemed to be highly and generally desirable, the human subject might well find herself longing wistfully for some quality time with family or friends. Quality time is the time of intimacy, of analogue, face-to-face, intersubjective attention: *How are you, my friend? What's going on with you?* It can only occur in disconnection from wider social and technological networks. And yet, of course, the magic circle of interpersonal intimacy is not completely secure, not for long, not typically. Considered in context, quality time is not just time for intimacy; it is also the time of intimacy-at-risk-of-distraction. It wouldn't exist as a term without a background assumption of its quantitative scarcity. "Oh shit," groans Jesse (Ethan Hawke) near the end of the movie *Before Sunrise*, as the nightlong parenthesis in their respective travels that he has been occupying with Céline (Julie Delpy) begins to close upon the break of day. "We're back in real time." "Yeah I hate that," she responds, announcing the imminent conclusion of what we can now see has been a filmic embodiment of quality time, a time for talking and touching, for being close, atypical only in that the two lovers began the night as strangers.[9] The enjoyment of quality time usually takes place against a backdrop of alienated familiarity.

Historically, the idea of "quality time" originates in the subjugation of mothers to the developmental needs of their children. It dates from the early 1970s, where it emerged as the social-scientific description of an ideal form of childrearing practice. The stage for the appearance of the term was set when, in *Interactions between Mothers and Their Young Children*, Allison Clarke-Stewart declared there to be an important distinction to be drawn between merely spending time with children and really interacting with them. "In many homes," she concluded, generalizing from the data she had collected in formal field observation, "mothers are often available to their children but seldom interact with them in playful, affectionate, and stimulating ways, nor is their behavior consistently responsive to the child's expressions."[10] In other words, it is not enough to spend large quantities of time with one's child—say, cooking and cleaning while the child plays nearby. To develop his maximum cognitive potential, the child must be taken as an object of attention in his own right and acknowledged as a fully functional human interlocutor. For this purpose a certain physical posture and type of activity were to be recommended: the child should be addressed face-to-face, with mother and child engaged in a common task. The unspoken implication here was that the ideal form of childcare must take the form of a certain kind of compulsory sociality and even auto-infantilization on the part of mothers. (The question of the father's role in the cognitive development of the child barely comes up.) Role-modeling, setting an example of efficacy in the world—these would not be enough. The ideal mother would have to meet her child halfway or more on the road to adult social intercourse, acting as though the stream of relative nonsense emanating from the child were valuable contributions to a conversation. This is the origin of the concept of quality time.

In the wake of this study, the distinction between quality and quantity in childcare caught on, and it was solidified when Clarke-Stewart returned to the issue in *Child Care in the Family* (1977), a book-length review of then-recent trends in child development research. As Clarke-Stewart summarized, the "most recent trend in research on child care is to study more closely the *quality* of stimulation provided by the parents or the environment."[11] It is perhaps no surprise that as specialists sought the truth of childcare in the qualities of this interaction, quality was discovered to be more important than quantity in the parent-child bond:

> Unlike motor development . . . cognitive development seems to be significantly shaped by the mother's behavior. . . . *Quality* of stimulation is also important. Measures of the mother's adaptation of stimulation . . . are even more closely related to infants' IQ than are measures of the sheer amount of maternal stimulation. . . . It is not related to the amount of time spent in caretaking activities—during which mother and infant typically are at a 45 degree orientation to one another—but is related to the time spent playing together, since this activity is almost invariably face-to-face. Once eye-to-eye contact is established, it has ramifications for both participants. . . . The adult now perceives the infant as "human" and becomes dramatically more involved with him or her.[12]

This is fascinating as much for its fine-grained attention to bodily postures as for its linking of literal faciality to the production and recognition of the human as such. Here is at least one scene of the closing of the human social circuit against the inhuman alterity of the outside. Whether there might be a reciprocal dehumanizing effect on the subjectivity of the mother in this process was not addressed.

Let us recall that, even as Clarke-Stewart was publishing these conclusions, the social historical context within which mother-child social interactions were assumed to be occurring was in the process of rapid transformation, if not disintegration. To the extent that it had ever been a norm, the 1950s image of the stay-at-home mother, the one-wage-earner family, and relative unthinkability of divorce was becoming ever more rarified as women joined the fulltime workforce in large numbers and families became more elastic in their definition and duration. This begins to explain how what began as a call for compulsory sociality and auto-infantilization on the scene of mothering evolved, over time, into a relative convenience to parents—and even more so to their employers. A woman with a full-time job typically spends a much lower quantity of time with her children than her "stay-at-home" counterpart, but this was less troubling than it might have been if what really mattered was the *quality* of that time. As such, quality time could be safely compartmentalized, even scheduled for a given time of day, whether in the morning or in the evening after work. In this manner, "quality time" now became a complex cipher in the rising call for women's liberation. Already by 1973, reports one student of the history of phraseology, it had taken on a distinctly feminist meaning as a defense against demands for large quantities of a woman's time: her life would instead be organized around segments of quality time, including time spent in self-fulfillment.[13]

Of course, the pursuit of self-fulfillment in the form of a career would prove to be fraught with complexities. As has been documented by Arlie Russell Hochschild in a series of bestselling studies of the transformed relation between the spheres of work and home in the United States, diminishing a woman's relative subjugation to the needs of her children is replaced, frequently enough, by a woman's increased subjugation to the needs of employers. In *The Second Shift: Working Families and the Revolution at Home*, Hochschild documented the disturbing extent to which the feminist project of "having it all" had, by the 1980s, devolved, for many American women, into *doing* it all—that is, remaining primarily responsible for child care and housework even as they held down full-time jobs.[14]

Hochschild followed up *The Second Shift* with *The Time Bind: When Work Becomes Home and Home Becomes Work* (1997), which pursued to its logical conclusions the transformation of home into a second (but unpaid) workplace for women. The transformation occurred, by her account, under the banner of "quality," and as part of the same general shift in management theory that brought us into the age of just-in-time production. In the late 1980s the American workplace modulated from a Taylorist conception of worker productivity dating from the late nineteenth century—with its demand for maximum worker efficiency and regularity down to the smallest physical movements—to the new, more emotionally rewarding ethos of Total Quality Management. TQM, as it soon came to be called, echoing the ironic shortening of "quality time" to QT, sought to distribute the spirit of the quality control inspector throughout the organization, to all levels of management and labor, measuring the success of the organization as much by the quality as by the quantity of its output. And yet TQM was not just focused on the literal production process. It also carried an associated set of values into the workplace, including a new attention to the quality of the social interactions that took place there, and to the "quality of life" of workers in general. It was, in short, with the rise of TQM that American corporations began to represent themselves to themselves as "families," with all of the (ominously nonnegotiable) emotional attachments that implies. This, it was said, was the only way to compete with Japan.[15]

Picking up the story in the 1990s, in a period of relative prosperity, Hochschild's ethnography documents how, under the influence of TQM, the workplace in many ways became more personally, and indeed emotionally, rewarding for women than life at home, even as the corporation demanded more and more of its workers' time. Hochschild's ingenious argument was that while the ideology of TQM had made the workplace more familylike, the bad old regime of Taylorization had simultaneously fully infiltrated the home, with family activities now scheduled down to the minute. Telling the story of Gwen and John Bell at the pseudonymous corporation Amerco, Hochschild describes a work-life balance grown increasingly out of whack, but with most of the resentment reserved for the nature of home life. For Gwen, she explained, "her workplace has a large, socially engineered heart while her home has gained a newly Taylorized feel."[16] Hence Gwen and John find themselves responding

> to their time bind at home by trying to value and protect "quality time." A concept unknown to Gwen's ancestors, quality time has become a powerful symbol of the struggle against the growing pressures on time at home.... Many Amerco families were fighting hard to preserve outposts of quality time, lest their relationships be stripped of meaningful time together.[17]

From its origins in research on childcare, the concept of quality time had come a long way, all the while retaining its primary meaning as time for intimacy. The new implication that had by now become obvious to everyone who used the phrase was one of severe scarcity. In a world of Total Quality Management—a world where employers, in return for caring for their employees in some new ways, now make ever-larger demands upon the time of their workers—the very idea of "quality time" seems the token of a lost childhood idyll, ill equipped to compete with the unceasing demands of the real-time market.

Perhaps the same could be said of literary fiction, which is nothing if not the virtualization of quality time. The novel finds its thematic substance in the myriad forms of human intimacy and intrigue, but its typical grammatical form—the past tense—indicates its at once alluring and dangerous irrelevancy, its removal from the real time of the reader's present. Even with remarkable innovations like stream-of-consciousness narration, which simulates the tracking of that stream in real time, fictional time is shifted time, but it can only be enjoyed in real time.

The problem with quality time, in other words, is that it happens in real time—which is to say, first of all, that it happens in a wider social surround geared to an ideal of ceaseless economic production and consumption hostile to the pleasant longueurs of human intimacy, let alone serious reading, or even sleep. Wreathed in fiber-optic cables and orbited by communications satellites, the contemporary world revolves in real time, 24/7, occasioning, as Jonathan Crary puts it, the "generalized inscription of human life into duration without breaks, defined by the principle of continuous functioning."[18] Real time is the time of organizational command, minimizing the lag between command and execution, as between desire and fulfillment, and feeding results back into the system without delay. It is the apogee of technocratic impatience, and as such has proved notoriously difficult to dispute, since arguments take time. Always rushing forward, ever privileging action over reflection, the force of real time is felt more in the form than in the content of contemporary systems, and is difficult to get a handle on. This is true of fiction, as we have seen, but it is perhaps even more poignantly true in film and television, which abound (as *Before Sunrise* or any old sitcom ensemble illustrates) in compensatory representations of people enjoying quality time, even as the rigorously timed, segmented, and commercially sponsored nature of that enjoyment fades into the background. Meanwhile, in the temporality of Twitter and other social networking platforms, this phenomenon takes a further step: the mediated social interactions it facilitates could in theory take place "in good time" but never do, the ideology of hurry now having become fully internalized and autonomized. Thus Twitter does not seem to function as a medium of deliberation, let alone of intimate conversation, only of a demand for instantaneous redress of the latest, but soon to be forgotten, outrage.

Thus it is clear that it is wrong to think of the enjoyment of quality time as an act of simple resistance to, let alone negation of, the real-time regime, which in fact contains and conditions it. Yet the relation of quality time to real time can be a positive one. After all, to spend quality time with family or friends implies bodily co-presence, the live-action sharing of experience without significant delay; it implies, that is, communication in real time. As James Gleick writes in his popular account of the contemporary culture of speed, "real-time oral communication is what used to be called *conversation*," and conversation is indeed at the heart of the idea of quality time.[19] Far from simply being the limiting context of quality time, then, real time ticks at its core, and neither one provides any real critical distance on the other. This becomes clear in the sequel to *Before Sunrise*, which, nine years later in the lives of both its protagonists and its lead actors, replays the earlier movie's twelve-or-so-hour parenthesis of quality time but compresses it into something approaching the film's running time of eighty minutes. In *Before Sunset* (2004), Jesse, now a bestselling author, has a plane to catch not the next morning but in a few hours, and if he is to make that flight the couple's second round of musings on life and love will have to be scheduled accordingly.[20] Seen from this perspective—and notwithstanding the deliberate doubt sown, as the credits roll, about whether or not he will actually make that plane—the film sequence betrays the intensifying inscription of quality time into and *as* real time.

By the same token, on at least one of its vectors, the ideal toward which real-time communications reach is the facilitation of quality time-at-a-distance, which is to say, in the product language of Apple Computer, "FaceTime." In the third installment of Linklater's trilogy, *Before Midnight* (2013)—which (surprise!) finds Jesse and Céline in their seventh year of marriage and the harried parents of twins—the two find themselves at dinner with a new young couple who triumph over long distance via Skype even as they take for granted that someday they will break up.[21] Whether as parents or as young lovers, the suggestion seems to be, we now live in real time all of the time, and quality time is just one of its modes.

Quantity Time

In the speculative space of cultural criticism, however, the claim upon the "real" asserted by real-time capitalism needn't simply be granted without struggle. To let that happen would be to capitulate to what Mark Fisher has called "capitalist realism," to submit to the way capitalism hustles us unthinkingly into attenuated futures.[22] As an antidote to such capitulation, and an attempt to reconstruct the terms of the deeper dialectical collision of which our clichéd terms might only be the visible linguistic residue, one might offer the term *quantity time*. This is not to be confused with "clock time," which bears the correct implication of inhuman inexorability but would appear to depend upon the existence of clocks. Quantity time is rather the absolute time of the physical world, of the second law of thermodynamics, of perpetual transformation in the quantitative disposition of matter and energy in the universe. As recently reintroduced to philosophy by Quentin Meillassoux, it is time utterly evacuated of qualia, the stuff of human experience, but useful for imagining the ultimate contours of the circuit of human intersubjectivity.[23] As such, it is even realer than real time and more truly inexorable, since it is not subject like the technical system to wear and tear and catastrophic failure.

Something like the latter must once have occurred in the ancient Greek culture whose ruins provide the backdrop, in *Before Midnight*, of Jesse and Céline's long conversation about the family life they have lived together, which, for want of quality time, and caught in Hochschild's time bind, seems like it might now be headed toward divorce. If these civilizational ruins are meant, in this sequence, to stand as admonitory analogies to the coming ruin of one relationship, they clearly exceed the individual real lifetimes innovatively monitored by the trilogy and point it toward something much larger and more inevitable, the disappearance of any given civilizational form, and eventually of everything human.

Real time, meanwhile, filling out the semiotic structure generated by our initial opposition, is ironically revealed as taking place in *artificial time*—that is, time as "subjectively" shaped by and for the ends of humanity, whether cognitively or otherwise. Indeed, we can now see that quality time and real time represent competing but thoroughly intertwined orientations, in artificial time, to the fact of quantity time. While the first exerts whatever drag it can upon time's entropic flow by way of the construction of intersubjective meaning, shared history, and memories, the second simply tries to keep up, to stay continuously, aggressively, and profitably informed. The dialectical collision of which they are the shards can now be seen as the one between artificial and quantitative or, as they are more familiarly known, subjective and objective time, which relation is placed under a new kind of stress, in late capitalism, at the hinge between the "real" and the Real. As the realest real time of all, quantity time will, in time, run both "real time" and "quality time" into ruin, as we see imagined not just in *Before Midnight* but, less subtly, in any one of the seemingly hundred massively popular apocalyptic visions—in print, video, and video game—available to us for purchase in real time. We eagerly await the arrival of artists capable of envisioning the future in credibly better, credibly positive terms. In the meantime, as an unstoppable engine of negation, as the guarantor of historical change, quantity time supplies the form if not the content of hope.

NOTES

1 Paul Virilio, *Speed and Politics*, trans. Mark Polizzotti (Cambridge, MA: Semiotext(e), 2006), 149–67.
2 See Stephen Kern, *The Culture of Time and Space, 1880–1918* (Cambridge, MA: Harvard University Press, 1983), 65–88.
3 Benedict Anderson, *Imagined Communities: Reflections on the Origin and Spread of Nationalism* (London: Verso, 1983).
4 Manuel Castells, *The Rise of the Network Society*, 2d ed. (Malden, MA: Wiley-Blackwell, 1996), 465.
5 Eric Norden with Marshall McLuhan, "The Playboy Interview: Marshall McLuhan," *Playboy* (March 1969), quoted in Bob Hanke, "McLuhan, Virilio and Speed," in *Transforming McLuhan*, ed. Paul Grosswiler (New York: Peter Lang, 2010), 211.
6 *Michigan Technic* 72, no. 4 (January, 1954): 38.
7 *Michigan Technic*, 38.
8 Virilio, *Speed and Politics*, 156.
9 Richard Linklater and Kim Krizan, *Before Sunrise*, directed by Richard Linklater (1995; Turner Home Entertainment, 1999), DVD.
10 K. Allison Clarke-Stewart, *Interactions between Mothers and Their Young Children: Characteristics and Consequences* (Chicago: Society for Research in Child Development, 1973), 94.
11 Alison Clarke-Stewart, *Child Care in the Family: A Review of Research and Some Propositions for Policy* (New York: Academic Press, 1977), 5–6.
12 Clarke-Stewart, *Child Care in the Family*, 15, 19.

13 The Phrase Finder, "The Meaning and Origin of the Expression: Quality Time," http://www.phrases.org.uk/meanings/297250.html. The Phrase Finder finds the first instance of "quality time" in a January 1973 edition of the Maryland newspaper *The Capital*, in an article titled "How to Be Liberated."
14 Arlie Russell Hochschild with Anne Machung, *The Second Shift: Working Families and the Revolution at Home* (New York: Penguin, 1989).
15 Arlie Russell Hochschild, *The Time Bind: When Work Becomes Home and Home Becomes Work* (New York: Holt, 1997), 18–20.
16 Hochschild, *The Time Bind*, 49.
17 Hochschild, *The Time Bind*, 50.
18 Jonathan Crary, *24/7* (London and New York: Verso, 2013), 8.
19 James Gleick, *Faster: The Acceleration of Just About Everything* (New York: Vintage, 1999), 67.
20 Ethan Hawke, Julie Delpy, and Richard Linklater, *Before Sunset*, directed by Richard Linklater (Warner Home Entertainment, 2004), DVD.
21 Richard Linklater, *Before Midnight*, directed by Richard Linklater (Sony Pictures Home Entertainment, 2013), DVD.
22 Mark Fisher, *Capitalist Realism: Is There No Alternative?* (Winchester, UK: Zero Books, 2009).
23 Quentin Meillassoux, *After Finitude: An Essay on the Necessity of Contingency*, trans. Ray Brassier (London: Bloomsbury, 2010).

PART III

Time as Culture

Mediating Time

13

Aesthetic / Prosthetic

JESSE MATZ

Wired to a tank lies the mutilated corpse of Captain Colter Stevens, an army pilot killed in action in Afghanistan. His body is gone below the ribcage. But his brain lives still, patched into a computer system with a vital purpose: time travel. Sent into the past for information about a terrorist attack, Stevens ultimately saves Chicago from a massive dirty bomb. That such a radically disabled body could have such ability—to travel in time, to save a city—is the central premise of Duncan Jones's 2011 film *Source Code*.[1] The film demonstrates a remarkable form of prosthesis by which technology not only aids the disabled body but gives it a superhuman temporality. Technological prosthesis has been the premise of much science fiction, but in *Source Code* it entails a special proposition: that film itself might be a salvific form of temporal prosthesis, one that makes time an aesthetic achievement.

It takes nine tries for Captain Stevens to get the information he needs. The "source code" system enables only a brief interval of time in the past, and Stevens must repeat his efforts to make it time enough to find the bomb. In the process, he also meets a girl, finds love, and achieves personal redemption. *Source Code* is about time travel, but it also has another story to tell, one about second (or, rather, ninth) chances, in which redoing the past saves the soul even as it saves the world. A number of contemporary films tell a similar story. Harold Ramis's *Groundhog Day* is a foundational example, for its sense of the personal salvation to be had through screen time. More generally, this linkage of filmic temporality and human betterment is the basis for films including George Nolfi's *The Adjustment Bureau*, Gaspar Noé's *Irreversible*, Christopher Nolan's *Inception*, Neil Burger's *Limitless*, Doug Liman's *Edge of Tomorrow*, and even Woody Allen's *Midnight in Paris*, all of which feature prosthetic means through which to master time, and, in the process, to cultivate enhanced humanity. But what really distinguishes them from other films and fictions of time travel and technological transformation is their sense of the relationship between temporal prosthesis and filmic art. Enamored with forms of filmic practice through which temporal mastery might be represented, these films boast a new function for aesthetic engagement. They show how film might actually intervene to transform our temporal capacities, much the way common forms of prosthetic action transform the practice of daily life. Joining temporalities that are aesthetic and prosthetic, these films represent a contemporary sense that a prosthetic art of time might cultivate a new humanity.

Aesthetic

Aesthetic time is that which the work of art creates. It is not just the temporality of aesthetic engagement or art's ontological time; it asserts the further possibility that art makes time meaningful. Aesthetic time motivates Romantic poets and champions of photography alike, for whom art is an opportunity to capture lasting moments and also, importantly, to make a difference to time itself. Phenomenological philosophy, too, used this form of explanation and advocacy, which can be said to culminate in the work of Paul Ricoeur, for whom time becomes human time to the extent that fictive narration reconfigures it. Ricoeur's examples are the "tales about time" invented by Virginia Woolf, Thomas Mann, and Marcel Proust, who use narrative to wrest meaningful concordance from time's confusion.[2] Ricoeur credits them with special knowledge of art's signature power to configure time's aporias in such a way as to make human time emerge. And that power has been observed elsewhere—in other theories of literature, in other forms of art. Theorists as diverse as Frank Kermode and Paul de Man also see this relationship between aesthetic engagement and temporal possibility—Kermode in his account of how and why we "make objects" of temporal concordance, de Man in his reading of the effectual rhetorics of temporality.[3] Other arts of time include cubist painting (which invents a new form of serial simultaneity) and film, which, for theorists ranging from Gilles Deleuze to Garrett Stewart, frames time in such a way as to reinvent it for human use.[4]

As Ricoeur implies, it is in modernist art that aesthetic time asserts itself most fully. Whereas the art of time is practiced wherever art is in play, its aesthetic ambition is a product of modernist ideology. That is, time is shaped by art of all kinds—in Romantic spots of time and postmodern "chronoschisms" as much as any modernist epiphany—but in modernism it becomes subject to an aspirational aesthetic.[5] From Impressionism to cubism, in Woolf and in Proust, in the first efforts at cinematic montage, there is a reflexive wish to create human time. The crisis in time caused by modernity would be resolved in art: what is elsewhere a matter of correlation becomes, in modernism, a causation in which aesthetic engagement repairs or even reinvents temporalities jeopardized by the pace, priorities, and sheer novelty of modern life.[6] To say this, however, is not to consign aesthetic time to the past—not to claim that it was really active only in modernism's historical moment. For this is a modernism that persists, continuing to inspire fantasies of temporal redress.

It is very much active in *Source Code*. If Captain Stevens gains power over time, it is represented as an essentially filmic style of repetition, a virtue of the film aesthetic. Each time Stevens returns to the past for his eight-minute chance to save the world, the scene is just the same: he sits on a train opposite a girl named Christina, coffee spills on his shoe, a conductor asks for his ticket, and other passengers pursue their same morning routines. In each replay, changes to the same scenario are decisive. Sameness enables Stevens to identify potentially significant points of departure for his investigations. When the final replay goes like clockwork, it is still essentially the same as the first, but Stevens's mastery of it produces the successful outcome. Such an experience might seem to derive from the structure of interactive video-gaming, but it is actually that of a filmic mode that has incorporated virtual interactivity into its *mise-en-scène*. In *Source Code*, film shows how it can conjoin mere repetition to human agency and thereby model a form of temporal proficiency. That proficiency is essentially filmic because it exploits film's notoriously ineluctable indexicality: film can show us the same thing again and again because its way of indicating its objects seems to show them untransformed. Each time Stevens returns to the past, it is that same past awaiting his greater mastery. As Mary Ann Doane might say, film has "archived" the past moment in its fullest contingency, readying it for true repetition as only film can do.[7] But *Source Code* entails a contemporary revision to the classic way film has been seen to structure time—by Doane, and by Deleuze and Bergson before her. Here, cinematic time becomes a reflexive object of thematic self-importance.[8] Its aesthetic value is something *Source Code* promotes.

But if *Source Code* asserts aesthetic time, it also boasts another kind of temporal mastery. If it is aesthetic, it is also prosthetic, offering to supplement our natural sense of time with technological enhancement. The prosthetic motive has a similar ambition but a very different relationship to modernity, to art, and to humanity.

Prosthetic

Human time has always been prosthetic, known through external objects, even before there was any possibility of what we now consider technological enhancement. Time as we know it has always been distributed among the vast range of physical objects and measures through which time is made and told. Always already posthuman, time has long presented the challenge we now associate with prosthesis: how to take on the abilities it confers without conceding human disability—without allowing dehumanization. In the longer history of temporal objects, prosthesis has seemed dehumanizing, for the more we have deferred temporal agency to clocks, railroad timetables, and the prerogatives of industrial mechanization, the more we have seemed to disable temporalities fundamental to human flourishing. In this view, efficient chronology detracts from some idealized fullness of time and our freedom within it. Such has often been the response to prosthetic time and its consequences; the sense that time has become prosthetic but that prosthesis destroys time is a fundamental paradox in time's longer history. But if it has always been a source of anxiety and doubt, it came to a crisis in modernism. Like aesthetic time, prosthetic time comes to cultural consciousness most fully with modernist ideology, and indeed the two define each other by opposition: aesthetic time reacts against time's prosthetic objectification, in a bid to restore human agency in time. This reaction is definitive for Stephen Kern's influential account of the modernist "culture of time and space," in which time-culture often asserts a humane alternative to time's new public instrumentality.[9] And for Tim Armstrong, the modernist artwork is often a response against technological prosthesis more generally, an aesthetic resistance we might also see at work in the recuperation of human time.[10]

After modernism, however, prosthesis gets rethought and revalued. Once technological enhancement becomes the essential condition of modern humanity—that is, in the moment of the posthuman—prosthesis ceases to align simply with a bad kind of reification. It inspires an array of responses, ranging from uncritical enthusiasm for technological enhancement to a new kind of alarmism about technological takeover, from grudging acceptance of the prosthetics of human subjectivity to celebrations of cyborg humanity. This mixed array of responses defines prosthetic time.[11] For although *Source Code* would seem to operate with something like Donna Haraway's enthusiasm about cyborg enrichment, and even if Captain Stevens seems to represent the kind of "cyborg acceptance" that Scott Bukatman attributes to fictions of technological capitulation, the film stops short of any cyborg outcome.[12] Stevens's final replay does not end, but extends into an alternate reality in which he and Christina walk off together into a perfect Chicago afternoon. Not only has he changed the very course of time, he has a nondisabled body, if a virtual one. Prosthetic time, it seems, repairs itself away, and this new version of the old paradox indicates a need at once to have and to deny prosthetic temporality.

It is an ambivalence that plays upon the strange supplementary logic of prosthesis itself. In his foundational work on the relationship between bodies and machines, Mark Seltzer observes a "double logic" by which prosthetics extend the human body but thereby undo human agency, not just by absorbing it into technological practice, but by reducing humanity to a state of debility.[13] This, of course, is the logic of the supplement, and for Seltzer it sets up a problematic back-formation in which prosthesis redefines the human (for all humans) not as a technologically enhanceable capacity but as a cancelled category. Of course, this view of human disability is problematic, and it is more common to stress "ambiguity": Sara Brill, for example, following Elizabeth Grosz, argues that the "ambiguous" duality of prosthesis functions "not only to supplement a lost capacity . . . but also to generate new capacities and experiment with new fields of action."[14] But supplementary prosthesis remains a problem, as Sarah Jain observes, when its new capacities are subject to old social constructions. Jain writes, "prosthesis can fill a gap, but it can also diminish the body and create a need for itself"—but then she argues that this need must be read in terms of the sociocultural affordances for which prosthesis is but a metaphor, the "economic-discursive apparatuses" that characterize reparation.[15] Prosthesis might construct the body that needs it, naturalizing an artificial addition, but it takes a social construction to make that naturalization work, and to produce disability as a fact of identity. It is important not to ignore the contexts necessary for human identity to come to reside in "bodies that are never considered whole enough," to reflect "normative values of wholeness" that drive the prosthetic imagination to ever greater normativity.[16] Jain questions the tropic disavowal in play here, noting that the tendency to imagine all kinds of technological supplementation in terms of prosthetics applies the language of medical enhancement to what are really sociocultural conditions and, as a result, naturalizes disabilities that have been constructed socioeconomically.

But if the "prosthetic imagination" exploits a certain "metaphorical opportunism," the metaphor in question is one that might have a range of effects. Seltzer and Jain are right to question its dehumanizing tendency—the way prosthetics imply their own necessity in such a way as to make the human animal dependent upon their completion, often in order to totalize an economic system of production—but if such a metaphor defers humanity to mechanical means, it can also defer it to means that are, inversely, always already human. Sometimes, that is, the machines are aesthetic ones, and therefore themselves constructed according to humane prerogatives. The result is not necessarily better for actual people. Rather, it involves a "language of displacement" that tries to convert disavowal to fantasy, recovering human agency in the less instrumental—the merely purposive—quality of the aesthetic. Bernard Stiegler has most notably questioned this variety of prosthesis in his account of the "cinema of consciousness."[17] In their special issue of *New Formations* dedicated to the question of the "prosthetic aesthetic," Joanne Morra and Marquard Smith have collected an array of essays on the overlapping impulses of these two modes of reparation, many of which explain how aesthetic objects bolster the prosthetic metaphor.[18] What prosthesis fails to secure, aesthetics supply. That is, one way to meet the challenge of prosthesis—to take on its abilities without conceding human disability, without allowing deficiency to characterize human time—is the aesthetic, which humanizes prosthetic time by making it a matter of human art rather than inhuman technology. Elizabeth Grosz has asked if "prostheses [should] be understood more in terms of aesthetic reorganization and proliferation, the consequence of an inventiveness that functions beyond and perhaps in defiance of pragmatic need."[19] But such an aesthetic might be decisively pragmatic. Indeed, this dynamic explains how the conflict between aesthetic and prosthetic time has now become an important collaboration: in prosthetic-aesthetic time, art humanizes time with all the power of prosthesis, and prosthesis augments the human with all the power of art.

Prosthetic-Aesthetic

Prosthetic-aesthetic time actually begins in modernism, when the art of time first begins to present itself reflexively as an object. That we have no "organ for our sense of time" is the famous lament of Thomas Mann's *The Magic Mountain*, a novel all about temporal disability and its cultural sources and effects.[20] But the lament is insincere, because the novel itself provides such an organ, albeit prosthetically. Although Hans Castorp has nothing like eyes or ears for time, Mann's narration compensates, embodying all the acuity necessary to make up for temporal incapacity. Whenever Castorp is bewildered by a sense of timelessness, his novel, by contrast, flaunts temporal proficiency. This is the greatest irony of *The Magic Mountain*—that it is itself the "cure" its protagonist cannot find. It is also the irony whereby many modernist texts propose their own forms as temporal prosthetics. William Faulkner's *The Sound and the Fury* famously centers on the smashing of a watch but does not therefore forsake prosthetic means of temporal perception. Rather, the novel designates itself as the better timepiece, showing in its formal structure how to remedy the temporal disabilities of its characters. Marcel Proust even makes this objective explicit. He writes of his readers, "it seemed to me that they would not be 'my' readers but the readers of their own selves, my book being merely a sort of magnifying glass like those which the opticians at Combray used to offer his customers—it would be my book, but with its help I would furnish them with the means of reading what may lie inside themselves."[21] Modernist time more generally may be defined in terms of this nascent prosthetic-aesthetic impulse, which is what sets it apart from earlier efforts to develop an art of time. Thematizing the practical use of their work, these writers assert a new ambition to make the aesthetic prosthetic.

If this prosthetic aesthetic comes into its own with the postmodern arts, which engage so much more actively with technology and its potential inhumanity, we should note that postmodern prosthesis is much less likely to align with aesthetic redress. Almost by definition, postmodernism rejects the salvific aspirations often essential to the modernist arts, and its prosthetic temporalities therefore tend to disclaim aesthetic mastery. Whereas *The Magic Mountain* and *The Sound and the Fury* aspire to remedy temporal disabilities, postmodern texts like Kurt Vonnegut's *Slaughterhouse-Five* or Harold Pinter's *Betrayal* only suggest skepticism about any aesthetic version of prosthetic time. By contrast, the prosthetic impulse at work in a film like *Source Code* is a provocative revival of a modernist intervention into time, a "remodernist" aesthetic, and one that ought to draw the kind of skeptical attention postmodernism paid to modernist aestheticism in general.[22] Moreover, as a remodernist phenomenon, the prosthetic temporality of contemporary film coincides with other contemporary developments that seem to share its impulse rather than that of postmodern temporality. Whereas the classic time-theorists of the postmodern moment (Jameson, Lyotard, Harvey) stress the crisis by which postmodernity has irreversibly diminished "time today," a newer generation of theorists pursue conceptual and practical means to redress this crisis.[23] Most recently, for example, Jonathan Crary has rehearsed the standard critique of the compression, emptiness, and "time without time" symptomatic of our moment, only to advocate a model for resistance or even redemption.[24] In *24/7: Late Capitalism and the Ends of Sleep*, Crary proposes that sleep, "unexploitable and unassimilable," is a "remission" within 24/7 culture that could inspire broader forms of refusal.[25] Arguing that "the imaginings of a future without capitalism begin as dreams of sleep," Crary theorizes something very much like a prosthetic aesthetic of time, insofar as he tries to reclaim temporal experience for a reconstructed humanity.[26] And this argument has counterparts in other contemporary cultural movements aimed at temporal redress—movements that similarly contribute to an explanatory context for the ambitions of films like *Source Code*. Efforts at "time ecology," "slow" movements, and the Long Now Foundation have a positive, practical impulse different in character from postmodernism's temporal ontologies, and it matches that of the prosthetic aesthetic of contemporary film in such a way as to help characterize—for better and for worse—the motives behind it.[27]

Two critics are alert to the specific prosthetic-aesthetic claims of film. Lev Manovich uses the term "cognitive prosthesis" to name a range of efforts to "externalize the mind," from the photographs of Sir Francis Galton to the films of Sergei Eisenstein and beyond. The argument naturally draws upon Marshall McLuhan's work on the "extensions of man," and, by entering the cognitive arena, jumps to a higher level: Manovich argues that information technologies serve a "metaprosthetic" function, augmenting mental acts that precede and govern any action in the material world.[28] The special metaprosthetic function of art forms was, according to Manovich, the utopian object of Eisenstein's theory of filmic montage. Montage created the possibility of "filmic reasoning," which transferred to the human mind differential logics once found only beyond it.[29] Here, any disability naturalized by prosthesis becomes an object of political restitution, since the inculcation of filmic reasoning is really a form of consciousness-raising, an association enabled by the aesthetic quality of the prosthetic object.

By contrast, Alison Landsberg observes a less utopian outcome in her account of prosthetic memory in late-twentieth-century American film. Discussing superhuman memory in *Total Recall* and *Blade Runner*, Landsberg describes a pernicious correlation between these films' thematic content and cinema as a cultural form: "The cinema . . . as an institution which makes available images for mass consumption, has long been aware of its ability to generate experiences and to install memories of them," and so "we might then read these films which thematize prosthetic memories as an allegory for the power of mass media to create experiences and to implant memories."[30] In other words, cinema has become aware of its own potential role in "cognitive prosthesis," and, in certain cases, has come to make it—to make Eisenstein's "filmic reasoning"—an explicit subject of thematic attention. For Landsberg, however, the result is dystopian. Like Stiegler, she sees film to be complicit in culture-industry deception at the "metaprosthetic" level, even—or especially—if it knows it.

Amid these interpretations of what happens when art becomes metaprosthetic—in the longer history that proceeds from modernism's utopian prosthetics to dystopian visions of the naturalization of mass-media experience—films like *Source Code* represent a new variation on an old theme. A reconstructed modernist utopianism meets a certain metaphorical insouciance to suggest that film really can make better human beings—or, more specifically, people with a more humane sense of time. Although *Source Code* would seem to represent a posthuman future in which humanity becomes its technological enhancement and thereby transcends any humanist psychology, sociability, or ethos, it actually promotes the prosthetic impulse in which humanist prerogatives are enhanced and undiminished by their technological remediation. What might otherwise be a merely supplementary relationship is rendered utopian by the apparatus special to art—that in which interests are to be suspended in favor of an autonomous form of attention. But we need not accept this alternative on its own terms. *Source Code* may suggest that its filmic form is a prosthetic means to improve human time without harming human beings, but its sentimentality, its nostalgia, its autonomy, and its heroic aspirations also suggest ideological deception, a false technological perfection in which time is no better than the "whole" human body. As Captain Stevens and Christina depart into an alternate reality achieved aesthetically, as they triumph over time, government, and terror alike, where are they actually going? If we accede to the enhancements of filmic repetition, making it a real way to imagine fuller present attention, do we just end up fantasizing about better worlds?

Such questions are raised more pointedly by another aesthetically charismatic enactment of filmic prosthesis: Neil Burger's *Limitless*.[31] If the feel-good message of *Source Code* shields it somewhat from our postmodernist skepticism—who would begrudge a disabled war veteran his happiness, even if it means accepting the ambitions of the filmic time-aesthetic?—*Limitless* lays bare the device more brazenly. It sets up scandalous links among film technology, pharmaceutical excess, and political power. Here, aesthetic prosthesis is literally a means of mastery, overstating film's claim upon time.

Eddie Morra is a failed writer and a failed person. He cannot even begin the novel he is writing—a utopian novel, naturally—and he hits rock bottom before reconnecting with his ex-wife's brother, Vernon, a drug dealer who claims he has gone straight and is now working for a pharmaceutical company. Hearing about Eddie's "creative problems," Vernon offers him NZT-48, an experimental drug that confers full access to the brain.[32] It gives Eddie total recall, perfect powers of prediction, and slow-motion control over the present. "What I could do with my day was limitless," he says. "Enhanced Eddie" is superhuman, and he soon draws the attention of Wall Street in the form of Karl van Loon (played with truly cinematic reflexivity by Robert De Niro), who becomes his mentor, making Eddie central to the merger of the century.[33] But of course there are side effects. Eddie starts losing time, skipping forward minutes or hours, jumping ahead and finding himself in the future. Further complications ensue, but Eddie does work things out. He perfects the drug, gets the better of his mentor, and is finally poised to become a U.S. senator.

Ultimately, Eddie is the poster child for the film's bold proposition: that the limitlessness conferred by the drug is safely conferred by film itself. In "enhanced Eddie," we get the model of a prosthetized temporal cyborg gifted with control over time; in his film, we're offered that gift ourselves, in a technique modeled for our use. For NZT-48 has a correlate in a film technique customized for *Limitless*: "fractal zoom," or, as it was refit for *Limitless*, "infinite zoom." Eddie can project himself into any future, and "infinite zoom" shows how: it represents the present moment opening onto an impossibly deep-focus sequence of future events. The visual-effects supervisor responsible for the technique, Dan Schrecker of Look Effects, says he tried to simulate something like the Mandelbrot fractal pattern, in which a closing-in yields to the iteration of a new variation.[34] This pattern inspired a way to do continuous zoom through time and space, a ceaseless motion that simulates limitless perception in both dimensions. It involves a three-camera rig—three high-resolution cameras with different focal lengths zooming at once through a scene. The footage from the camera with the longest lens is embedded in the footage from the two others, and that keeps the image consistent as it zooms forward.[35] Infinite zoom is the motion-picture of a limitless time-sense. It is the visual but also the aesthetic correlate of NZT-48— the film's effort to do what the drug does. The technique moves through space-time in summary style, one that shares Eddie's gift for temporal compression and inference. And just as we are shown how it works, we also see what inheres in this form of temporal proficiency, and what it produces. In Eddie's life, we see the "economic-discursive apparatus" involved in his prosthetic ability, and that metaphorical context may indeed make us wish to emulate it, given his high level of achievement.

Whether or not we could do so, his "infinite zoom" dramatizes a certain temporal ambition, showing us how film art now wishes to act, through us, upon temporal possibility.

The infinite zoom might be a sign of technological mastery or a symptom of cyborg disability. Is film's prosthetic project a dangerous thing—something to guard against? Or is the zoom really something film might offer as a kind of cognitive metaphor that could actually, redemptively supplement the human mind? *Limitless* seems at once to aspire to temporal enhancement and to warn against it, and this ambivalence is essential to the prosthetic-aesthetic ambition active in film today. If *Limitless* does warn against the fantasy that technological forms might enhance the human, it might usefully reconfirm a certain strain of ideology critique. But if it promotes this fantasy—giving it new re-modernist, aesthetic justification—we might also wish to take seriously what the film proposes, to wonder if the fantasy might not be a reality. If film today offers up prosthetic aesthetics for enhancing our sense of time, there is a chance that its products are useful, and should be studied or even adopted for their enhancements of the human. As much as a film like *Limitless* seems to encourage delusions of temporal grandeur, it might also be the better sort of "cognitive fiction," to use Joseph Tabbi's term for those fictional forms that embed and distribute useful cognitive resources.[36]

Does the filmic context only mask a dehumanizing technological imperative, or does it actually make temporal enhancement a posthuman possibility? In other words, if the prosthetic metaphor has tended to reinforce disability, distinguishing it from a normative wholeness no one can securely claim in actuality, do these films only aid in that supplementation, designating technological heroism as the unachievable norm for human time? Or does aesthetic mediation—filmic form specifically—actually serve the human purpose originally sought by modernist temporalities—not that of heroic mastery but rather the humane moderation of modern technology? For there is still the possibility that filmic prosthesis does conform to the wayward human will, demonstrating how we might make the most of present attention (as Stevens does in *Source Code*) or improve at foreseeing what will matter to us (like Eddie Morra in *Limitless*). That possibility is an old one, but new again today, and it circulates with new charismatic appeal in these films of time regained.

NOTES

1 *Source Code*, directed by Duncan Jones (2011; Universal City, CA: Summit Entertainment, 2011), DVD.
2 Paul Ricoeur, *Time and Narrative*, trans. Kathleen McLaughlin and David Pellauer, vol. 2 (Chicago: University of Chicago Press, 1985), 101.
3 Frank Kermode, *The Sense of an Ending: Studies in the Theory of Fiction* (Oxford: Oxford University Press, 1966), 4; Paul de Man, "The Rhetoric of Temporality," in *Blindness and Insight: Essays in the Rhetoric of Contemporary Criticism*, 2d rev. ed. (Minneapolis: University of Minnesota Press, 1983), 222.
4 Gilles Deleuze, *Cinema 2: the Time-Image*, trans. Hugh Tomlinson and Robert Galeta (Minneapolis: University of Minnesota Press, 1989); Garrett Stewart, *Framed Time: Toward a Postfilmic Cinema* (Chicago: University of Chicago Press, 2007).
5 Ursula Heise, *Chronoschisms: Time, Narrative, and Postmodernism* (Cambridge, UK: Cambridge University Press, 1997). See Heise's claim that postmodern strategies "preclude moments of epiphany and privileged insight," as well as her general discussion of the difference between modernist and postmodern temporalities (58, 50–53), which does not focus on the question of aesthetic objectives.
6 Stephen Kern's *The Culture of Time and Space, 1880–1918* (Cambridge, MA: Harvard University Press, 1984) remains the leading account of modernist time-crisis, and texts by Proust and Woolf are primary (though contested) examples of the reparative response to it.
7 Mary Ann Doane, *The Emergence of Cinematic Time: Modernity, Contingency, the Archive* (Cambridge, MA: Harvard University Press, 2002). I refer specifically to Doane's account of film's foundational "drive to fix and make repeatable the ephemeral" and the "pathos of archival desire" at work in that drive (22–23), but more generally I believe that the aesthetic impulse in contemporary film represents a latter-day, reflexive version of the "representability of time" in film as Doane explains it.
8 For a compelling account of this critical tradition, see the chapter on "Temportation" in Garrett Stewart's *Framed Time*, which makes an argument that complements mine by theorizing a postmodern tendency to disclaim something very much like the function I identify here. For Stewart, "framed time" is largely a postmodern, postfilmic tendency toward cinematic skepticism (122–63).
9 Kern 34. See also Heise's claim that modernist novels aim to "create a social 'soft-clock' temporality against the 'hard' clocks that divide public from private" temporalities (*Chronoschisms*, 51).
10 See Armstrong's chapter on "prosthetic modernism" in *Modernism, Technology, and the Body: A Cultural Study* (Cambridge, UK: Cambridge University Press, 1998), especially his work on Ezra Pound and other artists for whom "the status of art as an extension of the body [could] take on a sloganizing unreality in the face of its ideological construction" and his inspired rereading of the paranoia at work in William Carlos Williams's slogan "No ideas but in things" (105).

11 Marquard Smith and Joanne Morra provide a helpful survey of theory on prosthesis in their introduction to *The Prosthetic Impulse: From a Posthuman Present to a Biocultural Future* (Cambridge, MA: MIT Press, 2006), 1–16. A comprehensive sense of the range of theoretical views on the subject is given by N. Katherine Hayles, *How We Became Posthuman: Virtual Bodies in Cybernetics, Literature, and Informatics* (Chicago: University of Chicago Press, 1999). See Hayles's account of the way the "posthuman view thinks of the body as the original prosthesis we all learn to manipulate, so that extending or replacing the body with other prostheses becomes a continuation of a process that began before we were born" (3).
12 See Haraway's account of how "cyborg politics" would have us "rejoicing in the illegitimate fusions of the animal and machine" in her "Cyborg Manifesto" in *Simians, Cyborgs and Women: The Reinvention of Nature* (New York: Routledge, 1991), 176. Scott Bukatman, *Terminal Identity: The Virtual Subject in Postmodern Science Fiction* (Durham, NC: Duke University Press), 321.
13 Mark Seltzer, *Bodies and Machines* (New York and London: Routledge, 1992), 157.
14 Sara Brill, "The Prosthetic Cosmos: Elizabeth Grosz's Ecology of the Future," *Philosophy Today* 55 (2011 supplement): 248.
15 Sarah Jain, "The Prosthetic Imagination: Enabling and Disabling the Prosthesis Trope," *Technology & Human Values* 24, no. 1 (1999): 44, 46.
16 Jain, "The Prosthetic Imagination," 45, 46.
17 Bernard Stiegler, *Technics and Time, 3: Cinematic Time and the Question of Malaise*, trans. Stephen Barker (Stanford, CA: Stanford University Press, 2010), 64.
18 See Marquard Smith and Joanne Morra, "Introduction: The Prosthetic Aesthetic," *New Formations* no. 46 (Spring 2002): 5–6.
19 Elizabeth Grosz, *Time Travels: Feminism, Nature, Power* (Durham, NC: Duke University Press, 2005), 147.
20 Thomas Mann, *The Magic Mountain*, trans. John E. Woods (New York: Vintage, 1996), 64.
21 Marcel Proust, *In Search of Lost Time*, trans. Andreas Mayor and Terence Gilmartin, vol. 6, *Time Regained* (New York: Modern Library, 1993), 39.
22 Billy Childish and Charles Thomson, "Remodernism: Toward a New Spirituality in Art," Stuckism International, last modified March 1, 2000, http://www.stuckism.com/remod.html. This is the Stuckist "remodernist" manifesto.
23 Jean-François Lyotard, "Time Today," in *The Inhuman: Reflections on Time*, trans. Geoffrey Bennington and Rachel Bowlby (Stanford, CA: Stanford University Press, 1988), 58. For a fuller version of this characterization of postmodern time-crisis theory, see Ursula Heise, *Chronoschisms*, 16–28, and the opening of Jesse Matz, "The Art of Time, Theory to Practice," *Narrative* 19, no. 3 (October 2011): 273.
24 Jonathan Crary, *24/7: Late Capitalism and the Ends of Sleep* (New York: Verso, 2013), 29.
25 Crary, 24/7, 126.
26 Crary, 24/7, 128.

27 For a survey of this trend, see Jesse Matz, "Modernist Time Ecology," *Modernist Cultures* 6, no. 2 (2011): 245–69.
28 Lev Manovich, "Visual Technologies as Cognitive Prostheses: A Short History of the Externalization of the Mind," in *The Prosthetic Impulse: From a Posthuman Present to a Biocultural Future*, ed. Marquard Smith and Joanne Morra (Cambridge, MA: MIT Press, 2007), 204.
29 Manovich, "Visual Technologies," 207.
30 Alison Landsberg, "Prosthetic Memory: *Total Recall* and *Blade Runner*," *Body and Society* 1, nos. 3–4 (1995): 176.
31 *Limitless*, unrated extended cut, directed by Neil Burger (2011; Beverly Hills, CA: Relativity Media, 2011), DVD.
32 *Limitless*, 8:51.
33 *Limitless*, 27:37.
34 Tim Moynihan, "Dan Schrecker Q&A: Behind the Visual Effects in *Black Swan* and *Limitless*," *PC World/TechHive* (April 29, 2011, 6:00 PM), http://www.techhive.com/article/225669/dan_schrecker_interview_black_swan_limitless.html.
35 Moynihan, "Dan Schrecker Q&A"; Renee Dunlop, "Looking into Infinity: The VFX of *Limitless*," *CG Society* (7 April 2011), http://www.cgsociety.org/index.php/CGSFeatures/CGSFeatureSpecial/limitless.
36 See Joseph Tabbi, *Cognitive Fictions* (Minneapolis: University of Minnesota Press, 2002).

14

Analepsis / Prolepsis

JAMES PHELAN

I love you more today than yesterday. / But only half as much as tomorrow.
—Spiral Starecase, "More Today than Yesterday" (1969)

Analepsis and prolepsis are rough synonyms for flashback and flashforward. They were proposed by Gérard Genette as part of his theoretical and interpretive analysis of Marcel Proust's *À la recherche du temps perdu*, but they have since proved to be extremely valuable for thinking about the nature of narrative.[1] Observers of the "narrative turn," the burgeoning interest across multiple disciplines in the explanatory power of stories, often explain its emergence by noting that narrative is, in the emphatic words of H. Porter Abbott, "*the principal way in which our species organizes its understanding of time.*"[2] Paul Ricoeur makes a similar point by emphasizing the reciprocal relation between time and narrative: "time becomes human . . . to the extent that it is organized after the manner of a narrative; narrative, in turn, is meaningful to the extent that it portrays the features of temporal experience."[3] Thus to understand how we humans engage with time in the contemporary period, we should examine the links between the nature of narrative and its ways of handling time—and consider whether and how those ways have changed since the 1960s. As I have argued elsewhere, we turn to narrative to make sense of the world and our experiences of it because (a) it is necessarily selective (it sorts and abstracts those experiences into intelligible wholes) and (b) it allows for thick descriptions of what it selects.[4] Narrative's ways of handling time go hand in hand with both aspects of its nature, and analepsis and prolepsis are two of its most valuable means for such handling. Analepsis depends on the principle that in order to understand the present one needs to understand the (relevant) past. And I contend that prolepsis partly depends on the principle that in order to understand the present one needs to project a future.

My epigraph, the two-line chorus from Spiral Starecase's one-hit wonder, illustrates these points.[5] The male singer makes sense of the ongoing rush of his feelings by organizing them in time. Rooted in the present ("I love"), he testifies to his lover ("you") that his ever-deepening devotion can best be understood by reference to the immediate past ("more today than yesterday") and the immediate future ("but only half as much as tomorrow"). Furthermore, this move allows him to identify a constant amidst ongoing change: his feelings in the unfolding present always have the same relation to that past and, thus, to that future. More generally, the singer uses the analepsis and the prolepsis to construct a mini-narrative in the service of lyric ends: the swift movement from today-tops-yesterday to tomorrow-tops-today succinctly, wittily, and hyperbolically expresses his consistently expanding love.

To be sure, not all narratives employ analepsis and prolepsis, but these techniques have been a part of the storyteller's repertoire at least since Homer began the *Iliad* with an invocation to the Muse containing a proleptic abstract of the epic's action, and then turned to narrating that action *in medias res*. Not surprisingly, storytellers from historians to fantasists and from sportswriters to politicians have found countless uses for these techniques. Given the centrality of analepsis and prolepsis to narrative itself, and given their extensive use throughout its history, we should not be surprised to find that contemporary storytellers often use analepsis and prolepsis in ways that their predecessors did. But given the multiple historical traumas, widespread technological changes, and other cultural shifts that have altered the experience of time in the contemporary period, we should also not be surprised that storytellers have employed these techniques in new ways.[6] I turn to literary fiction to support these claims, because it typically both affirms its culture's fundamental commitments and reveals that culture's grappling with new aspects of human experience, such as new relations to time. I deliberately choose examples—Thomas Pynchon's *The Crying of Lot 49*, Toni Morrison's *Beloved*, and Ian McEwan's *Atonement*—in which the storyteller's grappling with time is not the central point of the fiction but rather plays an integral role in the treatment of other thematic concerns. Looking at how storytellers "do analepsis and prolepsis while doing something else" offers valuable insight into how contemporary storytelling has both perpetuated and challenged earlier ideas about time and narrative, especially in relation to each other.[7]

Analepsis

Gérard Genette's important discussion of analepsis and prolepsis is part of his larger treatment of narrative temporality, which he divides into three parts: order, duration, and frequency. In all cases, he focuses on the relation between temporality in the telling (the discourse) and temporality in the told (the story). Analepsis and prolepsis refer to mismatches between the order of events in the telling and their order in the told. In a narrative without analepsis or prolepsis, these two orders match ("I loved you very much yesterday, I love you twice as much today, and I will love you twice as much as that tomorrow"). In a narrative with an analepsis, as the telling moves forward in time, it deviates from the chronology of the told by going back to recount prior events ("I love you more today than I loved you last week"). In a narrative with a prolepsis, as the telling moves forward in time, it deviates from the chronology of the told by jumping ahead to tell of future events ("but only a fraction as much as I'll love you next year").

Genette defines analepsis this way: "any evocation after the fact of an event that took place earlier than the point in the story where we are at any given moment."[8] Crucial to Genette's account is the additional concept of "first narrative" (which for the sake of clarity I will amend to "primary narrative"): the events whose progress constitutes the narrative now, the continually unfolding narrative present. For example, in *The Crying of Lot 49* the primary narrative recounts Oedipa Maas's efforts to carry out her duties as "executrix" of Pierce Inverarity's will.[9] Consequently, the narration in the novel's first chapter about Oedipa's history with Inverarity—their affair and some of its aftermath—is analeptic.

Genette's conception of analepsis as a relation between telling and told differentiates it from the broader concept of retrospection. Retrospection is the default stance—and the past the default tense—of narrative, because, as Dorrit Cohn puts it, we live now and tell later.[10] We make sense of experience (our own and others') by looking back on it. Thus, analepsis typically occurs within the retrospection of narration, as the very first sentence of *Lot 49* illustrates:

> One summer afternoon Mrs Oedipa Maas *came home* from a Tupperware party whose hostess had put perhaps too much kirsch in the fondue *to find* that she, Oedipa, had been named executor, or she supposed executrix, of the estate of one Pierce Inverarity, a California real estate mogul who had once lost two million dollars in his spare time but still had assets numerous and tangled enough to make the job of sorting it all out more than honorary.[11]

The sentence establishes the retrospective primary narrative: Oedipa came home and learned of her new status. That primary narrative in turn helps us recognize three brief analepses, each signaled by the past perfect tense: (1) "whose hostess had put perhaps too much kirsch in the fondue"; (2) "she, Oedipa, had been named executor"; and (3) "Inverarity, a . . . mogul who had once lost two million dollars."

Notice that *what is retrospective for the teller is part of an unfolding present for the reader*. As the teller (in the default mode of narrative) looks back on what has happened, the reader looks forward to what will happen next. Analepsis from the perspective of the teller is a retrospection within a retrospection, a circumscribed looking back within the global looking back of the primary narrative. Analepsis from the perspective of the reader, however, is both a looking back from the unfolding present of the primary narrative and part of the relentlessly forward movement of the act of reading.

Foregrounding this paradox helps highlight the utility of *narrative progression* as a model to describe both the construction and consumption of narrative—and to deal with the functions of analepsis and prolepsis. Progression is the synthesis of *textual dynamics* (by which I mean the logic of the movement of narrative from beginning through middle to ending), and *readerly dynamics* (by which I mean the logic of the multilayered—cognitive, affective, ethical, thematic—trajectory of the audience's responses to the textual dynamics). Progression is the synthesis of textual and readerly dynamics because they are interdependent: while textual dynamics obviously trigger readerly responses, a storyteller who wants to generate a particular response will shape his textual dynamics accordingly. Textual dynamics in narrative typically proceed through the introduction, complication, and resolution (often only partial) of instabilities (within or between characters) and tensions (within or between authors, narrators, and audiences). Thus, as the primary narrative moves forward, readers develop hypotheses and expectations about its shape and direction, even as they form ethical judgments about and affective relationships with characters—and all this activity feeds into their understanding of a narrative's thematizing of its characters and events.[12] Analepsis affects textual dynamics by delaying the forward movement of the primary narrative, and it affects readerly dynamics in one or more of the following ways: building suspense; filling in gaps; providing information that influences readers' interpretive and ethical judgments of characters and events in the primary narrative; and affecting readers' expectations and desires about the developing trajectory of the narrative.

In the case of Pynchon's first sentence, the analeptic delays are minimal but they significantly influence the audience's understanding of this first event in the primary narrative. The delays highlight the event as an instability, indicating that Oedipa suddenly moves from a realm of Tupperware parties to an unfamiliar realm that promises to be much more complicated. Furthermore, the delays introduce tensions of unequal knowledge between teller and audience that fuel our interest in the primary narrative: why did Inverarity name her in his will? What kind of relationship did they have? Did he name her before or after she became Mrs. Maas? We read on to see how the instabilities will get complicated and to find out whether these tensions will be resolved.

Pynchon's sentence also illustrates two other important features of analepsis identified by Genette. *Reach* refers to the temporal distance between the events in the primary narrative and the events in the analepsis. Pynchon's first analepsis has the shortest reach (a matter of hours), while the reach of the second and third are not specified and so we cannot determine their temporal relation to each other. *Extent* refers to the amount of time represented in the analepsis, that is, the period from its beginning point to its end point (e.g., an hour, a week, six months). In Pynchon's first sentence we again cannot identify exact time spans, but we can infer that the extents of the first two analepses are brief (pouring kirsch into a fondue or naming someone in a will can be done in minutes) and that of the third is considerably longer (losing the money occurred during Inverarity's "spare time" rather than, say, "a single reckless wager"). Together, the reach and extent of these analepses initiate

Pynchon's audience into a narrative where just about every report of something definite is shadowed by something else that is uncertain.

Genette also uses the concept of reach to distinguish between external and internal analepses and the concept of extent to identify mixed analepses. *External analepses* reach back to points that remain outside the temporal borders of the primary narrative. Thus, in chapter 1 of *The Crying of Lot 49*, the narrator's account of Oedipa's trip to Mexico City with Inverarity is an external analepsis. There is a temporal gap between this trip, during which she realized that he "had taken her away from nothing, there'd been no escape,"[13] and her taking on the role of executor of Inverarity's will. This gap, characteristically for Pynchon's fiction, is never filled, and it renders unanswerable readers' questions about how Oedipa went from Inverarity to her husband, Mucho, and why Inverarity named her as his executor.

In contrast, *internal analepses* reach back to points within the temporal borders of the primary narrative. One of the most famous lines of Pynchon's novel comes in an internal analepsis: after narrating in chapter 3 the events of Oedipa's visit to The Scope, the bar where she first sees the post-horn symbol whose meaning she will pursue for the rest of the primary narrative, the narration moves forward in time. Then, early in chapter 4, it looks back to Oedipa's discovery of that image: "Under the symbol she'd copied off the latrine wall of The Scope into her memo book, she wrote *Shall I project a world?*"[14] By delaying this report, Pynchon highlights it and helps make it central to Oedipa's activity in the primary narrative. Finally, *mixed analepses*, amalgams of external and internal analepsis, reach back to points before the primary narrative but extend all the way to a point after the time at which the primary narrative paused for the analepsis.

Genette also uses the concept of extent to distinguish between *complete* and *partial* analepses. If the analepsis does not extend from a point in the past to the first event of the primary narrative, it is partial. If the analepsis does extend from a point in the past either to the first event of the primary narrative or to the point in the primary narrative at which the analeptic narration began, it is complete. Thus, all external analepses will be partial, and internal analepses will be partial or complete depending on whether they extend to the time of the primary narrative. In Morrison's *Beloved*, for example, the three different accounts of Sethe's taking the life of her own child (from the perspectives of the slave catchers, Stamp Paid, and Sethe herself) are three distinct partial analepses. If the narration continued to track Sethe's experiences from this point forward until it caught up to the moment in the primary narrative at which she tells Paul D her version, it would be complete.

To this point, my use of *Lot 49* and (more briefly) *Beloved* as illustrative examples provides evidence for my earlier claim that sometimes the functions of analepsis in the contemporary period are consistent with its functions throughout the long history of Western narrative. In the passages quoted here, both Pynchon and Morrison reinforce the principle that one can't know the present unless one also knows the relevant past. But I hasten to add that many writers since the 1960s, including Pynchon and Morrison at other moments, question how well one can know the past.[15] I will take up the consequences of this questioning after I look at analepsis's partner in time, prolepsis.

Prolepsis

Prolepsis initially appears to be temporal partner as mirror image: the evocation *before* the fact of an event that will take place *later than* the point in the story we have currently reached. From the perspective of the teller and of textual dynamics, this mirroring is evident. Where analepsis in the default mode of narrative is a retrospection within a retrospection, prolepsis is an anticipation within that retrospection. For example, in chapter 13 of McEwan's *Atonement*, the novel's (apparent) primary narrative about Briony Tallis's error and her efforts at expiation reaches its first major turning point: Briony mistakenly concludes that the man who raped her cousin Lola was Robbie Turner, the new paramour of Briony's older sister, Cecilia. Lola knows that Robbie is not her assailant but she does not correct Briony. The narrator then remarks: "And so their respective positions, which were to find public expression in the weeks and months to come, and then be pursued as demons in private for many years afterward, were established in those moments by the lake."[16] All the tenses are past here, but the reference to "the weeks and months to come" and to "many years afterward" evokes events that take place later than the unfolding present of the primary narrative.

As Genette shows, the concepts that help unpack the workings of analepsis can do the same for prolepsis. Both the reach and the extent in the example from *Atonement* are "many years." The prolepsis is both internal and partial because the primary narrative eventually takes Briony many years past the point at which she no longer clings to the position that Robbie was Lola's assailant.

When we shift to the perspective of the reader and the concerns of readerly dynamics, however, the metaphor of the mirror image becomes insufficient. If the metaphor were apt, we would simply conclude that just as analepsis depends on the principle that an understanding of the past is necessary to an understanding of the (unfolding) present, so too does prolepsis depend on the principle that an understanding of the future is necessary to an understanding of that (unfolding) present. To be sure, that principle generally applies and in some cases it is the most important one. But the narration from chapter 13 of *Atonement* does much more than shed light on the unfolding present, because its revelations are oriented at least as much to the future as the present. In this respect, the prolepsis works by a principle of bonus revelation about the future: "out of all the possible futures that you might project from this present, here is the one that occurred." Although McEwan's narration hints at, rather than fully reveals, Briony's and Lola's futures, it clearly reports that they will live in a gap between their public positions and their private knowledge. The effect of the proleptic bonus is to shift the emphasis of readerly concerns from "what" will happen to "how" and "why" it will happen.

Another way that the mirror metaphor proves inadequate can be illustrated by a scene at the end of Robert Zemeckis's *Back to the Future* trilogy.[17] In this scene Doc Brown (Christopher Lloyd) tells Marty McFly (Michael J. Fox) to stop worrying about what will happen in the future, because "the future is not yet written." True enough in life, but, as Mark Currie would say, not true in *Back to the Future* or in any completed narrative—because any story with an ending has its future already written.[18] Currie's point helps illuminate a paradox of readerly dynamics in any completed narrative: while we are in the midst of reading the unfolding narrative we have *the illusion* that the future is open-ended (we wonder, what will happen next?) even as we tacitly understand that the future has been decided—and has been written up in the remaining pages of the narrative. We can extend Currie's point by noting that analepsis allows readers to preserve Doc Brown's illusion and that prolepsis works to puncture it. While analepsis typically functions to increase the reader's interest in the question of what happens next (with the corresponding assumption that many things are possible), prolepsis converts possibility into actuality. In addition to shifting attention from "what" happens to "how" and "why" it happens, the bonus revelation will often affect our emotional responses to, and ethical judgments about, the characters in the present moment and across the reach of the prolepsis. Thus, for example, the three-page prolepsis in chapter 13 of *Atonement* increases our sympathy for Briony (in those weeks and months she wants to qualify her testimony but can never succeed) even as it underlines the magnitude of her mistake.

McEwan employs another significant variation of prolepsis earlier in *Atonement*, at the end of chapter 3, just after Briony achieves the insight that stories do not have to judge their characters as good or evil.

> Six decades later she would describe how at the age of thirteen she had written her way through a whole history of literature . . . to arrive at an impartial psychological realism which she had discovered for herself, one special morning during a heat wave in 1935. . . . Her fiction was known for its amorality, and like all authors pressed by a repeated question, she felt obliged to produce a story line, a plot of her development that contained the moment when she became recognizably herself.[19]

This variation is an analeptic prolepsis, a moving forward to a temporal point from which the character recalls the present moment of the unfolding primary narrative. In this kind of variation the reach of the prolepsis and the corresponding analepsis will be the same, as it is here ("six decades later"). (Other variations are also possible, such as an embedded prolepsis within an analeptic prolepsis, for example, "six decades later she would recall the moment in her thirties when she first articulated the significance of this moment," but these variations can be analyzed with the tools presented thus far.) McEwan also shows that the analeptic prolepsis can also include substantial new information about the future that has already been written, as his narrator mentions that "her fiction was known for its amorality." The larger effect is both on our understanding of the significance of the present moment and on our configuration of the (apparent) primary narrative: we are reading a portrait of the amoral novelist as a young girl.

Innovations

I have taken my examples of analepsis and prolepsis from well-known examples of post-1960s fiction in order to support my claim that contemporary storytellers have continued to use these techniques in ways that would be recognizable to their forebears. But these novels also point to a continuum along which we can place degrees of innovation in the techniques. This continuum tracks the degree of disruption the techniques cause to the stability of a primary narrative, the reference point upon which Genette's whole analysis is built. Without a stable primary narrative, such matters as reach and extent become much more difficult to identify, and so too do such matters as whether an analepsis or prolepsis is internal, external, or mixed, and whether it is partial or complete. Consequently, innovation happens when a storyteller destabilizes the primacy of the primary narrative and thus employs a "fuzzy temporality."[20] The greater frequency of such destabilizations in the contemporary period is part of the new relation between temporality and narrative form that Ursula Heise persuasively identifies in postmodern fiction.[21]

Highlighting the frequency of this destabilization raises other questions: does that frequency signify some fundamental shift in the understanding of time that is shared by contemporary storytellers? Is there one overarching concept that captures this shared understanding? While I do not want to rule out an affirmative answer, I also hesitate to offer one. My caution stems from my perception that "fuzzy temporality" designates a heterogeneous set of temporal representations that also have a wide range of effects and purposes across different narrative projects. Sometimes the destabilization of the primary narrative is in the service of epistemological uncertainty or even skepticism about how well we can know past, present, or future; at other times it seeks to replace linear temporal sequence with a conception of time as layered; at still others, it seeks to conceive of the same events as capable of fitting into multiple narrative configurations. And of course other effects are also possible.

Consequently, I propose a continuum of innovations based on the degree to which the temporality of the primary narrative gets destabilized, while also emphasizing that any destabilization can serve a variety of functions. I locate *Lot 49* on one end of that continuum and *Atonement* closer to the other end, with *Beloved* somewhere in the middle. Among the functions of temporal destabilization we often find the foregrounding of ontological questions, the phenomenon that Brian McHale identifies as the dominant concern of postmodernist fiction.[22] This function points to the link between temporality and ontology: destabilizing temporality often leads to destabilizing ontology.

As noted above, even as Pynchon sustains a primary narrative of Oedipa's experiences as executor of Inverarity's will, he includes many analepses to events whose temporal order we cannot fully specify. Furthermore, the analepses themselves often introduce tensions of unequal knowledge between the narrator and the audience. As a result, Pynchon makes his "fuzzy temporality" part and parcel of the uncertainty that pervades the primary narrative and the readerly dynamics. In this way it contributes to his purpose of conveying the nature of paranoia in the American 1960s: neither Oedipa nor the audience can pin things down, can sort out cause and effect, and so doubt and fear increase. Moreover, the relation of paranoia (as a psychic and as a social condition) to the temporality of prolepsis needs further investigation. Indeed, inserting a two-word phrase into the definition of prolepsis makes it an apt definition of paranoia as well: "the *perhaps delusional* evocation *before* the fact of an event that will take place *later than* the point in the story we have currently reached."

Like *Lot 49*, *Beloved* never abandons its commitment to an overarching primary narrative, but Morrison goes further than Pynchon in her efforts to destabilize the notion of linear time that underlies distinctions among the primary narrative, analepsis, and prolepsis.[23] First, Morrison introduces the concept of "rememory" to break down the usual border between past and present: a rememory is an event from the past that persists in the present. The concept is part of Morrison's larger vision of time as a palimpsest: the present does not simply succeed the past but rather overlays it. This conception underlies her construction of the character Beloved as having multiple possible identities with different connections to the past (she is Sethe's murdered daughter returned, she is an abused young woman who has recently escaped from her white abuser, she is a slave who has endured the Middle Passage). This view also underlies the present-tense monologue in Part Two that begins "I am Beloved and she is mine" and that melds experiences of the Middle Passage with the experience of Beloved's arrival at 124 Bluestone Rd.[24] While it is possible to naturalize this narration by declaring it the historical present tense, the concept of rememory runs counter to that move. Respecting that concept means taking the narration as an eternal present, one that renders distinctions among analepsis, unfolding present, and prolepsis moot. This vision of layered time—and the experience of reading it generates—ultimately serves Morrison's larger purpose in her historical novel. She challenges her audience to recognize the persistent effects of slavery in the contemporary period (for to her, slavery is a rememory) and to do something about them. Indeed, her narrator's present-tense insistence in the novel's conclusion that "this is not a story to pass on" refuses any proleptic gesture toward a period when slavery's effects belong only to the past.[25]

Atonement represents a more radical disruption of the concept of primary narrative because it deliberately misleads its audience about which narrative is primary. McEwan uses the final section of the novel, "London 1999," to reveal that the primary narrative is not about Briony's error and efforts to atone for it, but rather is about her struggle to tell the story of her crime and its consequences. This narrative concludes with her decision to give Robbie and Cecilia a happier ending than they had in life. McEwan's delayed revelation about the actual primary narrative has many effects on the readerly dynamics. For example, since the first event of the new primary narrative can occur only after Briony realizes that Robbie is innocent, the passage from chapter 13 analyzed above as prolepsis must now be reconfigured as a complete analepsis: it is at the end of those "many years" that Briony began her struggle. More generally, the revelation means that what much of what we initially took to be the primary narrative (the unfolding present of all of Part I, and some of the unfolding present of Parts II and III) needs to be understood as analepsis, that is, as a looking back at events that made the telling of the story such a struggle. In addition, the revelation means that we should read Parts I, II, and III as a version that shows not only the reasons for but also the signs of the struggle, especially in its treatment of Robbie's illness at the end of Part II and in the phantasmatic quality of Briony's meeting with Robbie and Cecilia at the end of Part III.

Nevertheless, these reconfigurations of the events and narration of Parts I, II, and III do not—indeed, cannot—erase the experience of reading the story of Briony's error and her efforts to atone for it as the primary narrative. Consequently, even though we have more certainty about the temporal order of events in *Atonement* than we do in *Lot 49*, the progression destabilizes the relationships among past, present, and future far more than does the progression of Pynchon's novel. Indeed, through his handling of the issue of the primary narrative, McEwan layers time in ways that have affinities with Morrison's practice. In *Atonement*, a single stretch of narration can be both prolepsis and analepsis, an event can be both part of an unfolding present and part of an analepsis. For McEwan these effects all contribute to his purpose of interrogating the powers and limits of storytelling as a means for apprehending and misapprehending the world—and as a way of atoning for misapprehensions that have serious consequences.

Of course, three examples of storytellers "doing analepsis and prolepsis while also doing something else" constitute a small sample size, but I offer my analysis of that sample as a warrant for further testing of this essay's main points: (1) contemporary storytelling has not completely transformed these techniques; (2) nevertheless, contemporary fiction does show that many storytellers feel a need to challenge the assumption that sorting experience into clearly demarcated timelines of past, present, and future is the best way to come to terms with that experience; and (3) the challenges to that assumption take a variety of forms as contemporary storytellers pursue a variety of narrative projects.

NOTES

1. See Gérard Genette, *Narrative Discourse: An Essay in Method*, trans. Jane E. Lewin (1972; repr. Ithaca, NY: Cornell University Press, 1980).
2. H. Porter Abbott, *The Cambridge Introduction to Narrative*, 2d ed. (New York: Cambridge University Press, 2008), 3.
3. Paul Ricoeur, *Time and Narrative*, vol. 1., trans. Kathleen McLaughlin and David Pellauer (Chicago: University of Chicago Press, 1984), 3. Since Ricoeur treats analepsis and prolepsis only briefly in volume 2 of *Time and Narrative* (trans. Kathleen McLaughlin and David Pellauer, Chicago: University of Chicago Press, 1985), I focus here on Genette's more extended treatment of the concepts.

4   See James Phelan, "Narratives in Contest; or, Another Twist in the Narrative Turn," *PMLA* 123, no. 1 (2008): 166–75.
5   Spiral Starecase, musical and vocal performance of "More Today than Yesterday" by Pat Upton, 1969, Columbia Records.
6   More common and familiar, prolepsis has received more recent attention than analepsis. In "Thinking Ahead: A Cognitive Approach to Prolepsis," *Narrative* 13.2 (2005): 125–59, Teresa Bridgeman offers a helpful cognitive analysis of how we process prolepsis. In "Many Years Later: Prolepsis in Deep Time," *Henry James Review* 33, no. 3 (2012): 191–204, Bruce Robbins ties the technique to a specific ethico-political function of "throw[ing] attention onto the boundaries and uncertainties of the community of fate" (199) and offers good insights into the novels he analyzes, but this generalization about the function of prolepsis seems unwarranted, since prolepsis, like any other technique, can be used for a wide range of effects. Mark Currie in *About Time: Narrative, Fiction and the Philosophy of Time* (Edinburgh: Edinburgh University Press, 2007) and *The Unexpected: Narrative Temporality and the Philosophy of Surprise* (Edinburgh: Edinburgh University Press, 2013) offers valuable discussions of prolepsis that influence this essay. But in *About Time*, he defines prolepsis to include rhetorical prolepsis, that is, the anticipation by a rhetor of an audience's response (such as an objection) and uses this expanded view as part of his case that in contemporary culture we now experience the present in relation to our future memory of it. I find the argument more intriguing than persuasive.
7   I also do not take up cases in which temporality is so difficult to pin down that it becomes impossible to decide what is an analepsis and what a prolepsis. Such cases are worthy of attention but are the subject for another essay, one whose interest is in the rejection of any fixed temporality in narrative rather than in innovations in the specific phenomena of analepsis and prolepsis.
8   Genette, *Narrative Discourse*, 40.
9   Thomas Pynchon, *The Crying of Lot 49* (1966; repr. New York: Harper Collins, 2009), 3.
10 Dorrit Cohn, *The Distinction of Fiction* (Baltimore: Johns Hopkins University Press, 1999), 96.
11 Pynchon, *Lot 49*, 3, my emphasis.
12 For more on progression, see James Phelan, *Reading People, Reading Plots: Character, Progression, and the Interpretation of Narrative* (Chicago: University of Chicago Press, 1989), and *Experiencing Fiction: Judgments, Progressions, and the Rhetorical Theory of Narrative* (Columbus: Ohio State University Press, 2007).
13 Pynchon, *Lot 49*, 13.
14 Pynchon, *Lot 49*, 75–76.
15 For more on this questioning that goes beyond the specific issue of analepsis, see Ursula Heise, *Chronoschisms: Time, Narrative, and Postmodernism* (New York: Cambridge University Press, 1997), with its discussion of "posthistory"; Fredric Jameson, *Postmodernism, or, The Cultural Logic of Late Capitalism* (Durham,

NC: Duke University Press, 1991), with its claim that the contemporary period is marked by a historyless present; and Hayden White, *The Content of the Form: Narrative Discourse and Historical Representation* (Baltimore: Johns Hopkins University Press, 1987), with its emphasis on how choices of narrative form shape historians' accounts of the past.

16 Ian McEwan, *Atonement* (New York: Doubleday, 2001), 157.
17 *Back to the Future Part III*, directed by Robert Zemeckis (1990; Universal City, CA: Universal Studios, 2009). DVD.
18 Currie, *About Time*, 33.
19 McEwan, *Atonement*, 38–39.
20 To his credit, Genette admits that his account engages in "excessive schematization" (79) and he demonstrates that not every mismatch between order of telling and order of told in Proust can be captured by his schema. Genette labels mismatches that fall outside the schema "achronic" narration (84n). David Herman, in *Story Logic: Problems and Possibilities of Narrative* (Lincoln: University of Nebraska Press, 2002), extends Genette's insight by drawing on philosophical work in fuzzy logic and noting that many narratives employ a "fuzzy temporality," one in which it is not possible to locate every event in relation to every other event. My analysis is very much in the spirit of Herman's discussion.
21 See Heise, *Chronoschisms*, 47–68.
22 Brian McHale, *Postmodernist Fiction* (New York: Methuen, 1987).
23 See Toni Morrison, *Beloved* (New York: Knopf, 1987).
24 Morrison, *Beloved*, 210–13.
25 Morrison, *Beloved*, 275. In *Sublime Desire: History and Post-1960s Fiction* (Baltimore: Johns Hopkins University Press, 2001), Amy J. Elias makes a similar point about Ishmael Reed's *Flight to Canada* and postmodernist historical fiction more generally. She writes that Reed's novel depicts trauma as a kind of psychic time that conflates past and present, and that post-1960s metahistorical romance is based in this posttraumatic approach to historical time.

15

Embodied / Disembodied

MICHELLE STEPHENS AND SANDRA STEPHENS

People Revisited, an interactive video installation by Sandra Stephens, presents life-size projections of people, walking in one main scene.[1] As viewers watch the video, they can use a trackball set up on a stand to control the movements of people within the projection (see figures 15.1 and 15.2).

Figure 15.1. *People Revisited* (2013), video installation, Sandra Stephens

As a viewer rolls the trackball over a particular person in the video, that person freezes, while at the same time a reflection of him or her (a semitransparent, ghostlike image) continues on its original trajectory, moving along with the other people in the scene (see figure 15.3).

Figure 15.2. *People Revisited* (2013), video installation, Sandra Stephens

Figure 15.3. *People Revisited* (2013), video installation, Sandra Stephens

Stephens's installation is a Conceptual artwork, participating in a post-1960s turn to interactive and idea-oriented art. The Euro-American Conceptual art movement of the 1960s was a reaction against formalism, where the ideas, content, and process behind the work took precedence over formal aspects such as line, color, shape, and texture.[2] Stephens's video piece visualizes a number of issues relevant to how we have come to think of temporal embodiment, both as art and as the subject of art. More specifically, it raises questions pertaining to why we have subordinated some modes of perceiving the bodied self that involve an experience of self as temporal rather than physical, "disembodied" rather than embodied (spatially).

We believe embodiment has come to be understood primarily as a spatial mode of experience rather than a temporal one: the body is often represented in physiognomic terms rather than in phenomenological ones. To correct this, we propose thinking about the body's temporalities, and we focus on two temporalities in particular. First there is what we call the "embodied body": an ideological construct that can be historicized. Second, alongside this, is a "disembodied body," a phenomenological entity that evades ideology and whose movements in time are much more difficult to signify. We claim that today's video installation art, new media art, interactive and performance art, and new visual technologies help to reveal this disembodied body, opening it to representation in this post-1960s ("postracial," posthuman, post-postmodern) moment. We would like to focus on approaches to Conceptual visual art as the locus, after the 1960s, for the movement from a universalized scopic regime of embodied space to a bifurcated view: a more historicized spatial embodiment and a disembodied proprioceptive regime of time. We see, in the work of many contemporary visual artists, a complex notion of the body that brings these together as reembodied, relational *timespace*. In the sections that follow, by placing both the embodied and the disembodied body back in time, historically and phenomenologically, we also arrive at this different notion of the body as a timespace.

Embodied

The installation *People Revisited* interrogates notions of the body that have been dominant in Western philosophy since at least the eighteenth century. In Anglo-European ontology and philosophical ethics, the human body generally has been understood as fixed, bounded, and materialized—to use Brian Massumi's term, "concrete."[3] By "concrete" he means a perception of the body both as an autonomous, bounded entity immobile in space and time and as an impermeable container, a hard membrane that separates inside from outside, an internalized and externalized body. While this body may have been seen, and naturalized, in these spatial terms, by historicizing this body one also places it back in a temporal frame.

Such a statement assumes that the way we have understood our bodies is shaped by a particular scopic regime. For example, Claudia Benthien describes the body of the Middle Ages as a "grotesque body" completely at odds with a more modern, post-Baroque conception of "an entirely finished, completed, strictly limited body" from which nothing "protrudes, bulges, sprouts, or branches off."[4] To the degree that this grotesque body was always uncontained—spilling and leaking and emanating beyond its spatial contours—it was also a sensory body continually in flux. In pre-modern conceptions of anatomy, the skin itself was seen as permeable and malleable. As such, it could not be the ground for difference, even in relation to the sex organs: Europeans believed in a "one-sex model" in which female genitalia were seen as merely a grotesque inversion of the male's.[5]

In contrast, by the Enlightenment, the skin and the body began to "harden," to be seen as less and less permeable, and difference began to be tied to the epidermal and physiognomic. The bodily surface, including the sex organs, became an impermeable container of difference. The body became less a cavity of temporally invaginated layers folding back in on themselves than a spatial site consisting of merely two distinct layers: a hard, impermeable outside covering a softer, organic interior. Sylvia Wynter observes that somewhere between the early modern era of colonial encounter and the nineteenth-century development of the racial sciences that attached human difference indelibly to the epidermal surface, a profound substitution occurred between anatomy and physiognomy.[6] When racial difference began to replace gender difference as the structuring division understood to define the human, a much deeper epistemic shift occurred in European understandings of the human body. In the intercultural context of colonial modernity, physiognomy—

especially the color of the skin—became the crucial marker of an essentialized or naturalized difference between peoples. Wynter's and others' observation—that the shift from malleable anatomy to a hardened physiognomy partly situates how Europeans thought about the *difference* of the racialized body—bears quite strongly, also, on our modern perceptions of the body as a spatially fixed, atemporal entity.

If one specific consequence of the foregrounding of physiognomic approaches to the body is precisely the emphasis on the body's atemporality, then, as Richard Twine argues, what is at stake also in this portrayal is the denial of mortality.[7] As Twine says, "The use of death masks as physiognomic tools [by eighteenth- and nineteenth-century scientists and phrenologists] provides the clue. May it be that the denial of temporality to the body is unconsciously a denial of physical change and ultimately one's own demise?"[8] Temporality, mortality, and motility together produce a body with subtle and minute changes in the present, and a body that changes in more gradual, larger sweeps over the course of a human life. Hence Twine asserts a further claim: "The role of physiognomic representationalism in ... visuality is most clearly challenged by introducing a sense of temporality into our conception of embodiment."[9] The death mask, in its effort to reproduce a stable and unchanging body, represents the antithesis of a body recognized as having its own temporalities—in history, and in the micro-movements of the present.

How we have thought about the body visually is structured by a physiognomic representationalism that "disciplines and freezes the inherent flux involved in the emotionality and materiality of the body."[10] This creates an atemporal view of the body, as Patrizia Magli describes: "Such a symbolising process introduces us into a new time: no longer is it the non-time of an actual face, lost in the interrupted fluctuation of lights and shadows. Rather, it is the time of a 'measure' that stills things, develops a formal image and locks it into an absolute fixidity, wherein it then interprets proportions, defines outlines, and attempts to establish essential traits."[11] What Magli is calling "time" here, and contrasting with a more bodily non-time, is a time that is more easily measured—rational, linear time, with past, present, future, minutes, seconds, and hours all easily sliced up and differentiated from each other. Physiognomy—the physiognomic representationalism of portraits, death masks, the photographs of a face—attempts to freeze the face in a rational conception of time. Consequentially, its manifestations offer one defining "look" to represent the person, and this often idealizing look is a smoothening out of all furrowed brows, winks and tics, distortions of the lips, animations of the face, that actually occur when a face is temporally in motion—what Magli describes as "the non-time of an actual face, lost in the interrupted fluctuation" of life.

In contrast, then, to a body in non-time is the understanding of spatial embodiment as being complete and closed unto itself, a state of being "composed."[12] Our bodies are now defined as always already dimensionally composed—frozen into one position and moment like the "snapshot" of a person caught in an idealized face. This body is then categorized through static dualities that eliminate a sense of the body's movement across and in and out of categories—young, old, dark and light, even male and female. Instead, embodied subjects are coded as if placed architecturally on a grid, in an "oppositional framework of cultural constructed significations: male versus female, black versus white, gay versus straight."[13] These identity terms make it easier to plot the body, easier to pin it down through a spatializing conceptualization.

The scopic regime thus constructs an atemporal, "embodied body." This notion of the embodied body in fact "effectively works to structure and contain" a body actually based in time and movement, one existing in reality as a composition in process.[14] According to Massumi, this "process body," what Magli also describes as a body in "non-time" (because bodily temporalities are not so easily measured, not as they are primarily experienced) can never actually be seen: "It always moves to the edge" of consciousness "or recedes infinitely into the shadows. It isn't an outline or boundary, but an indeterminate fringing," much like the figures merging and morphing in from the edges of *People Revisited*'s screen.[15] The temporal body that cannot be seen, that evades the scopic regime, is the incorporeal self as physical action-in-process, as an "interaction in the making" rather than an action with a terminated end.[16] This is also the bodily or bodied self that visual processes and epistemological discourses of embodiment seek to pin down, to concretize and fix, in atemporal space and in rational, measured time.

So powerful is the notion of the atemporal embodied body that even when certain artists aim to deconstruct and interrogate that spatial representation of the body, they often cannot find alternative (more temporally attuned) modes of representation to oppose to it. In Gunther von Hagens's controversial *Body Worlds* exhibits the artist presents whole human bodies plastinated in lifelike poses and dissected to show various, now frozen, structures and systems of human anatomy.[17] With each figure's muscles and tendons—the body's insides—now visible as surface—the body's outside—anatomy becomes physiognomy. Organic, vital interior systems are "stilled" to become startling, composed exteriors as von Hagens's *écorché*-like figures pose in stiff imitations of humans engaging in everyday actions (dancing, riding a horse, playing cards). Like the death masks of previous centuries, *Body Worlds* reproduces the notion of the atemporal, nonmortal, embodied body. Collapsing the distinction between represented and real to gruesome effect, von Hagens's body-figures are corpses caught between two deaths, petrified and kept alive in the symbolic order to represent a vision of humanity that is sealed up and closed.

In contrast, artists who are aware of prior representations of the embodied bodies of colonial subjects, where the logic of physiognomy reigned supreme, offer a more explicit counter to the spatially fixed "embodied body." Striving for a *decolonial*, rather than a postcolonial, aesthetics, they express an artistic methodology in which the impulse toward deconstruction has shifted to one of "*Aesthesis or Aiesthesis* . . . an unelaborated elementary awareness of stimulation, a 'sensation of touch' [that] is related to . . . sense experience and sense expression, and [that] is closely connected to the processes of perception."[18] One such artist who contests these fixed and spatial constructions of the "embodied body" is Camille Chedda, whose peeling and decomposing portraits on plastic bags provide not only a sense of the haptic but also a sense of the cultural uncertainty of fixing the black body. In *Built-In Obsolescence*, for example, shards of paint peel from twenty-eight tiny self-portraits painted on transparent plastic sandwich bags, the shards collecting in pieces at the bottom of the bags.[19] Although her work shows a keen interest in physiognomy (like the death masks of the past), it is an uncertain and in-flux physiognomy, one that is in stark contrast to the death mask images and that is cognizant of the debates on the (postracial, posthuman, post-postmodern) moment. Chedda uses nontraditional materials to create an image that will disappear in time, unlike the wax or plaster casts that remain stable over the ages. Chedda and other contemporary artists, with their use of more ephemeral materials and/or new media technologies, attempt to reveal the "process body," the embodied and disembodied body in time (see figure 15.4).[20]

Disembodied

The *People Revisited* video installation aims to move us toward the *disembodied* and the counter-tradition of process within Western philosophical ontology and aesthetics. First, there is the spectral figure who keeps moving, beyond the viewer's control, once he or she freezes a figure by moving a hand on the trackpad (see figures 15.2 and 15.3 above). This movement, both prefiguring and prefigured, represents a temporal movement, a movement in a non-time that has already gone on to happen, and will always go on to happen, without the viewer's consent. This act, independent of the viewer's hand, in fact asserts the human body's capacities as a line of flight or a movement through time rather than a thing in space.[21] It exists in time and space, but as a body without boundaries, or a body in excess of the image of its concreteness.

Second, as the moving spectre of the running man in *People Revisited* reconnects with his more solid image-form, the doubled motion or loop represented in the seven or so seconds of his appearance, presence, motion, and disappearance mirrors the estranging, disembodied ways in which bodies move in time through space—both projecting ahead unthinkingly, based on their memory of their previous proprioceptive motions, and orienting themselves spatially by aligning their geographical sense of the space around them with their motion memory. Massumi describes this joint activity as a folding in on each other of two different modes of perception, one projecting ahead, the other recurring after the fact. The first is based on the expectation of a certain bodily rhythm based on sensory experience, the other on more spatialized, visual coordinates, but both involve complex processes of negotiating the body's different (but also simultaneous) temporalities: past-present, present, present-future. It is as if this male figure in *People Revisited* knows where he is going before he gets there, has already moved himself there mentally before his body actually arrives to join him. He knows already, within a different register of the bodied self (in his muscle memory, for

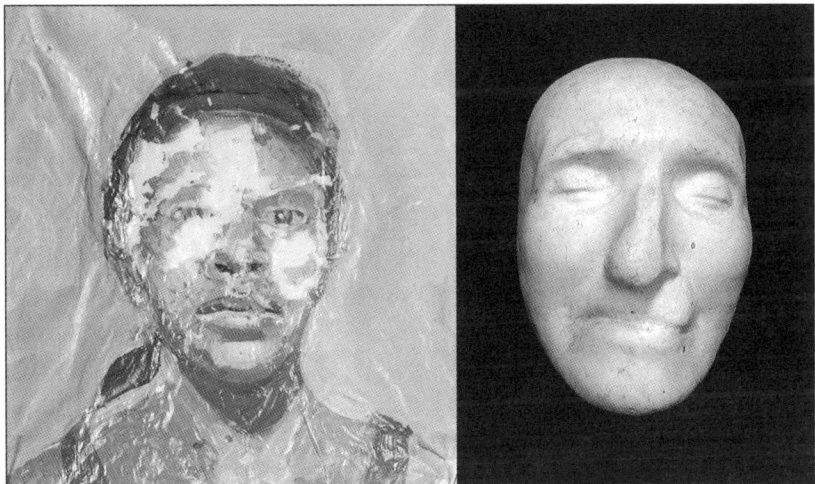

Figure 15.4. Left: Untitled, from the *Built-In Obsolescence* series (2011), acrylic on sandwich bags, Camille Chedda; right: *Thomas Paine's Death Mask*. (Source: Flickr user Ben Ledbetter, architect)

example), the spectral moves he will make to get from one place to the next. This all happens before he, retrospectively and in a matter of seconds, reconnects his body's movement with a more static sense of himself as fixed at a new and particular point in concrete space and time (see figure 15.5).

This is the body that Brian Massumi calls "abstract."[22] Massumi's project has been to reorient our thinking away from a more familiar, political, spatialized sense of what the body *is* to a less perceptible, somatic remembering of what the body *does in time*, an aim he captures with the telling mantra "Concrete Is as Concrete Doesn't."[23] As Massumi notes,

> Tactility is the sensibility of the skin as surface of contact between the perceiving subject and the perceived object. Proprioception folds tactility into the body, enveloping the skin's contact with the external world in a dimension of medium depth: between epidermis and viscera. . . . Proprioception translates [movement] into a muscular memory of relationality.[24]

Massumi uses the trope of the flesh to characterize this body that escapes both the image and the signifier, a quasi-corporeal or incorporeal body,

Figure 15.5. *People Revisited* (2013), video installation, Sandra Stephens

"the body without an image," which we come to know through the linked sensory modes of the proprioceptive (or muscular), the tactile (or haptic), and the visceral. Massumi would agree that even within the spatial grid of embodiment, people must move from point to point both mentally and physically to define the body, its boundaries, and its relation to that which is around it.[25] The body will not be still; it will not be contained in space and time.

In the post-1960s visual arts, these new ideas about the art object and changes in perception countered previous histories of how the "image" was used to fix the body. In her groundbreaking essay on the "dematerialization of the art object," for example, Lucy Lippard suggests that Conceptual art emerged from two directions—the art object as idea and as action.[26] Tracing the influences of "process art," "minimalism," and "Conceptual art" on contemporary art practice, she reveals how the artists in these movements were much more interested in the energy used to create and view a work of art than in the object itself. She also reveals how Conceptual visual artists, interested primarily in restructuring perception, were also more engaged with the environments and systems behind those perceptions than the object perceived.[27] Along these lines,

Lippard describes 1960s Conceptual artists' fascination with systems, especially spatial systems, and the objects within them:

> Donald Burgy's 1968 Rock series combined this impetus with the notion of context and took it to an almost absurd extreme, documenting "selected physical aspects of a rock; its location in, and its conditions of, time and space," including weather maps, electron microscopy, X-ray photographs, spectrographic and petrographic analysis. "The scale of this information extends, in time," said Burgy, "from the geologic to the present moment; and, in size of matter, from the continental to the atomic."[28]

Burgy's focus on subatomics, demonstrating how a rock can no longer be seen simply as a fixed element but must rather be regarded as fluid, composed of subatomic particles that are more energy and motion than anything else, rests on Bergsonian philosophical theories of matter.[29] Both Lippard and Burgy—though in different ways—moved an art of the body from a location in space to process and movement in time.

Post-1960s Conceptual artists aim to bring attention to the body's experience and to how traditional ways of understanding the body can be displaced by new ways of thinking about space in time. Conceptual artists today, such as Jacob François, reveal how "concrete is as concrete DOES." François's *Body in Space* (2012)—a sculpture made from volcanic stone contained by a fabric skin and suspended in the middle of six wooden rings—changes over time to reveal its parts.[30] François builds a relationship between this piece and his understanding of the decomposition of bodies: "Like the permeability of our own body, the fabric skin allowed the pulverized stone to slowly leach out of the body ultimately resulting in a pile of wood, fabric, string, and dust."[31] The sculptural body in space *is*, therefore, simultaneously a body evolving and changing over time (see figure 15.6).

Much post-1960s Conceptual art likewise aims to make us better able to see the *composing* (and decomposing) of bodies in the world: it exposes the bodied and bodily as experiential processes rather than stable, constituted entities that can only be represented, visually, in fixed, physiognomic terms. Such art plugs into trends within contemporary aesthetics and critical theory (such as affect theory and discussions of the body

Figure 15.6. *Body in Space*, (2012), wood, fabric mesh, string, stone dust, Jacob François

as sensational skin) that aim to bring us to a deeper understanding of our bodies in time as nonprivileged points, nonprivileged bodies with relational (and hence operational) consciousnesses.

To be "embodied," these artists claim, is also to be placed in history, as a body interpellated and constructed and, subsequently, marked as different. Massumi describes the act of body construction as taking "already extracted variables and recombin[ing] them," and Silverman confirms that, in Lacanian theories of the body, "a unified bodily existence comes into existence only as the result of a laborious stitching together of disparate parts."[32] This stitching—the activity of traditional efforts to construct the physiognomic, embodied body discussed above—seeks to override the awareness that the process body, the "proprioceptive ego" or the bodily ego implicitly shaped by the experience of the body in motion, "is *always* initially disjunctive with the visual image" of the body that we hold inside our heads.[33]

Not surprisingly, then, in the post-decolonization decades, feminist artists, performance artists, and writers of color spent much of their efforts taking back apart the impermeable, static body and the atemporal form of physiognomic embodiment that had become the container for essentializing notions of human difference. If the corpse as a sealed and plastinated container stands as the very opposite of the organicism and vitalism of a life lived in time and motion, as von Hagens's *Body Worlds* makes clear, a renewed focus on sensation swings back in the other direction, as artists and scholars find ways to express other experiences of the body, and particularly the decolonial body, in time as well as in space.

For example, in *Span II*, the famous performance piece by South African artist Tracey Rose, the artist deconstructs what has already been constructed to point to the effects of colonial history in manipulating bodies of color. Rose sat naked inside a life-sized glass cabinet, on a television turned sideways and displaying a classic image from European art history, the reclining nude. With her head shaved and bent, Rose ignored the gazes of viewers around her, occupying herself instead with knotting strands of her own shaven hair. In this act she referenced an earlier piece in which she used surveillance cameras to film herself shaving off her body hair, an act she described then as purposefully "demasculating and de-feminising my body [in a] kind of desexualization [that] carries with it a certain kind of violence."[34] Rose's stripping of her body not just of clothes, but more significantly, of her own hair, represents an attempt to deconstruct and undress the symbolic body of the woman of color as a body of difference, dressed in colonial languages of race, gender, and desire. Similarly to Chedda, she rejects the embodied spatial body to do things with it that mark its evolution *in time* as a body that sheds (skin, hair, fluids); as a body that decomposes second by second as it is posing; and as a live body that grows and is in flux, that will replace itself, over time, with new hair and new skin.

Reembodied

In *People Revisited*, bodies morph and merge into each other; they become spectral and resolidify at different paces. The interactive video installation images a world of permeable, moving parts that brush up against each other, flow together, and part, like atoms moving in space (see figures 15.7 and 15.8).

Figure 15.7. *People Revisited* (2013), video installation, Sandra Stephens

Figure 15.8. *People Revisited* (2013), video installation, Sandra Stephens

The phenomenologist Maurice Merleau-Ponty called this an experience of the "flesh of the world," whereby "every vision takes place somewhere in the tactile space" and so the touch and the gaze interact in a reversible, reflecting relationship to each other.[35] In *People Revisited*, the appearance of the couple at this moment is apt, for as their corporeal boundaries merge and draw apart, they illustrate the ways that the interpersonal relationship can reflect two consciousnesses in an interactive relationship and in motion in relation to one another, rather than in a closed or static relation to each other as entities fixed, separate, and apart.

People Revisited thus strives to capture modes of bodily being that break down the divide between the embodied (the point in time) and the disembodied (the movement through time), and it acknowledges the intimately intertwined nature of space-time, matter-energy, and our bodies. In the conversation taking place between the viewer and viewed, the final piece represents both continuous change and a snapshot created by the viewer of that particular moment as he/she interacts with the work. The video shifts our focus from the *appearances* of people fixed by the eye in space—a scopic, spatializing gaze—to the *appearing and disappearing* of people as spectral entities temporally in flux and physically in motion. As they participate in and interact with the work, viewers also become intimately connected to the scopic systems used to create it. The micro-action of freezing a person in time on the video screen reveals ways that philosophy has concretized our bodies in the past. However, the fact that viewers cannot fully stop the figures emphasizes the disembodied nature of those video actors in a very visceral and tactile way, and reminds viewers of their own disembodiments.

The body moves in time and in a way that the gaze can never fully capture, and how it moves constitutes a different order of perception. But bodies in relation to one another reveal the body's dual conditions as both embodied and disembodied entity, as relationality both asserts and breaks down the hard membrane of the embodied body. As noted above, Massumi specifies that "proprioception translates [movement] into a muscular memory of relationality."[36] This dual condition of the body is based in the synesthetic *interfusion* of the two regimes of perception, the scopic (linked above to the embodied body) and the proprioceptive (linked above to the disembodied body). In *People Revisited*, for example, the two regimes interpenetrate as the movement being seen and the movement being touched. The viewer's bodily memory follows each spectre with a tactile memory of movement—there we see the synesthetic combination of proprioception and vision. At the same time, the hand facilitates the gaze's freezing of individual figures—the haptic enabling the scopic. Naomi Segal calls this "consensuality": the way synesthesia implies a permeability between the senses.[37] Synesthetic stimulation turns inward and simultaneously reopens the body in a mode of non-time, "lost in the interrupted fluctuation of lights and shadows,"[38] while consciousness is subtractive, literally capturing a frozen or reified snapshot of the body in a slice of perceptual time.

For instance, Sandra Stephens's *Skin* (2012) in a similar way deconstructs the embodied body's scopic regime: tactile, haptic, proprioceptive pathways to disembodiment move to the forefront as the piece presents body without an image (see figures 15.9 and 15.10).[39]

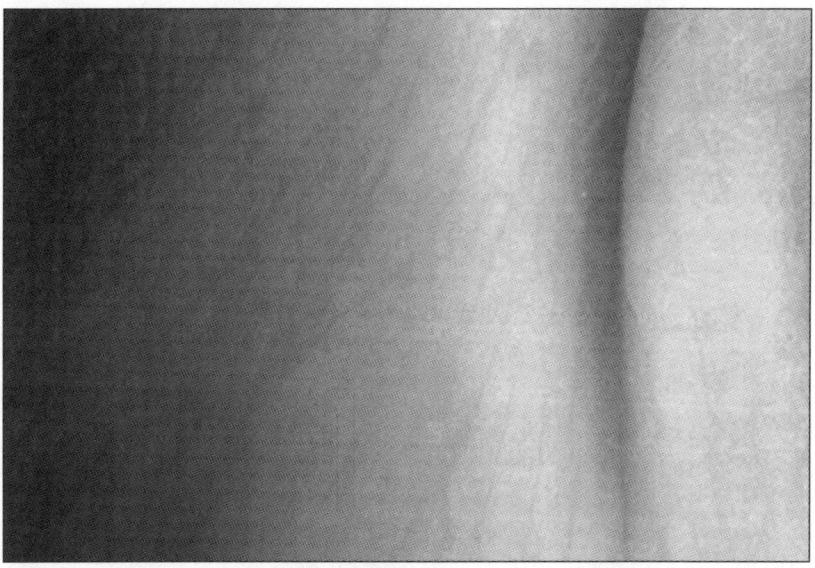

Figure 15.9. *Skin* (2012), close-up of video installation, Sandra Stephens

Skin visualizes skin as a disembodied environment for making both symbolic and sensory meaning. Shown as both a projected video installation (2012) and single-channel video (2013) of an image created by merging scanned sections of different peoples' skins, *Skin* employed

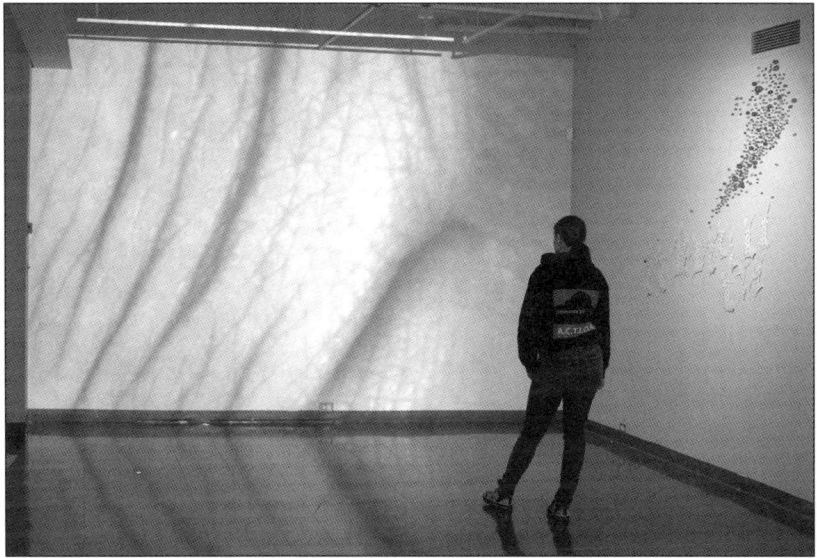

Figure 15.10. *Skin* (2012), video installation, Sandra Stephens

samples of various kinds of textures and complexions. The image created has discrete elements but also constructs a (dis)unified whole as the different areas of skin merge in a visual field. Scanning and montaging the skins of different subjects, putting various shades of color in motion across the screen and merging and blending them, Stephens combines the visual, the tactile, and the kinesthetic to bring us closer to a haptic appreciation of the skin, less as epidermis and more as invaginated flesh. The motion of the video across skins conveys a very different aesthetic than the physiognomic. Instead the skin becomes a milieu in flux, an environment with depth and texture evoking both insides and outsides, boundaries and planes, organic surfaces alongside invaginated folds and layers.

By coloring skin and juxtaposing those colors to one another not on a hierarchical ladder nor on a scopic grid but in a relational flow, Stephens's work implicitly comments on the racial politics embedded in Western philosophy's construction of the embodied body. But Stephens also reminds us that while people's skins can be culturally marked differently and as different, the skin is a synesthetic projection into the present that involves all of the senses and thus undermines this very perspective. Skin-based or skin-linked knowledges in fact have the capacity to bring the gaze back into relation with other sensuous forms of knowing the disembodied, proprioceptive body, a body in its fluid and changing temporal states. Both Laura U. Marks and Jennifer Barker, for example, describe how certain developing-world artists' and filmmakers' photographs and cinematic shots emphasize or foreground more haptic, bodily forms of knowing, and they contrast such techniques with traditional Western understandings of embodiment as a composed, impermeable surface moving on a spatialized grid.[40] They in fact suggest a rather different, more contemporary artistic response to an earlier effort by artists in the Global South to deconstruct symbolic constructions of the Other as a body epidermalized by difference through language.

Working in a similar way to these artists and filmmakers, Stephens in *Skin* harnesses a different perceptual regime, one that uses visual cues to evoke touch beyond sight, what Barker calls the "tactile eye." Rather than focusing on the segmented representations of the symbolic grid, Stephens aims to focus our attention as viewers on the more disembodied, more sensory and phenomenological experience of the skin as the largest organ of the body (see figure 15.11).[41]

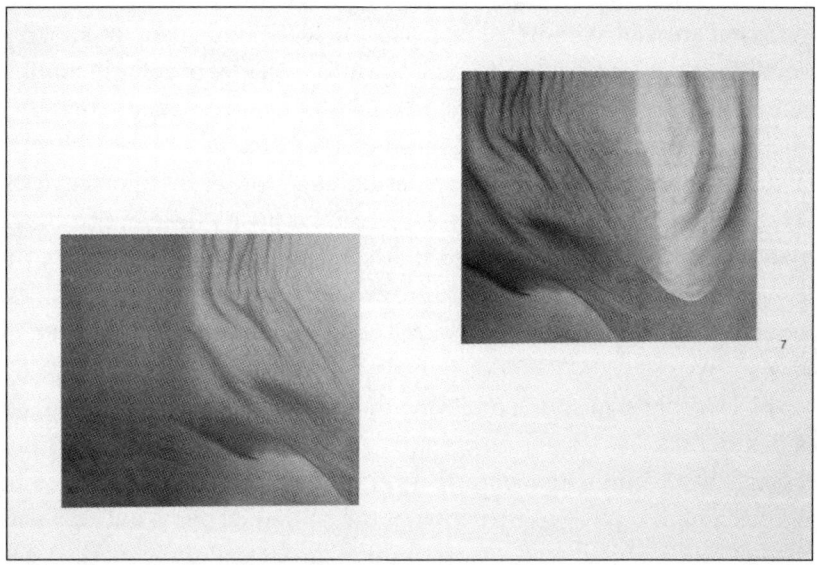

Figure 15.11. *Skin* (2013), metal prints, Sandra Stephens

Skin reminds us that seeing the bodily surface *only* as the shadowy afterimage of a distancing, othering, colonial gaze keeps us confined to the skin's limited symbolic and semiotic meanings.[42]

The skin—as both an ever-changing bodily environment in time and a hardened, epidermal surface and container—is itself the perfect milieu for describing and defining the body in timespace. The skin in timespace represents a combined corporeal, experiential medium, one in which the embodied experience of space—also the experience of the body as viewed by the Other, from the outside, from the surface of the skin—interacts and intersects with a proprioceptive experience of the body from within. The body in this latter sense is experienced as a fleshy, tactile, kinesthetic entity in a kind of non-time that can be measured only once it is frozen and made still. Erin Manning describes this as making "incipient movement felt" or experiencing the "sensing body in movement."[43] The skin as a milieu, an environment, and a context is constantly in motion and in process. This body-of-skin, or body-in-its-skin, is only ever evident in a constantly flickering view, representing the vicissitudes of two simultaneous views of the body.[44] To the degree that this simultaneous, dual perspective on embodiment allows us to think about timespace as relationality, and the body as a "relationscape" as Manning further defines it, we move away from visual and epistemological regimes of difference that have bounded and framed the human body in relation to the world and others.

Since bodily composition "involves the unfolding of an absolutely singular worlding relational whole," which can never actually be seen as whole, expressing the experience of bodily composition can be near impossible.[45] Contemporary artists thus face "the nagging problem of how to add movement back into the picture," how to capture the body's temporality while somehow avoiding the mortifying effects of physiognomic logics of representation.[46] In *Spatialization of Time* (2008), for example, Jacob François brings attention to timespace relations by showing how space both connects and separates bodies, and how this space is broken down by connections people make with each other through their senses (figure 15.12).[47]

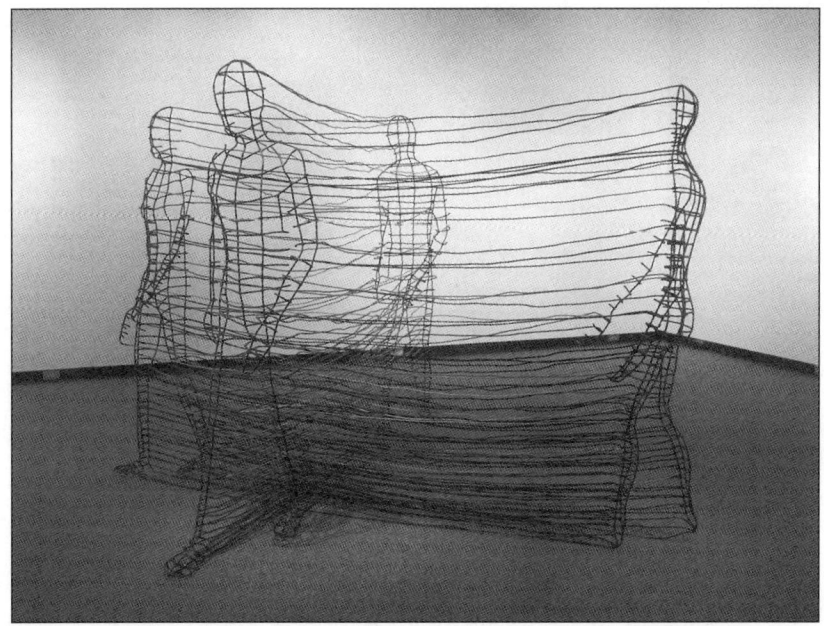

Figure 15.12. *Spatialization of Time* (2008), steel, rubber, wire, Jacob François

François creates two sculptural figures that appear to bifurcate each other with their movements through space. His figures' movements are visualized with motion lines that have a direct relationship to the space around them. The motion can be seen not only to relate to each other but to morph and change the space around them.[48] The figures also move across the timespace of past-present, present, and present-future, appearing at two points simultaneously, like Stephens's spectral video people.

If representing the body in timespace motion represents one set of efforts to capture bodily temporality, another can be seen in contemporary decolonial work that articulates—that is, embodies—disembodiment as a *decomposing* of the body of matter. Unlike the deconstructionists, authors of these works focus less on the body's symbolic fracturing and more on the effects of organic processes connected to and shaped by the vicissitudes of time.[49] Aging, evolving, mortal bodies replace fixed, impermeable, static containers, as in Camille Chedda's paintings that disintegrate in time.

Another contemporary artist whose work offers a useful contrast to that of Gunther von Hagens, for example, is Jason deCaires Taylor, whose underwater sculpture *Vicissitudes* (2012) captures precisely this dialectic between stillness and motion, impermeability and leakage, inorganic and organic, rigidity and fluidity, timelessness and history (see figure 15.13).[50]

Figure 15.13. *Vicissitudes* (2012), PH-neutral concrete (a mix of marine-grade cement, sand, and micro-silica), reinforced with special fiberglass rebar, Jason deCaires Taylor

Taylor's pieces are large-scale underwater sculptures that develop over time into artificial coral reefs. In these works, the environment is as important as the art object and the subject of the piece becomes the decomposition of the figurative body by the environment. In time, these statues will become less like static stone and, *à la* Burgy, more like rocky bodies created and recreated in dynamic tension with the environment, coralline bodies in perpetual composition/decomposition. The movement of water in between each figure makes visible the substance, the fluid environment, that moves between bodies and is a part of the work. The organic sea creatures that attach and adapt to each figure's presence also literally reenact the symbiotic relationship between different bodies as permeable, relational beings in space (see figures 15.14 and 15.15).

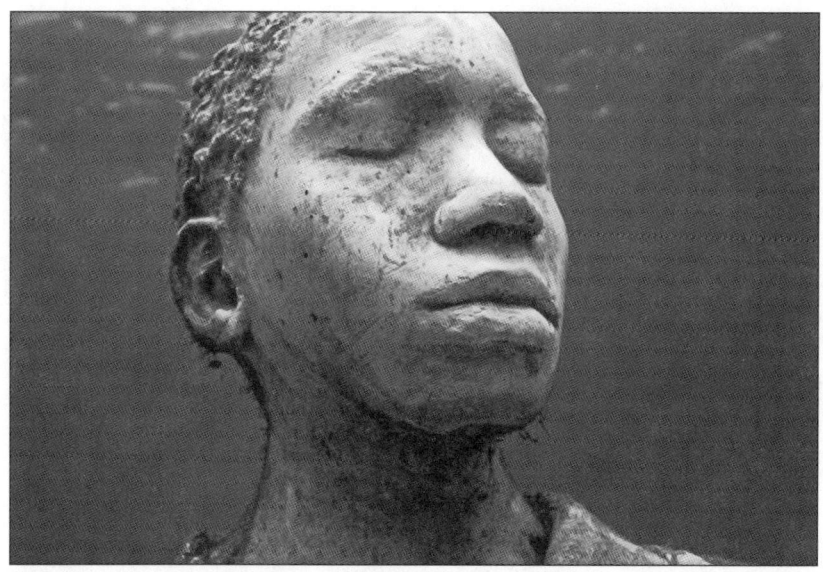

Figure 15.14. *Vicissitudes* (2012), PH-neutral concrete reinforced with special fiberglass rebar, Jason deCaires Taylor

Figure 15.15. *Vicissitudes* (2012), PH-neutral concrete reinforced with special fiberglass rebar, Jason deCaires Taylor

In *Vicissitudes*, Taylor also plays with our certainties about the body in space. He places the bodies in a circle looking outward, with no privileged position or viewpoint for the gaze to locate itself or fixate on. Instead, our focus is drawn to the piece's overall evolving texture.[51] Even more to the point, in terms of thwarting our desire for spatial fixity and locational certainty, there is a mystery to Taylor's piece in its inaccessibility. We are unable to see the work easily in its natural habitat, buried as it is five meters deep under the sea off the coast of Grenada in the West Indies. However, in the photographs of its evolving status as an artwork (perpetually changed by and interacting with its environment), the sculpture offers a broader metaphor for the buffeting of our bodies by the tides and waves of history. Just as the aquatic environment directly transforms Taylor's sculptures, the cultural milieu of the contemporary moment creates the context for a more complex, open view of bodies in time and in history. Taylor's spatially embodied bodies decompose in such a way as to reveal, and mimic, the human body's multiple temporalities—in the non-time of a process body whose external motions and internal fluctuations can never be fully measured or seen; in the mortality of a fleshy, haptic body, whose physiognomy erodes into a decomposing face; and in the historicity of a body that is ideologically constructed, but also exists in a fluid timespace representable in the new artistic modes, conceptual methods, and visual technologies of time-based arts.

NOTES

1 Sandra Stephens, *People Revisited*, video installation for *Rationalize & Perpetuate* solo show, *ArtRage*, March 2013, Syracuse, New York. Posted at Vimeo by Sandra Stephens, June 15, 2013, http://vimeo.com/68461329.

2 *Global Conceptualism: Points of Origin, 1950s–1990s* is a key exhibition organized by the Queens Museum of Art and curated by Terry Smith et al. that questions the primary association of Conceptual art with Europe and North America and presents work to show the movement's connection with "social, political and economic revolution throughout the world" (see http://www.walkerart.org/archive/8/A9739147A35626FC6162.htm). See also Smith's association of Conceptual art with "Euro-Americanism" at the Revisiting Conceptual Art: Russian Case in an International Context symposium (2011) organized by Boris Groys with the Stella Art Foundation, which posed the question "Is there only one or are there many Conceptualist traditions?" (http://ncca.ru/innovation/en/shortlistitem?slid=127&contest=23&nom=3).

3 Brian Massumi, *Parables for the Virtual: Movement, Affect, Sensation* (Durham, NC: Duke University Press, 2002), 1.
4 Claudia Benthien, *Skin: On the Cultural Border between Self and World* (New York: Columbia University Press, 2004), 38.
5 Vanita Seth, *Europe's Indians: Producing Racial Difference, 1500–1900* (Durham, NC: Duke University Press, 2010), 183.
6 Sylvia Wynter, "Beyond Miranda's Meanings: Un/silencing the 'Demonic Ground' of Caliban's 'Woman,'" in *Out of the Kumbla: Caribbean Women and Literature*, ed. Carole Boyce Davies and Elaine Savory Fido (Trenton, NJ: Africa World Press, 1990), 355–71.
7 Richard Twine, "Physiognomy, Phrenology and the Temporality of the Body," *Body and Society* 8, no. 1 (2002): 67–88.
8 Twine, "Physiognomy, Phrenology and the Temporality of the Body," 84.
9 Twine, "Physiognomy, Phrenology and the Temporality of the Body," 83.
10 Twine, "Physiognomy, Phrenology and the Temporality of the Body," 84.
11 Patrizia Magli, "The Face and the Soul," in *Fragments for a History of the Human Body*, ed. Michel Fehrer et. al., (London: MIT Press, 1989), 90. Quoted by Twine.
12 Kaja Silverman, *The Threshold of the Visible World* (New York: Routledge, 1996), 11.
13 Massumi, *Parables for the Virtual*, 2.
14 Massumi, *Parables for the Virtual*, 11.
15 Massumi, *Parables for the Virtual*, 174; Magli, "The Face and the Soul," 90.
16 Massumi, *Parables for the Virtual*, 9.
17 See more on Gunther von Hagens's *Body Worlds* exhibits at http://www.bodyworlds.com/en/exhibitions/current_exhibitions.html.
18 See the statement by Alanna Lockward et al., "Decolonial Aesthetics (I)," Transnational Decolonial Institute (May 22, 2011), http://transnationaldecolonialinstitute.wordpress.com/decolonial-aesthetics/.
19 Camille Chedda, *Built-In Obsolescence* series (2011), in *conversationXchange: The Universal Black Body*, a two-person show with Andrés Montalván Cuéllar, PrattMWP School of Art Gallery, March 2015, Utica, New York.
20 Thomas Paine's death mask in figure 15.4 is accessible at http://flickr.com/photos/benledbetter-architect/2624231911/.
21 See Gilles Deleuze and Félix Guattari on deterritorialization and lines of flight in *A Thousand Plateaus: Capitalism and Schizophrenia* (Minneapolis: University of Minnesota Press, 1987), 3, 10.
22 Massumi, *Parables for the Virtual*, 31.
23 Massumi, *Parables for the Virtual*, 1.
24 Massumi, *Parables for the Virtual*, 58–59.
25 In *The Grid Book* (Cambridge, MA: MIT Press, 2009), Hannah B. Higgins does note that even grids can be "open."
26 Lucy R. Lippard, *Six Years: The Dematerialization of the Art Object from 1966 to 1972* (Berkeley: University of California Press, 1997).

27 Lippard, *Six Years*, xv.
28 Lippard, *Six Years*, xvi. Here Lippard quotes from a passage from Burgy with information on the piece, which she presents under the heading "September 4, 1968, Bradford, Mass., Donald Burgy, *Rock #5 (Rep.)*" (51).
29 See Marta Braun's explanation in *Picturing Time: The Work of Etienne-Jules Marey* (Chicago: University of Chicago Press, 1992), 278–79: "Matter, for Bergson, is best conceived as energy, and energy is the ultimate form of motion; thus the shapes of material objects are not properties of the objects but are 'snapshots taken by the mind of the continuity of becoming.'" Christian de Quincey, in *Radical Nature: The Soul of Matter* (2002; repr. Rochester, VT: Park Street Press, 2010), notes that artists also reference Einstein's theory of a space-time matrix (22).
30 Jacob François, "Body in Space," wood, string, fabric, stone dust, 2012 (*Contrary Behaviors* group exhibition, Parallax, Lincoln, Nebraska).
31 Email conversation between Sandra Stephens and the artist, October 10, 2013.
32 Massumi, *Parables for the Virtual*, 174; Silverman, *The Threshold of the Visible World*, 17.
33 Silverman, *The Threshold of the Visible World*, 17.
34 Sue Williamson, "Tracey Rose," *Artthrob* 43 (March 2001), http://www.artthrob.co.za/01mar/artbio.html.
35 Maurice Merleau-Ponty, *The Visible and the Invisible*, ed. Claude Lefort, trans. Alphonso Lingis (Evanston, IL: Northwestern University Press, 1968), 134.
36 Massumi, *Parables for the Virtual*, 59.
37 Naomi Segal, *Consensuality: Didier Anzieu, Gender and the Sense of Touch* (Amsterdam and New York: Rodopi B. V., 2009).
38 Magli, "The Face and The Soul," 90.
39 Sandra Stephens, *Skin*, video installation for *conversationXchange: Shifting Representations of Color*, a two-person show with Oneika Russell, PrattMWP School of Art Gallery, March 2012, Utica, New York.
40 Laura U. Marks, *The Skin of the Film: Intercultural Cinema, Embodiment, and the Senses* (Durham, NC: Duke University Press, 2000); Jennifer Barker, *The Tactile Eye: Touch and the Cinematic Experience* (Berkeley, CA: University of California Press, 2009).
41 Sandra Stephens, *Skin*, metal prints for *Rationalize & Perpetuate* solo show, ArtRage, March 2013, Syracuse, New York.
42 As Claudia Benthien states further: "The body is not only a cultural sign but also an entity with sensation and perception" (*Skin*, 12).
43 Erin Manning, *Relationscapes: Movement, Art, Philosophy* (Cambridge, MA: MIT Press, 2012), 9, 6.
44 For Manning, the dancer is the figure that best captures the body as an entity of timespace. In her words, "A dancing body is a sensing body in movement" (*Relationscapes*, 70).
45 Massumi, *Parables for the Virtual*, 174.
46 Massumi, *Parables for the Virtual*, 3.

47 Jacob François, "Spatialization of Time," steel, rubber, 2008 (*Reassessing*, University Gallery, Southern Illinois University Edwardsville, Edwardsville, Illinois).
48 François's piece also visualizes work being done by Manning, who founded Sense Lab as a "laboratory for thought in motion," and her collaborator Brian Massumi.
49 Massumi, *Parables for the Virtual*, 3. One might note here the art of Maria Driscoll McMahon, whose street performances and videos decompose and make abject the bodies of less privileged, lower-class rural subjects. See Maria Driscoll McMahon, *This Mad Attachment: The Burdocks Project*, documentation of performance, interviews, and sculpture, at http://www.mariadriscollmcmahon.com/mariadriscollmcmahon/Street.html.
50 Jason deCaires Taylor, *Viccisitudes*, underwater sculpture, Grenada, West Indies.
51 Massumi, *Parables for the Virtual*, 30.

16

Theological / Worldly

STANLEY HAUERWAS

Christianity is a faith that roots, in time, those who would be Christian. But those same Christians are bound by those temporal roots, haunted by the knowledge that they cannot escape time. I want to explore that paradox in an effort to make what I hope will be some useful observations about how Christians understand, as well as tell, time by briefly examining time as God's time (theological time), time as the secular time of modernity (worldly time), and the time of lived reality for the Christian, which mediates these other senses of time and imbricates them in one another (other-worldly time). In particular, I will challenge the oft-made presumption today that Christianity is a technology designed to aid those who are Christians to escape temporality. I contend that if time is but a passage to the eternal, Christians risk turning Christianity into another community identified by adherence to a set of ideas rather than into a people who believe a revelation in history has made them possible.

Theological

"Out of all the peoples of the world I have called you Israel to be my promised people" is the claim that makes time unavoidable for Christians. We are of a faith that depends on the presumption that events have occurred that have changed the world. Accordingly, Christians are a people who have staked their lives, quite literally their living and their dying, on their conviction that God has been present to us "in time."

Addressing this point in his now classic book, *Christ and Time: The Primitive Christian Conception of History*, Oscar Cullmann argued that it is "the cosmic extension of the historical line" that makes Christianity so offensive in modernity. Cullmann explains that "all Christian theology in its innermost essence is Biblical history," which means Christians believe that the ordinary processes of history constitute a straight line of time through which God reveals himself as the Lord of history and nature.[1] In short, Christians believe a story, a story that invites as well as requires ongoing revision, a story that can be told about all that is and is not. Few phrases are more significant to the notion of a theologically determined, eternal time than "In the beginning."

According to Cullmann, this unique character of the Christian conception of time as the scene of redemptive history has a twofold character. In the first place, Christians believe that salvation is bound to a continuous time process that has a past, present, and future. Cullmann contrasts this understanding of history with the Greek cyclical conception of history in which salvation is always available in the "beyond." Secondly, Christians believe that this redemptive process in time is determined by one historical fact that constitutes the midpoint of time. That historical fact has an unrepeatable character, but as such it "marks" all other historical events. That fact is the death and resurrection of Jesus Christ. On that event all of history and nature turns.[2]

Yet do not Christians believe that God is eternal? Do not Christians believe that God is before all time yet still present fully in time in the person of Jesus? Such questions have led many Christian theologians to rethink what they mean by "eternity." For example, John Howard Yoder argues that in biblical thought, the eternal is not atemporal. Rather, according to Yoder, the eternal "is not less like time, but more like time. It is like time to a higher degree. If real events are the center of history—certainly the cross was a real event, certainly the resurrection is testified to as in some sense a real event—then the fulfillment and culmination of God's purposes must also be really historical. The God of the Bible is not timeless."[3]

Yoder's position on time and eternity is not unique. No less a theologian than Karl Barth seems to have views on theological temporality that are very similar to Yoder's understanding of time. Barth asserts that "even the eternal God does not live without time. He is supremely temporal. For his eternity is authentic temporality, and therefore the source of all time. But in His eternity, in the uncreated self-subsistent time which is one of the perfections of His divine nature, present, past, and future, yesterday, to-day, and to-morrow, are not successive, but simultaneous."[4] This is not a casual remark by Barth. In point of fact, he intensifies the claim that God lives in time by asserting that God's time *is* eternity. But eternity is not time without beginning or end. For to so identify eternity, Barth suggests, would be to attribute to it an idealized form of creaturely existence. That would be wrong, because to say "eternity" is to say "God," and God is not an idealized form of creaturely existence but a relational Other to that existence. God is not the reflection of our desire to live immortally, in a manner that refuses to believe we will ever die. Rather God is the ground and form of His existence, which means in His eternity He is Creator of time in which He Himself is beginning, middle, and end. Accordingly Barth maintains, like Yoder, that "eternity is not timelessness."[5] It is time that is other to the time of humanity—even to our conception of eternal time—and the laws of the physical universe occupied by that humanity.

According to Barth, that means that the time in which we live as part of God's creation is "an allotted time." As creatures of time, we have the task of resisting the temptation to make the time in which we live something other than a time we have been given. Since we are human and not God, we must learn to live in created time. Created time has a beginning, a middle, and an end, which means that there are boundaries in the time we have been allotted. To learn to live in the time we have been allotted, from Barth's perspective, is to learn that we are participants in a storied time that makes our lives more than simply one damn thing after another. We are embedded in histories we have not chosen, but by having our lives storied by God, fate can be transformed into destiny.[6]

Worldly

Another way to put these matters is to consider the significance of death for making possible the storied character of our lives. As *memento mori* traditions taught, death makes life valuable. If we did not die, our stories would have no limit or endpoint, and our relationships would have no trajectory, no narrative arc. Moreover, endings are always important to revelations of meaning in stories. For Barth, death creates an economy that makes me who I am but would not be if I could live as if time had no limit.[7] We are, therefore, haunted by the time that makes possible that our lives can be "storied," but that same condition that makes a narrative of our lives possible threatens that which we have learned to call "my life."

Christians are haunted by time because one of the forms time takes, particularly in our time, is our consciousness of our own historicity. Such consciousness, Cullmann argues, is personified in the figure of the modern historian who seeks neutrality toward that which she studies in an effort to tell time "objectively." While post-1960s historiography has radically called such objectivity into question, Cullmann argues a different point in relation to conceptions of historical time: namely, that secular history's claim to neutrality is constituted by a philosophical view of history that cannot help but make problematic the claim that all history turns on the cross and resurrection of Christ.[8] Cullmann's historian is a professionalized, worldly figure who represents the secular consciousness of "most people" in modernity. I am not referring only to a general distrust modern people have about claims that a human being was resurrected to sit at the right hand of God. Rather, I am trying to make articulate the stance born of our sense that as people overwhelmed by the vastness of time, we simply cannot fathom how all time can be determined by the crucifixion of an obscure Jewish man in an even more obscure outpost of the Roman Empire. To the extent that Christians created a consciousness of time that can be storied, they gave birth to the possibility that the story they believe to be true is just one story among others.

The time consciousness of modernity that I am trying to suggest is aligned with a modern cosmopolitanism that is inclusive of difference in a leveling gesture and thus is aligned with a kind of worldly time-sense. Cosmopolitans try to enact a generalized humanism that values universal understanding through intercultural contact; moving through worldly space without boundaries, they also extend time (of contact, of movement) infinitely. Time and space both are implicitly unlimited within a cosmopolitan ethos. Accordingly, moderns advocate and live within unstoried space—a space populated by many stories but with no frame that would permit a coherent narrative arc internal or external to them. They believe it is possible to rise above all time in an effort to avoid being limited by any time. From a theological perspective, they are a timeless people who have been made possible by a time when time no longer has the character of a time.

Of course, within theological circles there have been many attempts to accommodate the timeless time of cosmopolitan modernity to the rooted time of Christianity. Christian theologians have tried to re-present Christianity in philosophical modes that do not commit Christians to the view that the cross and resurrection of Jesus constitute time. The problem with that strategy, which has often been impressively executed, is that there is an element of throwing the baby out with the bathwater in these attempts to exorcise the haunted Christian. It is not clear, that is, why one needs the Christ if Christianity is but the exemplification of ideas that are accessible to anyone without reference to Christ. When the cross is made merely a "symbol" of the possibility of redemptive suffering, then there seems to be no reason, accept perhaps as a historical "fact," why you need Jesus at all.

It is equally unclear why you need a church. The church is necessary if Christianity is about a fact of time—a fact of time otherwise not available unless people who have experienced a sense of such time become witnesses to other people, who become witnesses in their turn across time. So the church is a correlative of the Christian conviction that at the appointed time in Palestine a savior was born. It too embodies time, a time that is other to worldly time and cannot be fully enunciated within the terms of modern rationalistic thought. The very existence of a people who are gathered to worship the God they believe to be the Lord of history is an indication that Christianity is a faith that cannot desire to escape time.

From an ecclesial point of view, it is not accidental that John Howard Yoder and Karl Barth emphasize that the God Christians worship is "timeful." For both Yoder and Barth are the sworn enemies of that form of Constantinian Christianity in which the church, in order to make itself intelligible in advanced social orders, becomes a representative of worldly logic and ideals. That is why so much of contemporary Christianity is Gnosticism in disguise. Charles Taylor has in fact characterized the dominant understanding of time today as "empty," and Taylor's empty time bears similarities to a Gnostic understanding of time as that from which we should recoil, that which is itself "worldly." To characterize time as empty, to assume that time is indifferent to whatever is used to fill it, in a complex manner reproduces the Gnostic presumption that the task is to escape time.⁹ Yet as I have tried to suggest, the Christian knows there is no escape. Put even more strongly, the Christian has no desire to escape time.

Taylor suggests it is Augustine whose account of time provides the most illuminating alternative to the empty time in which we now live.[10] Therefore I think it well worth our time to attend to Augustine on time, because by doing so I hope to make clear some of the claims about time I have just made as well as to introduce new considerations concerning the politics of time.

Other-Worldly

Augustine's extended account of time is in Book Eleven of the *Confessions*.[11] Yet it would be a mistake not to regard the whole of the *Confessions* as a meditation on time. It is always tempting for readers of the *Confessions* to stop reading after Augustine stops telling the story of his life. For some readers, Book Ten on memory and Book Eleven on time seem unrelated to his telling of God's refusal to let him be, but memory and time are at the heart of the story Augustine has to tell. For that is exactly what he discovers memory and time make possible: a narrative of his life. In Books Ten and Eleven, Augustine explores memory and time as crucial for his ability to tell how he has learned to tell time in the light of Christ.[12]

Earlier in Book Four of the *Confessions*, Augustine reflects on the death of his close friend and observes that "time takes no holiday. It does not roll idly by, but through our senses works its own wonders in the mind. Time came and went from one day to the next; in its coming and its passing it brought me other hopes and other memories, and little by little patched me up again."[13] Augustine's reflections on his developing feelings about the death of his friend indicate that early in his life journey, he began to understand the essential relation between time and death as well as how tempting it is for us to use our memories to construct continuous time between past and present—a timeless time—as false comfort.

It is, therefore, not accidental that in Book Ten, Augustine explores the nature of memory and what it entails, because he is aware that the *Confessions* are themselves an exercise in memory. Memory for Augustine is a "huge court" containing sky, sea, and earth, but memory also contains the self he thinks himself to be. For memory "weaves into the past endless new likenesses of things either experienced by me or believed on the strength of things experienced, from these again I can picture actions and events and hopes for the future, and upon them all can meditate as if they were present."[14] That memory is capable of such mediation is a sign of its power. This power has made possible the story he has told of his life, but through that very telling he has discovered that the "mind is not large enough to contain itself."[15] For example, how, he wonders, is he able to remember, as he is, what he has forgotten? By the very fact he has remembered what he had forgotten, his mental process is no longer "forgetfulness."[16] There seems something magical about memory, not only in that it can conjure the past but that it can cross logical categories and present what should be unpresentable.

Yet Augustine is well aware that his memory, like his life, is filled with scattered and seemingly unrelated images. He doubts, therefore, that he can tell the story of his life unless God makes such a telling possible. Strewn with irrelevant or distracting details, his memory cannot make story-time cohere. The ability to narrate his life is, then, but a microcosm of what it means to tell time. But then he, thus, must ask: "But where in memory do You abide, Lord, where in my memory do You abide?"[17] He answers that God abides in the one alone who is the Mediator between God and men, the man Christ Jesus, who "appeared between sinful mortals and the immortal Just One, for like men He was moral, like God He was Just; so that, the wages of justice being life and peace. . . . As man, He is Mediator; but as Word, He is not something in between, for He is equal to God, God with God, and together one God."[18]

That "answer" sets the agenda for Augustine's exploration of time in Book Eleven. In particular he addresses those who suggest that if something can be found in God that is not eternal but that can be said to be God's will that such a creature should exist from eternity, then why can we not say some creatures exist from eternity. These are the same kind of people, Augustine observes, who ask what God was doing before he made Heaven and Earth. Augustine is tempted to give the answer that God was preparing Hell to receive people who pry too deeply. Rather than to give that answer, however, he prefers to simply say, "I don't know." He prefers that answer to avoid having the one who asks the question win applause by being given a worthless answer.[19]

Augustine observes, however, that if there was no time before Heaven and Earth were made, the question "What were You doing *then*?" makes no sense. "Then" suggests temporality, embedding time in both language and universal substance.[20] Therefore the question is absurd: it cannot be that God was in time before all time. God cannot be in time and at the same time be the Maker of all time. So God must be before all time, which means there was a time when there was no time. But then Augustine finds he must ask, "What then is time?" His response may well be the most quoted sentences from the *Confessions*:

> What is this time? If no one asks me, I know, if I want to explain it to a questioner, I do not know. But at any rate this much I dare affirm I know: that if nothing passed there would be no past time; if nothing were approaching there would be no future time; if nothing were, there would be no present time. But the two times, past and future, how can they *be*, since the past is no more and the future is not yet? On the other hand, if the present were always present and never flowed away into the past, it would not be time at all, but eternity. But if the present is only time, because it flows into the past, how can we say it *is*? For it is, only because it will cease to be. Thus we can affirm that time is only in that it tends towards non-being.[21]

Here Augustine addresses several ontological issues related to the nature of time, but he also responds to his own worries about time by calling attention to our everyday expressions about it. For example, we speak of long or short times in order to describe the past or the future. Thus a hundred years ago can sound like a time long past, but Augustine takes this simple expression of time and derives from it a philosophical as well as a linguistic question—namely, in what sense can it be said that that which no longer exists is "short?" He concludes that we must not say that the past was "long" or that a day in the present is "long," for there is a category difference between time itself and our perception of it; there seems not to exist a time that was "long" or "short."[22] Augustine acknowledges that while time is passing it can be subject to measure, but once it is passed it cannot. Once passed, time is not. He continues his exploration of the past and future by asking not about their duration but about their location: he asks *where* they are. For example he calls attention to his boyhood, which no longer exists, but the likeness of his boyhood can be recalled in his memory. He notes that the same might be true of prophecies of things to come, because we consider future actions in advance in the present, but yet it remains the case that the future action does not exist. The mystery of past and future, our inability to characterize their being, leads Augustine to the judgment that there are few things we know how to characterize in thought and language, and that time seems to be one of those things.[23] Time seems other-worldly.

Augustine, therefore, ends by confessing that he does not know what time is, but he also wonders even how he knows that he does not know what time is. If he cannot know what he does not know, how then can he possibly think he knows how to measure time? He can only conclude he is able to measure time because it is in God that time has being. He says, "What I measure is the impress produced in you by things as they pass and abiding in you when they have passed and it is present. I do not measure the things themselves whose passage produced the impress; it is the impress that I measure when I measure time. Thus either that is what time is, or I am not measuring time at all."[24] He concludes with his prayer:

> Thou, O Lord, my eternal Father, art my only solace; but I am divided up in time, whose order I do not know, and my thoughts and the deepest places of my soul are torn with every kind of tumult until the day when I shall be purified and melted in the fire of Thy love and wholly joined to Thee.[25]

Augustine's interrogation of time is the philosophical and theological expression that makes the form of the *Confessions* explicit. The form of the *Confessions* is prayer, because Augustine is convinced that what makes possible the narration of his life is the time that God has provided. Augustine may finally be unable to provide an account of time *qua* time, but in the process of that investigation he has made explicit what his *Confessions* show—that is, that his life as well as all his life entails is storied by the God who is the creator of time.

The Christians that gave us the New Testament presumed (a presumption almost beyond belief given their insignificance) that what they had witnessed in Christ made possible an account of time that we now call "history." They called it "providence." Augustine is quite right to doubt that he knows what time may be, because time is not a "thing." Rather, in the theological context of Christianity, time is part of God's creation that makes possible the narrative arc of our lives.

Sheldon Wolin observes that Augustine's conception of time was not only one of the most original and significant contributions Christianity made, but also that it was one that had enormous political implications.[26] Wolin explains that Christianity broke the classical closed circle of cyclic time by conceiving of time as a series of irreversible moments extending along a line of progressive development. Accordingly, history was transformed into a drama of deliverance enacted under the shadow of apocalypse.[27] According to Wolin, Augustine's conception of time, with its emphatic distinction between church time and worldly time, meant that the classical quest for an ideal polity was now rendered problematic. After Augustine, eternity is understood to be the province of the City of God, making the church more "political" than Rome—thus Augustine's presumption that the church knows better than Rome itself how to tell the story of Rome. The church does so because those who make up the church know they have been storied by a gracious God who has made the story of time possible. No small feat.

* * *

Even the most sympathetic reader may be wondering where all this has gotten us. How, if at all, has it helped us locate who we are as well as where we are in time?[28] I cannot, of course, answer that question for anyone, but I do believe that Christians have some response as a people rooted in time. Of course, as should be clear from the above, it is not any time in which Christians exist. Christians believe they exist in the time God enacted in the Son. As a result, Christians tell time on the basis of that enactment. They may not know what time is, but they know it is Advent. Liturgical time turns out to be the fundamental way Christians have learned to tell time.

Such a sense of time has worldly implications. In his extraordinary book *The Vietnam War and Theologies of Memory*, Jonathan Tran argues that "time *as creation* is meant for worship." He then draws on Yoder's claim that cross and resurrection, not cause and effect, determine the direction of Christian history, and yet this suggests that Christians do not believe that history is a closed system determined by an economy of immanence.²⁹ Instead, Christianity seems to forward the proposition that history is the domain of God's time, which means that as creatures who worship God, Christians may be haunted by history but are never "boxed in" by it. As Tran memorably puts it, because Christians have learned to tell time doxologically, they are not allowed to do nothing, but rather they believe they can do anything.³⁰ In particular, Tran argues that the Eucharistic memory that shapes all Christian worship "makes possible the impossibility of forgiveness and that presages a politics of memory revolutionary in a world fixed on revenge/forgetting."³¹

Tran's book is unrelentingly theological, but that does not make it irrelevant for discussions of time by those who believe, as David Scott suggests in his *Omens of Adversity: Tragedy, Time, Memory, Justice*, that a new time consciousness is emerging everywhere in contemporary theory.³² Like Taylor, Scott draws on the work of Walter Benjamin to suggest that something has happened to our sense of time just to the extent we are no longer sure how the future makes the past possible. He observes that this development has even made Augustine's account of the intractable problems of time intensely interesting again. Augustine is back, according to Scott, because the enduring temporalities of past-present-future that animated Marxist historical reason no longer line up in the purposive fashion that once seemed possible: "The old consoling sense of temporal concordance is gone."³³

Thus Scott's judgment that Benjamin, like Hamlet, was right to describe our time as one that is out of joint. The time that is out of joint is nowhere better seen than in the disjunction between time and history—that is, the disjunction between our experience of time and the experience of history. Scott elaborates that claim by drawing on Agamben's suggestion that "embodied in the conception of history that we moderns have taken for granted, and by which we have ordered our experience of past-present-future, there is an unexamined conception of time."³⁴ Scott develops the sense of the present as a "ruined time" by calling attention to the current attention to the relation of memory and trauma. He observes that memory and trauma are at the center of discussions of reparatory justice, but such an understanding of justice carries little weight because the only truth today is that every human being has a right to a perspective on truth: "What now counts is each story of what is true." There is now no metatruth or metanarrative by which to judge the injustices laid upon past victims, yet they and their persecutors are enjoined to seek reconciliation and thus reconstruct the past as one they can share. We are living, he contends, in a stalled present that undermines a sense of social justice predicated on forward movement, action, future. He concludes with the judgment that "'forgiveness' is the name of a moral politics for an age characterized by being stranded in the present."[35]

No matter their differences, Scott and Tran are on the same page as they each seek to negotiate the present, a time paralyzed by a historical reality of a past that cannot be redeemed. Under such a condition, time is brought to a standstill, and history is devoid of hope. Yet Tran refuses that refusal of hope and the future. He does so in the name of a time that, contra Scott, is *made possible* by forgiveness. However, that type of forgiveness has a narrative grounding and a figure that give birth to a specific notion of time, a time that allows for a future. It is a time that has been redeemed by the cross, resurrection, and ascension of Jesus Christ, the Christian time of, ultimately, hope. The Church through its liturgy and its revelation of liturgical time creates narrative conditions enabling us to hear a call from the past as well as from the future: the Lord reminding us that we have been given all the time we need to be reconciled to one another and thus to God. Telling time for Christians means embodying the hope of the Gospel by loving our neighbors, who are sometimes our enemies. It means escaping the stalled present and entering time, yet again.

NOTES

1 Oscar Cullmann, *Christ and Time: The Primitive Christian Conception of Time and History*, trans. Floyd Filson (Philadelphia: Westminster Press, 1949), 23.
2 Cullmann, *Christ and Time*, 32–33.
3 John Howard Yoder, *Preface to Theology: Christology and Theological Method* (Grand Rapids, MI: Brazos Press, 2002), 276.
4 Karl Barth, *Church Dogmatics* 3, no. 2 (Edinburgh: T &T Clark, 1960), 437.
5 Barth, *Church Dogmatics*, 558.
6 This is the main theme that shapes Samuel Wells's extremely illuminating account of my work in his, *Transforming Fate into Destiny: The Theological Ethics of Stanley Hauerwas* (Carlisle: Paternoster Press, 1998). Wells provides a very helpful account of why time is so crucial for my account of the character of the Christian life in chapter 7 of his book, "From Space to Time" (141–63). Wells is not suggesting that I leave behind all considerations of space (I obviously think space and time are mutually implicated), but Wells is quite right that I think that space to be such must be "timed."

7 Barth, *Church Dogmatics*, 561–63.
8 Cullmann, *Christ and Time*, 21–22.
9 Charles Taylor, *A Secular Age* (Cambridge, MA: Harvard University Press, 2007), 58. Outside of ecclesial circles, Harold Bloom has proposed a similar linkage between Gnosticism, American religion, and time; see *The American Religion: The Emergence of the Post-Christian Nation* (New York: Simon & Schuster, 1992).
10 Taylor, *A Secular Age*, 56.
11 Augustine, *Confessions*, trans. F. J. Sheed (Indianapolis: Hackett, 1993), 211–31.
12 Augustine, *Confessions*, 10.5, 176.
13 Augustine, *Confessions*, 4.8, 57.
14 Augustine, *Confessions*, 10.8, 179.
15 Augustine, *Confessions*, 10.8, 180.
16 Augustine, *Confessions*, 10.16, 185–86.
17 Augustine, *Confessions*, 10.25, 191.
18 Augustine, *Confessions*, 10.53, 207–8.
19 Augustine, *Confessions*, 1l.12, 218.
20 Augustine, *Confessions*, 11.13, 218.
21 Augustine, *Confessions*, 11.14, 219.
22 Augustine, *Confessions*, 11.15, 219–21.
23 Augustine, *Confessions*, 11.19–20, 222–23. Augustine considers two other possibilities as well: (1) that time may be determined by the movement of the sun and/or (2) that time is measured by the movement of the body. Yet neither suggestion works because of the variety of the movement of the sun and the body.
24 Augustine, *Confessions*, 11.22, 229.
25 Augustine, *Confessions*, 11.28, 230.
26 Sheldon Wolin, *Politics and Vision* (Princeton, NJ: Princeton University Press, 2004), 112.
27 Wolin, *Politics and Vision*, 112.
28 For the best account of how those questions may be answered, see Nicholas Boyle, *Who Are We Now? Christian Humanism and the Global Market from Hegel to Heaney* (Notre Dame, IN: University of Notre Dame Press, 1998). Boyle provides an extremely important argument (154–55) that the consumptive economies that shape our sense of time in effect rob us of time.
29 Yoder's dramatic claim is in his *The Politics of Jesus: Vicit Agnus Noster* (Grand Rapids, MI: Eerdmans, 1995), 232.
30 Jonathan Tran, *The Vietnam War and Theologies of Memory* (Oxford: Wiley-Blackwell, 2010), 8.
31 Tran, *The Vietnam War*, 208.
32 David Scott, *Omens of Adversity: Tragedy, Time, Memory, Justice* (Durham, NC: Duke University Press, 2014), 1.
33 Scott, *Omens of Adversity*, 6.
34 Scott, *Omens of Adversity*, 12.
35 Scott, *Omens of Adversity*, 14.

17

Authentic / Artificial

ANTHONY REED

Questions of authenticity and artificiality haunt black culture's transition from an appropriated set of practices to a set of commodity objects and reveal along the way a racialized set of spatiotemporal assumptions that persist into the digital era. The possibility of counterfeit haunts the digitally produced cultural object in two specific ways: concern over the legality and provenance of the physical object, and concern over the legitimacy of the art as "expression" of a nebulous "folk-spirit." The latter is my primary concern in this discussion. For if the possibility of recording retrospectively creates the "live" performance, the possibility of the inauthentic creates authenticity. The techniques of reproducibility that Walter Benjamin discusses develop in the era of industrialization and high imperialism, both of which intensify the meanings of cultural distinctness—on the one hand, through the threat of competition in multicultural societies such as the U.S., or, on the other hand, as a justificatory premise of imperial domination.

What I am calling a "racialized spatiotemporal schema" names the ways that the ideology of cultural distinctness from the start conflates space and time. As historian Karl Hagstrom Miller observes, racialized music categories develop in the context of broader interest in folk culture, where spatial isolation translated easily into temporal remoteness from modernity that needed to be preserved. The Other having been denied what Johannes Fabian terms "coevalness," the existence of the Other at the expanding edge of modernity generated anxiety over his/her disappearance through movement or greater participation in modernity, which is a problem of human knowledge and of plausible alternative cultural formations to capital.[1] The remote, isolated Other— with "isolated" functioning as both adjective and verb—figures a living artifact who shares a common horizon of biological, but not cultural, time. This theoretically allows her to solve the problem of her disappearance (and justify her isolation) by means of recording technologies.[2] Nonwhite and nonincorporated Others (e.g., Native Americans, African Americans, "hillbillies" and cowboys living on the margins of modern culture) are valuable to the extent that their cultural practices can be appropriated—understood as discrete objects of knowledge that can then be transported to other discursive domains—and verified as "expressing" authentic folk-spirit.

The appropriation of African American spirituals in antislavery activism transformed those songs from noise to objects requiring specific hermeneutic practices to hear them as testimony against the "peculiar institution." That process marks the transition from what sociologist Jon Cruz terms "ethnosympathy," or "the new humanitarian pursuit of the inner world of distinctive and collectively classifiable subjects," to an ethnography, initially folkloric in nature, whose aims need not be directly political.[3] The introduction of sound reproduction technology to early ethnographic practice in the early twentieth century—which underwrites John and Alan Lomax's collection of Negro spirituals, work songs, and blues—eventually gives way to the commercialized circulation of ethnically marked sounds. Consider, for example, the "folk revival" of the 1960s, including performers Skip James, Son House, Mississippi John Hurt, and others: those sounds continue to be at least partially valued insofar as the spatial remoteness of the Mississippi Delta also signals for black and white listeners *temporal* distance from more racially mixed urban areas. The "primitive" of the earlier moment becomes the "authentic," and that distinction has salience to the degree that control over the racialized spatiotemporal schema continues to be reserved for the modern bourgeois subject, who has the prerogative of defining margin and center, present and past. I would go a step further: the racialized spatiotemporal schema, as a practice of race, produces the privileged perspective of the modern, while tacitly (at least) supporting Euro-American imperialism, domination and exploitation of nonwhites and "backward" rural whites, on the one hand, and supporting the fantasy of an authentic culture, somewhere, untouched by capitalist modernity. The notion, ultimately, encodes both containment and utopian liberation from the bonds of the present.

Authentic

Through this briefly sketched historical process, "authentic" comes to refer to the degree to which a particular performer or song under consideration hews to a supposed "folk-spirit," itself knowable only through specialized hermeneutics and the racialized spatiotemporal schema that converts social isolation and racial or ethnic difference into temporal difference. The very existence of such a hermeneutic and scholarly apparatus, however, reveals the degree to which the folk-spirit of *Volksgeist* precedes the "folk" who nominally "express" it.[4] Authenticity, then, is a name for the conversion of space into time, as well as the conversion of race into a set of appropriable and iterable social practices, texts, and performance styles. These are interchangeable in a consumerist framework and become marketing devices that promise consumers the experience of the real.[5]

Modernist art from the turn of the century and the Beats in the mid-twentieth century, for example, could draw on black music as a resource or model only insofar as they could imagine the blacks themselves as having maintained cultural purity, and themselves as having contact across the color line. In this context, allowing the possibility of "authentic Negro elements in jazz," Theodor Adorno's skepticism toward the white bourgeoisie taking up the music they heard in Weimar Germany as the genuine article critiques the naïveté of believing cultural objects could pass through the culture industry without bearing its imprint.[6] Instead, it was a "black mask" that offered an illusory freedom to the white bourgeoisie from the oppressive structures of modernity they correctly identified, without forcing recognition of the relationship between their oppression and black oppression. Through the bebop era (mid-1940s and 1950s), critics insistently labeled black musicians "primitive," "instinctual," "childlike," or otherwise uninhibited, and the musicians themselves used terms such as "soulful" to describe their practices even as they fought to claim an openness to nonblack sources, and thus to claim participation in modernity. From this perspective, the concept of authenticity more or less functions as a sign of property, allowing listeners and performers to police performance boundaries that may be in the interests of the culture industry, but also to express agency in ways that challenge as well as reproduce dominant ideologies.

Authenticity becomes an acutely temporal concern under the procedures of the culture industry, where, as economist Jacques Attali notes, *"people must devote their time to producing the means* [money, i.e., 'exchange-time'] *to buy recordings of other people's time,"* i.e., symbolically stockpiled "use-time."[7] Where appropriated culture once served as a kind of testimony for those who could not represent themselves, commodified culture—especially early recordings of work songs and blues—represents sounds used or stored by others that derive their value from the continued relations of exploitation. Music was (and in genres such as rap that are thought to be documentary, still is) valuable to the extent that it avowed and elaborated the genuine experiences of the producers, who represented the consciousness of "the folk." The same ideology, however, required the continued isolation of the folk in order for it to retain its authenticity. Hence one sees the relatively ineffectual nature of those projects geared toward increasing "awareness" or "visibility" through cultural texts: the value of those texts diminishes to the degree that those who produce them enter more fully into a coeval time.

A full discussion of the transition from culture in politics to culture *as* politics is beyond the scope of a short essay. One consequence of the racialized spatiotemporal schema relevant here is the contemporary policing of legitimate and illegitimate uses of cultural practices, even when divorced from the folkloric context within which it emerges. That transition has helped to make charges of "artificiality" or "cultural appropriation" especially acute within discussions of rap and other black music that positions itself (and has been positioned by record labels) as the voice of "the people," making realness an end in itself. The contemporary jargon of "appropriation" reproduces the older racialized spatiotemporal schema, but now with a heightened realization that the sense of time, and labor time, circulate in congealed form as commodities. The sociological notion of appropriation already names such a distillation of race, space and time into discrete objects. The jargon of appropriation recognizes the interaction of the concept with commodified authenticity, exacerbating the anxiety that always attends iterable practices: the counterfeit or artificial passing as the genuine article.

Artificial

Though all art may be artifice, an abiding requirement of "realness" serves to prevent the artwork from being "artificial" or mere technique. In the age of digital reproduction, the problem of resemblance—the ability of the artificial or secondary to pass for the authentic—intensifies. Likewise, the spatiotemporal schema begins to break down, thanks to im/migration patterns in the African diaspora following the legal dismantling of segregation and colonialism. Though others bring their own cultural practices (which themselves distill race, space, and time through notions of authenticity and tradition), the divisions of space-time within a given nation-state still shape understandings of culture. What happens, in other words, when travel or digital dissemination makes the porousness and arbitrariness of national boundaries—and the assumptions of the spatiotemporal schema—more visible? To work through this question, I will focus on hip-hop, specifically rap music, which places a rhetorical premium on authenticity and which, like other African American art, is often made to testify to the prominence of racially unmarked American culture as a sign of past ills overcome. Though discussions of hip-hop privilege and centralize the United States, the nation (as Tricia Rose, Richard Iton, and others have shown) tends not to be the most salient domain of the music itself, which focuses instead on the city, the block, the street, the corner. The distance of these locales—now urban ghettos within the borders of cultural centers—from the cultural spatiotemporal mainstream is key to this music's value. A fetishized verisimilitude communicates the "authenticity" of hip-hop's interconnected micro-locations and situations understood as a temporal genre linking life rhythms and a horizon of plausible expectations: the song is real to the extent that listeners accept it as speaking their experience.

While a thick historical account of the origins of hip-hop in the South Bronx requires an analytic that exceeds national boundaries, commentators often insist on naturalizing the form, which means understanding it exclusively through the lens of a certain U.S. history. In an exemplary article, Johan Kugelberg argues that the "many *consecutive* births of hip hop" include "the power of the Jamaican sound system" and "the Jamaican MC moves of Coke La Rock that *echoed* back to King Stitt on the island" that Jamaican-born Clive Campbell (Kool Herc) brought to the Bronx.[8] Jeff Chang's timeline within the same volume highlights the completion of the Cross Bronx Expressway in 1972, subsequent white flight and "the sound of Bugalu—featuring artists like [Filipino-American] Joe Bataan, [Puerto Rican] Joe Cuba, [Puerto Rican] Pete 'El Conde' Rodriguez, who blend Afro-American and Latino rhythms and styles—replac[ing] doo-wop as the sound of the Bronx."[9] The discussion of the artists charts a postcolonial confluence and articulation of Latin American, Jamaican, and African American people with different practices and traditions that at the very least raise questions about the interaction of the U.S. and other nations. To trace that line would be to understand culture as a set of acts in time rather than objects in space.

Kugelberg laments hip-hop's "transition from performance-based culture to a recorded culture," which allowed "every swinging dick with the means to release an independent record and get it distributed" to do so, especially after the Sugarhill Gang's "Rapper's Delight" brought rap music to "the collective mind of the five boroughs (and then the world)."[10] From that point on, in a narrative that traces a familiar trajectory of waning authenticity, the fake mingles with the real, and the culture becomes a set of appropriable objects without any necessary connection to a race-place-time nexus from which one could develop a hermeneutic or axiological scheme. The problem, as Kugelberg frames it, is a surplus of performances following "Rapper's Delight." This set of oppositions—legitimate/illegitimate, black culture/white mainstream, performance/inscription—reinscribes the necessary temporal distance of "authentic" folkish culture from the rhythms of capitalist modernity. In doing so, the racialized spatiotemporal schema obscures the ways rap emerged "as one of the primary ways Caribbean immigrants and their children negotiated their identities as both African Americans and West Indians,"[11] and obscures the South Bronx as a diasporic space where aesthetics and politics in formal and mundane senses co/operate.

My point is not to argue that hip-hop does not originate in the South Bronx with a party hosted by Kool Herc and his sister Cindy. Instead, I wish to draw attention to the ways the Bronx, as an "illusory but meaningful space between the national and the imperial,"[12] helps us understand a cultural form ill understood within the narrative prerogatives either of the nation-state or the racialized spatiotemporal schema by which the music is commodified as the newest moment of

African *American* culture. In different terms, the artificial—rap music not produced by African Americans, strictly speaking—precedes and is retrospectively naturalized by a jargon of authenticity and its cognates. Understanding hip-hop as an art of the African diaspora, where both bodies and cultures move, especially in our globalized, networked moment, raises new questions about the meaning of it as cultural *practice*. Returning to an earlier point, if the "folk-spirit" precedes any example of folk culture, then folk-spirit itself belongs to the general law of iterability, repetition, corruption.

Especially in the digital age, where spatial distances are easily overcome and the margins of culture are more intimate with the center, the former periphery is culturally simultaneous with—or produces the advanced culture of—the cultural center.[13] Prevailing theoretical models of adaptation, in which the postcolonial world adopts aesthetic techniques generated in the overdeveloped world to "local materials," rely on an implicit assumption that temporal distance follows from spatial distance. Getting out of that temporal schema, which is rooted in a pernicious cultural hierarchy made possible, then supported, by the social scientific appropriation of culture, requires more careful consideration of specific practices and generic negotiations of the authentic/artificial binary.

Resemblance

Semblance, however, becomes a way out of the real/fake, authentic/inauthentic binary if we accept the more radical implications of technical and then digital reproducibility. But first one must be attentive to the conceptual possibilities of thinking African American culture as participating in global black culture in the current age of migration. As Alexander Weheliye argues,

> black popular music, and hip-hop in particular, serves as a global forum through which different black diasporic subjects negotiate shifting meanings of blackness, as well as other forms of social and political identification. This negotiation, far from being uncomplicated or uncritical, entails facing the prominence of African American cultural practices such as hip-hop within the continuum of the African diaspora.[14]

I would stress, however, that "African American culture" is not the stable ground against which "*different* black diasporic subjects" form and measure their own identities, but is itself negotiated in and through cultural practices. I make this distinction in order to short-circuit an expressive chain—the record expresses (contains a performance thought to express) a culture understood as the "objective" form of the people of a given nation—that otherwise threatens and overdetermines the "content" of any record, artificially stabilizing identity into an ontology rather than letting it remain a process, a negotiation of past and present. In the history of hip-hop, the temporal relationship between the authentic and artificial is not based in succession (though one could argue that relation has never truly been consecutive and that the authentic is in fact an effect of the artificial). One need only listen to the famous opening Afro-Cuban *cascara* pattern reconstructed from Chic's "Good Times" on "Rapper's Delight," or look at the overwhelming number of hip-hop artists of Caribbean or African heritage. One would need to attend to—and listen for—the ways people construct and negotiate, rather than "express," blackness in a thickly historical context alert to nuanced interactions of culture and politics.

In cultural studies, hip-hop's individual lyrics and recordings—rather than performance techniques or the musical ensemble within which the lyrics are situated—remain the primary unit of analysis, which leaves analysis of "the unpopular behind so-called popular culture,"[15] the construction of "the popular," and "the popular's" relationship to broader movements too little addressed. Attention to both along with an analysis of dissemination (both literal transmission and the internal distribution of differences within a given song), alongside analysis of techniques such as sampling and interpolation, might unseat the power of the fake/real binary and uncover new ways of considering the political possibilities of rap, and cultural production more generally. With a differently attuned critical apparatus, the increasing presence of African-born hip-hop artists—e.g., Dizzee Rascal and Akan (Ghana), Afrikan Boy (Nigeria), Jean Grae (South Africa), and Oddisee (Sudan)—rising to prominence in the U.S. and Europe and distributing their music through record companies or digitally through SoundCloud, Bandcamp, and electronic "mixtapes" will be unsurprising.

Mixtapes, typically the work of a DJ or an MC, tend to have a looser format—and shorter time from conception to release—than the commercial album, and they can allow artists to release material that record companies refuse to release. They are a traditional means of hip-hop dissemination, distributed hand-to-hand in the cassette and CD era, allowing performers to release music without worrying about sample clearance or sales figures. Without merely celebrating the Internet as automatically "liberating" or "democratizing," I want to suggest that the mixtape and digital dissemination are dramatic examples of the nonclosure of culture, and the collapse of racialized and spatialized distance upon which many understandings of culture rest. As Iton argues, "with regard to the ways technology, popular culture, and politics intersect, it is important to consider the particular forms of sound systemization that dramatically reconfigured the boundaries between black public and private spaces" and nations in the past decades.[16] If in earlier moments one would have to wait for the newest records from the U.S. to arrive in Kingston, Lagos, London, or Cape Town, or vice versa, digital distribution makes culture more or less simultaneous. We can ask again: when—and from what directions—is the diaspora? To begin to answer this question requires a notion of diaspora that avoids a detemporalized Africa, that attends to the ways diasporas work within, against, and across state boundaries. We must understand it as a political field of contestation and negotiation, a counter-rhythm of the rhetoric of modernity and an articulation of coeval time that doesn't reduce to simple analogy behind which would lie familiar racializing spatiotemporal schema and priorities.

For the remainder of this essay, I will consider the collaboration of two artists—K'Naan, Somali-born and based in Canada, and Wale, born in the U.S. to Nigerian parents—on "Um Ricka" (a phonetic rendering of "America") from the latter's 2009 mixtape *Back to the Feature*. The song features the two rappers telling the stories of their family's migrations from Africa to "the land of the white man." At this level, "Um Ricka" participates in a musical lineage about rappers' upbringing and contemporary problems that includes Grandmaster Flash and the Furious Five's 1982 "The Message" through rap of the present, where the autobiographical "I" is now typically assumed to overlap with the narrating "I" and witnessing eye. "Um Ricka" ties Africa (e.g., "little nut jobs come swarming the village up") to postindustrial and inner-city crime in Washington D.C. ("no income, no bills getting paid / not to mention Wale is on the way"), before presenting a banal rags-to-riches story (e.g., "K'Naan and Wale got money in the bank") and model minority fantasy. Indeed, the success of recent African and Caribbean immigrants in the U.S. typically operates within ahistorical arguments about African American pathology that wish to ignore history and justify African American immiseration. My interest is the implicit comparison of Mogadishu and Washington, D.C. (with the New York and Los Angeles of gangsta rap lurking behind both) as two Afro-diasporic urban spaces that, despite spatial and geopolitical distance, share a temporal horizon of black vulnerability. Of course, stressing the danger of one's subsection in a larger metropolitan space is a common authenticating trope: one must "represent" one's home town in word and deed.[17] Like many mixtape songs, however, this song primarily displays the rappers' verbal and musical skills to generate interest, and that is where this song's play of temporalities gets the most interesting.

Produced by London-based Mark Ronson, "Um Ricka" is built around two samples. One comes from Amadou and Miriam's "Ce n'est pas bon," featured on the album *Welcome to Mali*, their first album to be released in the United States. (Released on Nonesuch Records in 2009, it had been released by Because Music in England in 2008.) "Ce n'est pas bon" features an antiphonal structure with Amadou Bagayoko calling out a line (e.g., "l'hypocrisie [or "la démagogie," "la dictature"] dans la politique"—hypocrisy/demagogy/dictatorships in politics) to which he and Miriam Doumbia sing-chant the response ("ce n'est pas bon, ce n'est pas bon, nous n'en voulons pas"—it is not good, it is not good, we don't want that). Their vocals are set against a midtempo Afrobeat-like groove that recalls Fela Kuti and Africa 70's "Water No Get Enemy," though it does not seem to be a sample. Those two interpolations signify "Africanness" as something other than a sign of purity: two rappers of African heritage rap over a track produced by a white Englishman that features a sample from a Malian group's first U.S.-released album and an interpolated groove inspired by Afrobeat, itself an amalgam of highlife, traditional Yoruba music, jazz, and 1970s funk styles. One could suspect here a lazy exoticism, keeping the Africans and African styles segregated, insisting on their otherness, but what is more compelling are the ways that sampling brings all of its sources into one common time.

The song features neither hook nor chorus: instead, the two rappers trade back and forth in the antiphonal style of the Amadou and Miriam sample, which becomes a structuring device rather than a signal of romantic, translingual pan-Africanism. His voice treated with reverb, Wale concludes his narrative of his family moving from dangerous Washington, D.C. ("that is not an ice cream truck, that's a coroner") and providing him with the discipline that "made me the man I am today." Prosodically and rhythmically compelling, the message is again banal. The song, however, takes an interesting turn when K'Naan begins his verse "I was similar" before rehearsing familiar tropes of wartorn Mogadishu as more compelling than the cinema. His verse pivots on the very "similarity" that would ostensibly yoke him, Wale and other black people in a common diasporan identity, obscuring important differences and incompatibilities in structures of the state and extra-state institutions, in history, and so on. One could perhaps link Washington, D.C., and Mogadishu historically through the Drug War or the Global War on Terror, but that linkage would imply a range of subordinate activities and governmentalities whose connection would require more analysis than a rap song typically provides. Instead, the rappers use similarity, the very disavowed point in notions of authenticity, to negotiate a shared diasporic identity around a shared horizon of time.

Mogadishu (metonymically the "village" in the line I cited above), whose life rhythms are structured around a similar-but-different adjacency to spectacular death that "makes boring the cinema," has historical and temporal links to Washington, D.C., as Somalia has with Nigeria, but no other immediate cultural links. Picking up K'Naan's stated refusal "to be less African and get amnesia," that is, to assimilate fully and lose the source of his authority, Wale's second verse begins by calling the very point of reference for similarity into question: "And I'm similar." The refusal to "get amnesia," his willingness to represent the known rather than the unknown Africa, makes critics "fall in love," lured by the draw of the imitation and its likeness—its superiority—to "the real thing." But that object—the real one—is elusive. The U.S.-born Wale acknowledges that his stories, so long as they are plausible, will satisfy critics invested in their own notions of the authentic global black experience. But attention to that "authentic" experience will miss the anxious negotiation one can see here in the excessive "Africanness" of sonic referent and the anxious repetition of similarity. Playing on, against, and within the prescriptive assumptions of "Africa," K'Naan and Wale advance a postauthentic, transgressive identity that calls into question spatiotemporal assumptions of center/margin, authentic/artificial. "Um'Ricka," like the technically reproducible recording in general, has origins irreducible to any single site.

The methodological privileging of a notion of culture "expressing," speaking for, the people of a nation—and the related assumptions of proper and improper uses of cultural forms—has consequences beyond rap or black music, as I hope is clear. I have gestured toward some of the effects of the privileging of the record as transparent re-presentation. The methodological question underpinning these comments on authenticity and culture might be a familiar question of the archive as commencement, commandment and "a house, a domicile, an address, the residence of the superior magistrates, the *archons*, those who commanded."[18] It is the home of the sovereign and, by metonymic extension, the home of sovereignty, the very sovereignty of the sovereign. In the case of hip-hop and other racially marked music, tracing the first or "legitimate" records requires, initially, some housecleaning, specifically to neutralize or *put in its place*—that is, outside—anything that cannot be retroactively claimed as belonging. The impure origin becomes prehistory, the space-time of the authentic, which nonetheless persists in cultural practices linked by race or nation without any consideration of the means through which traditions are handed down, and in that transmission, changed. Prior to this moment of origin, if such a "prior" can be imagined, there will have been a moment of nonidentity, of nonself that precedes the encounter, and persists within the formations that follow. This revenant trace of alterity shares the structure of the diaspora, gesturing to a relation more fundamental than any identity—it belongs nowhere. It has no proper belonging but disrupts and defers the proper.

Diaspora is a modern form of association, naming dispersal from some "original" site and the participation of that site in global forces and flows that overflow national borders. Understanding diaspora as a name for that which differs and defers the proper, the "final instance" of national belonging enables its conceptualization as a relation between differences, rather the being-together of the Same. The eschatological notion to which diaspora—like authenticity—is always proximate resurfaces, adding a different temporal dimension to the analysis of authenticity and artificiality

I have offered here. If one awaits return to a promised land, awaits the return of the voice of the dead or vanquished, what present and what future does one live in relation to now? To answer that question without affirming that one does not belong where one is requires abandoning prescriptive notions of time and sharing time, and more careful engagement with the legacies that yoke together race, place and time. One must also ask, given the physical, economic, political, and virtual interconnectedness of spaces, *when* is the diaspora, if one can speak, in all justice, of "the" African diaspora. Hip-hop, at least, does not refer to its origin in the singular, as a unidirectional monoculture that others define themselves within and against, marking those who are not African American as simply derivative. To think the diaspora is to conceive a culture that does not belong to the time of one nation or racial folk—that which suspends and reworks that distinction, without subordinating one to the other, or putting either under erasure. Hip-hop, as a diasporic form, disrupts simple narratives and assumptions of race and, indeed, can reveal something significant about the negotiations of blackness in a global context whose boundaries have shifted and become porous.

Diaspora, a framework that does not begin from the nation-state or the assumption of a state of affairs, is a situation, in the temporal sense I discussed above: a way of considering time's unfolding without privileging identities or nation-states. The hierarchical opposition modern/modernizing, or African American/global black, encodes the latter term, further away from the source, as less potent, even more corrupt, delayed, and, ultimately dependent. To see them as simultaneous, without immediate enclosure in questions of sovereignty, citizenship, legitimate participation, or the imperatives of authenticity, is to return again to the notions of modernity and the relations between culture and politics and see them as transactional and ongoing rather than a ceaseless reiteration of some prior "root" or unfolding of some transhistorical spirit. The ground of identity is nothing but this reverberation—one thing is similar to another because the other is similar to it. There is only this dispersal, this reverberation—both questions of temporality—and of likenesses without the priority of an original to which they refer, or an end to this play of differentiation: it is an impossible similarity rooted in shared time rather than speculative or mythic history. This impossibility may serve as one definition of modernity itself.

NOTES

1. See Johannes Fabian, *Time and the Other: How Anthropology Makes Its Object* (New York: Columbia University Press, 2002). For a specific account of the relationship between the folkloric paradigm and implicit support for segregation, see Jerrold Hirsch, "Modernity, Nostalgia, and Southern Folklore Studies: The Case of John Lomax," *Journal of American Folklore Study* 105, no. 419 (1992): 183–207.
2. Jonathan Sterne writes in *The Audible Past: Cultural Origins of Sound Reproduction* (Durham, NC: Duke University Press, 2003) about the process of sound recording coming to fulfill what he sees as a larger societal concern with preservation in the late nineteenth and early twentieth centuries; see especially his discussion on pages 287–334.
3. Jon Cruz, *Culture on the Margins: The Black Spiritual and the Rise of Cultural Interpretation* (Princeton, NJ: Princeton University Press, 1999), 3.
4. On this point, see John Hagstrom Miller, *Segregating Sound: Inventing Folk and Pop Music in the Age of Jim Crow* (Durham, NC: Duke University Press, 2010), and Robin D. G. Kelley, "Notes on Deconstructing 'The Folk,'" *American Historical Review* 97, no. 5 (1992): 1400–8.
5. Limiting myself to one example, one might think of the short-lived, black-owned record label Black Swan advertising itself as "the only *genuine* colored record."
6. Theodor W. Adorno, "The Perennial Fashion—Jazz," in *Prisms*, trans. Samuel M. Weber (London: Neville Spearman, 1967; repr. Cambridge, MA: MIT Press, 1981), 119–32. Adorno is the most relevant of a long line of writers who have theorized racialized spatiotemporal authenticity up to the present.
7. Jacques Attali, *Noise: The Political Economy of Music* (Minneapolis: University of Minnesota Press, 2011), 101.
8. Johan Kugelberg, *Born in the Bronx: A Visual Record of the Early Days of Hip Hop* (New York: Rizzoli International, 2007), 140, emphasis added.
9. Jeff Chang, "Timeline," in *Born in the Bronx: A Visual Record of the Early Days of Hip Hop*, by Johan Kugelberg (New York: Rizzoli International, 2007), 56.
10. Kugelberg, *Born in the Bronx*, 140–41.
11. Richard Iton, *In Search of the Black Fantastic: Politics and Popular Culture in the Post-Civil Rights Era* (Oxford: Oxford University Press, 2010), 250.
12. Iton, *In Search of the Black Fantastic*, 257.
13. Nicholas Brown persuasively argues that this is the case prior to the digital moment, and that scholars therefore should attend to the emergence of modernism as a response to empire expanding national borders. See his discussion in *Utopian Generations: The Political Horizon of Twentieth Century Literature* (Princeton, NJ: Princeton University Press, 2009).
14. Alexander G. Weheliye, *Phonographies: Grooves in Sonic Afro-Modernity* (Durham, NC: Duke University Press, 2005), 146. In Weheliye's account of the African diaspora as created and sustained through sonic transmission, throughout which he takes pains to undo the opposition between black culture and Western

modernity, he notes the often-suppressed non-U.S. origins of several prominent performers but does not raise the more general question of what it would mean to consider hip-hop itself as *originally* diasporic, despite the hegemony of the U.S. recording industry.

15 Cruz, *Culture on the Margins*, 15.
16 Iton, *In Search of the Black Fantastic*, 124.
17 This is especially true in K'Naan's oeuvre. He consistently compares the spectacle of wartorn Somalia with the images in U.S. rap songs and cinema of urban street violence to establish his superior "street cred." See Wale and K'Naan, "Um Ricka," *Back to the Feature*, Allido Records, 2009, compact disc. In the last lines of the song, K'Naan claims of one of hip-hop's many apostrophized, phantom haters and "sucker MCs": "Oops, I'm so sorry / just turned your street cred to whole wheat porridge," alluding to his hero Bob Marley ("No Woman No Cry") to establish his belonging to the African diaspora.
18 Jacques Derrida, *Archive Fever*, trans. Eric Prenowitz (Chicago: University of Chicago Press, 1996), 2.

18

Batch / Interactive

NICK MONTFORT

"Batch" and "interactive" are modes in which computer systems operate, the latter being a mode that is much more common and visible in the late twentieth and early twenty-first centuries, the former having dominated mainframe computing, beginning in the late 1950s. In early, archetypal batch processing, programs were written out, punched cards were prepared on keypunch machines, the deck was brought in and—when the computer could accommodate the job—the program was run. A tiny error would mean that the programmer would have to come back the next day and try again with the corrected program. Interactive computing, which began on a large scale with so-called time-sharing systems, turned the previous paradigm completely around so that a computer, rather than being guarded and inaccessible, was available to many people at once. By parceling out the computer's processing time among many users, it became cost-effective for many people to program and use computers on their terms, changing code and providing input during a session. This interactive time-sharing caused the computer to operate on a human time scale; the system took input and provided output directly to users, and while its time was initially split among many people, the computer was waiting on them, rather than the other way around.

The effects of this interactive revolution changed not only how computer scientists and programmers used the computer, but also how people related to computing more generally, through specific systems such as BASIC (a programming language for nonexperts, built for interactive use) and various types of creative computing, including "interactive fiction." Interactivity was an essential part of personal computing. This paradigm, arising after time-sharing, provided one computer per person. Today, time-sharing is still an important aspect of networked computing, but instead of there being many people to a single computer, a single person's computing environment usually involves many computers, local and remote. Today, programs run in both batch and interactive modes, with batch operations often automating what has first been developed interactively. The prevalence of interactivity has helped to shift the popular concept of the computer from one of a "big iron" billing system to that of a system for networked communication, creative production, and exploration. While interactive computing was not developed for artistic purposes, it provided new ways for programmers to explore the possibilities of computing and fostered a great deal of clever hacking, enjoyable game-making, and innovative textual and graphical production.

Even in today's environment, there are still many good reasons for batch jobs to run, and many of them are regularly carried out in ways that are invisible to most users. While batch and mainframe modes of computing have not been fully contained in entirely separate eras, and still coexist, the two paradigms do embody different attitudes toward time: batch jobs are set up beforehand by people with the computer's time in mind, while interactive systems are organized so that the computer's use of time serves people and fits the way they experience time.

Batch Time

In discussing the elaborate process by which early airline reservations were made in the 1940s and 1950s, Martin Cambell-Kelly and William Aspray describe a mode of batch processing that predated general-purpose automatic computing:

> The reason for this largely manual enterprise—it could have been run on almost identical lines in the 1890s—was that all existing data-processing technologies, whether punched-card machines or computers, operated in what was known as batch-processing mode. Transactions were first batched in hundreds or thousands, then sorted, then processed, and finally totaled, checked, and dispatched. The primary goal of batch processing was to reduce the cost of each transaction by completing each subtask for all transactions in the batch before the next subtask was begun.[1]

While batch processing is often strongly associated with mainframe computing, it's important to understand that digital computers did not prompt the invention of batch jobs—if anything, it is more sensible to think of computers as having been developed in their early form because of the existing practice of batch processing. Doing work in batches was the procedure not only for airline reservations but also in early data processing, which was done using tabulation machines rather than computers, as in the 1890 census, which was done on early punched cards, called Hollerith cards. Traditionally this mode was also part of accounting practices such as billing and payroll—business activities that were batched together long before the middle of the twentieth century. Some of these types of batch processing even predate industrialization: payroll, for instance, was developed around the same time as double-entry bookkeeping and other early accounting practices, in fourteenth-century Italy.

During World War II, the development of the significant early computer ENIAC began at the University of Pennsylvania. The ENIAC's first task was the computation of firing tables for artillery. This job was batch task. Such computations could in theory be done as needed by people in the field, but there was an obvious advantage to doing many of them at once, beforehand, so that soldiers could simply and quickly look up the result, a result that had been prepared and checked beforehand in a calmer context. The computation of tables of figures was important not only in the Army but also for business, scientific, nautical, and other purposes, and it was a classic early computer task, done in a batch mode.[2]

The standard way of interacting with a computer in this early era is described by Paul Ceruzzi: "a typical transaction began by submitting a deck of cards to an operator through a window (to preserve the climate control of the computer room)."[3] These cards would need to be prepared beforehand on a special keypunch machine; being able to operate this was not synonymous with being a computer programmer. As Ceruzzi notes:

> Sometime later the user went to a place where the printer output was delivered and retrieved the chunk of fanfold paper that contained the results of his or her job. The first few pages of the printout were devoted

to explaining how long the job took, how much memory it used ... and so on ... written cryptically enough to intimidate any user not initiated into the priesthood.[4]

There were rare cases (as with military applications) of interactive computing during the early years of computers, but most computer programmers faced this sort of batch-processing situation. And even though these people were among the few who had the skills and circumstances to program computers, they still had to submit their work to the priesthood in the climate-controlled room. In terms of time, this meant waiting in line to submit punched cards, waiting for the computer to process them, and then, often, waiting until the next day when the programmer/user would be allowed to resubmit the program with errors fixed.

As a mode of computing, batch processing helped to provide the cultural idea that computing was centralized, inaccessible, corporate and/or bureaucratic, and threatening. The nature of work had to be rearranged for the sake of the computer, as could be seen in the reconfiguration of architectural space to accommodate a massive computer in the 1957 film *Desk Set*, starring Spencer Tracy and Katherine Hepburn.[5] This film's perspective on computing is lighthearted; the mainframe here makes mistakes that can be corrected and is ultimately being put in place to help employees with their jobs. But the alien logic of the computer goes beyond threatening job security in Jean-Luc Godard's 1965 film *Alphaville*, which portrays the Alpha 60 as operating a dystopian dictatorship and ruling brainwashed citizens.[6] While there is an "interactive" sequence in this film (of Lemmy Caution's interrogation by Alpha 60), the system is clearly one that operates mainly in batch mode and, as much as it exerts a new technological influence, it also seems to grow out of existing bureaucratic governmentality. The computer's indifference to users' time and to the human experience of time more generally importantly contributed to this unpleasant perspective on early computing.

Interactive Time

There are various early visionary writings that describe some of the most important aspects of interactive computing. One of them, imbued with a dismal outlook, is E. M. Forster's "The Machine Stops," first published in 1909, many decades before general-purpose computing.[7] In it, the entire population of earth becomes completely dependent on a global machine that uses elements of batch and interactive computing. Well provided for (until, of course, the machine stops), people telecommunicate while being physically isolated, each using his or her own computing station. While Forster's is a provocative story, it did not provide a blueprint for the future but rather a vision of forthcoming dystopian catastrophe.

Vannevar Bush's vision of the Memex, laid out in his *Atlantic Monthly* article "How We May Think" (1945), offered a more positive example of how people and computers could interact. While mentioning aspects of digital computing, the technology Bush focused upon was microphotography, which he saw as a critical substrate for this type of information system.[8] The Memex concept can be seen to relate very directly to networked computing and the Web—and, indeed, it was influential in the development of both. His article described a device that would allow researchers to build trails through a body of existing scholarship, making connections in the associative way that the mind thinks. But Bush also recognized that operations that were more in the batch processing mode would continue to be necessary: "Such machines will have enormous appetites. One of them will take instructions and data from a whole roomful of girls armed with simple key board punches, and will deliver sheets of computed results every few minutes. There will always be plenty of things to compute in the detailed affairs of millions of people doing complicated things."[9]

Bush's Memex idea involved turning the wartime scientific effort to peaceful, progressive purposes. It considered the slower timeframe in which researchers could access many different library resources (which took longer than did a single conversation or the reading of a single article), noted that the body of work in humanity's "great record" was continually growing, and suggested a way to align the breadth and speed of information access with human thought. Although Forster and Bush came to radically different conclusions about where interactive computing would lead, Bush, too, thought that computing technology should be shaped to fit people, not the other way around. In this sense, at least, they would agree with Forster's character Kuno in "The Machine Stops," who declares, "Man is the measure."[10]

Following Bush's work, in 1960 a remarkable article appeared in the journal *IRE Transactions on Human Factors in Electronics*. It was by J. C. R. Licklider, who was known as "Lick" and who brought his background in both engineering and the behavioral sciences to bear on the future of computing. The article was "Man-Computer Symbiosis," and although it did not use the term "interactive" specifically, it did set out a plan for how people and computers could work together, a plan that became influential in the history of computing.[11] Licklider explained the reasoning behind having computers cooperate and collaborate with people, describing in detail the work day of a scientist and how interfaces (such as speech and handwriting recognition) and time-sharing could facilitate the human-computer symbiosis he imagined. Discussing time and the time scales upon which computers and humans operate, he noted, "Any present-day large-scale computer is too fast and too costly for real-time cooperative thinking with one man. Clearly, for the sake of efficiency and economy, the computer must divide its time among many users."[12] Although this seems an unusual idea from today's perspective, it was exactly the insight needed to shift costly, large computers from the mode of batch operation to interactive computing.

In one of his rare statements that wasn't prescient, Licklider, arguing that effort be spent on a speech recognition project, wrote that "one can hardly take a military commander or a corporation president away from his work to teach him to type. If computing machines are ever to be used directly by top-level decision makers, it may be worthwhile to provide communications via the most natural means, even at considerable cost."[13] Licklider, who was here being very respectful of the time of certain important people, didn't foresee that interaction with computers would change the definition of what constituted valuable executive and military-commander skills. Nonetheless, he set the direction for computing in many other ways. In a later article, he even discussed the networked computer as a communication device and foresaw (in 1968) that there could be a digital divide between those able to get online and those without access.[14]

Another early discussion of interactive computing is found in a March 1967 article by R. W. Taylor in the renamed *IEEE Transactions on Human Factors in Electronics*, one that directly references Licklider's and which introduces a special issue on interactive computing. Taylor noted the progress, both in terms of system-building and perspective, that had been made in seven years:

> there are perhaps several thousand people in the United States who are participating to various degrees in the man-computer symbiotic relationship anticipated by Licklider. The tasks accomplished by these people-computer partnerships have been impressive enough to suggest that we encourage the forming of millions of such partnerships. Today the question is no longer whether we should bring interactive computing to the sciences, engineering, law, publishing, libraries, the government, economics, banking and finance, manufacturing, management, and education—but how.[15]

Perhaps as notable as the use of the term "interactive" in Taylor's article is the phrase "man-computer." Although "people-computer" is also used, that other term appears repeatedly in the abstract, and of course it appeared in Licklider's article seven years earlier, as well. Before that, as noted above, Vannevar Bush described how "a room of girls" would be necessary to supply the Memex system "in the application of science to the needs and desires of man." Even the humanist E. M. Forster, who made a female character the most avid computer user in his short story, echoed old humanist and masculinst values when he had his male protagonist declare, "Man is the measure." Of course "man" was often *meant* to stand metonymically for humans in general, as was the conventional use of the time. Yet while the phrase may not have been consciously constructed by authors to exclude women, there *were* gender-neutral ways available to discuss interaction between people and computers in the 1960s, as the phrase "human factors" in this journal's title indicates. Bush himself uses "human" more often than "man" to indicate a person or to reference general humanity.

This usage of "man" might not stand out from other contemporary uses, except for the fact that in this case, it was "man" in the narrower sense who was the prototypical computer user, at least the ultimate user, the commander or executive who was the ideal user from the perspective of interactive computing. This was the case even though the first programmers of the ENIAC were women and there was some acceptance of mathematically trained women in programming roles. As computing became interactive, it was often imagined that women, or "girls," would work in the back office aiding with batch-processing tasks, while male decision makers did the interacting. This concept was in play in *Desk Set*, in which a male engineer brought a computer into a company's library—with the assumption that it would replace the female librarians and directly serve the company's executives.

The development of the BASIC programming language at Dartmouth, an important interactive computing project that was undertaken after Licklider's article was published and before Taylor's appeared in print, was in fact done by men in the context of an all-male college. BASIC co-creator John Kemeny later wrote a popular book about access to computing that was called *Man and the Computer*.[16] Yet high school students of both genders and female summer school students at Dartmouth did have access, via the network, to Dartmouth's time-sharing computers and to early versions of BASIC. Kemeny's book title, while not as equitable as it could have been, was a reference to the lecture series in which he began to develop the book, the "Man and Nature" series at the American Museum of Natural History, just as the title of Taylor's article is a reference to Licklider's. (Also worth noting is that as president of Dartmouth College in 1972, Kemeny led the school to become coeducational.)

Thus the development of interactive computing did not affect men and women equally, but neither did every aspect of this transformation privilege one gender over the other. Part of the issue was how men and women had used their time to gain different skills, such as the ability to type, and how the time of powerful men was too important to allow them to become skillful typists. As Licklider stated, interactive computing gave those who could type immediate access to programming and computation, advantaging them. Typists would not need the intermediary of a keypunch machine and its operators nor those operators who were handed cards and ran them through as batch jobs. Because of the gender norms prevalent for secretarial jobs, a large number of those who could type in the 1960s were women. In this way, at least, considering the distribution of office skills, interactive computing had the potential to offer some advantage to women. Kemeny himself was not a highly skilled typist, and there is evidence that his and Kurtz's BASIC programming language treats spaces as optional so that it can be more easily used by poor typists.[17] Of course, typing ability and a typewriterlike interface was not all that was needed; women would still have to be granted access to interactive systems in order to put their typing skills to use as programmers.

Any profound change in how computers and people relate to one another—for instance, because computing becomes much more accessible in many all-male contexts or because computers become more easily operated by typists, who are mostly women—is sure to have such differential effects. These depend, in part, on previously established abilities, environments, rhetorics, and stereotypes. While new ways of interacting with computers can be revolutionary, their revolutionary effects can also be contained or distorted, to a greater or lesser extent, by the way they are framed and the contexts in which they play out. Yet while the thinking that went into the project of interactive computing was often focused on how it could serve "man," interactive computing continued to explore its potential to bring computers onto a human time-scale.

For people and computers to work together at a similar pace, it was necessary to design interactive interface components that were different from keypunch machines and card readers. Doug Engelbart, who led the development of the groundbreaking NLS (oN Line System), developed not only the mouse but also a special chording keyboard that could be used one-handed while the other hand employed the mouse to point. Engelbart also developed videoconferencing; real-time online collaboration and collaborative editing; the first practical, working hypertext system, which was also a hypermedia system; word processing; dynamic file linking; and version control.[18] He did all of this as part of his program of furthering the computer augmentation of human intelligence and the bootstrapping of human intellectual processes into computerized systems. And, on December 9, 1968, he showed off NLS and these capabilities to a thousand onlookers at an extraordinary event in San Francisco, making him the inventor of the modern technology demo as well. The event came to be called "the mother of all demos," and it revealed that progress toward better interactive systems was made, at times, on many fronts at once. It also showed the computer working to facilitate human collaboration and real-time conversation—an even further orientation of the computer toward human time.

Ubiquitous

"Interactivity" has been valorized as a positive quality of computing systems and particular programs, as Espen Aarseth has pointed out:

> Interactivity ... has long been associated with the use of computers that accept user input while a program is running, as opposed to "batch" computers, which process only preloaded data without interruption. Interactive thus came to signify a modern, radically improved technology, usually in relation to an older one. The industrial rhetoric produced concepts such as interactive newspapers, interactive video, interactive television, and even interactive houses.[19]

Seen from this perspective, a term such as "interactive fiction" could appear as merely a marketing gesture, an attempt to get onto the interactive bandwagon, and Aarseth dismisses it as such: "The word *interactive* operates textually rather than analytically, as it connotes various vague ideas of computer screens, user freedom, and personalized media, while denoting nothing."[20]

Aarseth is uninterested in exploring the nuances of the term "interactive," and particularly its temporal nuances—probably because he is mainly working here to replace it with his own neologisms. In the framework of batch vs. interactive computing, however, the earlier term does clearly indicate that this is a program that allows user interaction, and when a phrase such as "interactive fiction" is coined, there is the indication that one's dialogue with the computer can influence a fictional world. Those who popularized the term (mainly at the company Infocom) were well aware of batch and interactive computing and, in addition to using a marketing term, were also making an appropriate distinction, calling attention to how the new form in which they were working differed from "batch" fiction such as the short story or novel. A noninteractive story generation program could also be considered, perhaps even more appropriately, as "batch fiction."

The term "interactive fiction" has survived any widespread commercial market for interactive fiction; it is used by a community of practice and play that reviews new (and old) releases at the Interactive Fiction Database, participates in the annual Interactive Fiction Competition, and gets together locally for meetings and events organized by clubs such as the People's Republic of Interactive Fiction in the Boston area.[21] While the scale of today's interactive fiction activity does not rival that of big-budget AAA games or Hollywood movie productions as an industry, it represents a thriving type of creative computing practice alongside the artgames movement, net art, and the demoscene. Interactive fiction certainly pertains to the way people read and respond, on the command line, to interactive computer systems. It also draws on how they read, how they study a text or environment carefully to try to understand it—a process that involves a different temporal engagement than does "surfing" or other modes of computer interaction. While the computer is responsive and waiting on the user in this interactive experience, interactive fiction invites significant engagement and attention and discourages multitasking, interruption, and play in short sessions.

Some believe this command-line interaction is very directly the basis of interactive fiction,[22] while I and others also see an important relationship in this art form to human dialogue, tabletop role-playing games such as *Dungeons and Dragons*, and the simulation and transformation of fictional worlds.[23] In any case, even when "interactive" *is* deployed as a marketing term, it can be useful to see what aspects of existing media (fiction, video, newspapers, etc.) are being referred to and what idea is in place for developing these existing media in new, dialogic ways.

A category or paradigm that has been proposed for computing, and which goes beyond batch or interactive but would not be possible without interactivity, is *ubiquitous*. Although this term indicates pervasiveness throughout space, the concept has implications for time as well. Devices such as the PDA, the smart phones and tablets that are now in widespread use, and increasingly other devices including smart watches and small health-monitoring units—all function to connect us to computing, and often to the network, at almost every moment. Consciously initiating an interactive session is still a possibility, but we are just as likely to be interrupted by an alarm that reminds us of an appointment on a networked calendar, letting the corporate model of time intrude on our own events and personal experience of it, or to glance down and check to see how many of our steps have been counted during the current day.

"Multitasking" is another term to take into account. It was originally a term that applied to computers that could implement time-sharing and spend processor time on several different jobs in turn. It was used by IBM in 1965 technical documentation; by about forty years later the popular press was decrying the way people, and particularly the young, were multitasking rather than exhibiting the types of concentration and focus that were expected. *Time*, for instance, ran a cover story on "The Multitasking Generation," and within a few years articles were frequently appearing about the dangers of dividing one's attention and slicing one's time into smaller segments.[24]

Whether this response represents a media panic or an appropriate corrective, the way "multitasking" has been applied over the decades shows that people identify and are identified by others as taking on computer-like behaviors and dividing up time the way that computers do. While "interactive" systems were an attempt to align the computer's abilities with human dialogue and human time, people who "multitask" align human abilities with those of computers: they are behaving like computers and taking the computer's perspective on the allocation of time and work. While on the one hand this looks like a return to computercentric time, just as batch processing, invented before computing, was an efficient and computerlike way to process large numbers of transactions, there could be good uses for certain types of computer-inspired multitasking.

Yet if this is not so, if we should hold that "man is the measure," then what is a healthy relationship for people and computers to have? Specifically, how should we negotiate our experience and use of time and the computer's ability to function, respond, and interrupt us in time? Among several detailed suggestions that have been proffered, Douglas Rushkoff offers ten ideas (in imperative form) in his *Program or Be Programmed*. His first commandment/chapter, "do not be always on," states that people should not "marry" their "time-based bodies and minds to technologies that are biased against time altogether."[25] This is a pointed example of how people believe that they can still make many of their own choices about how their work (and play) with computers influences their temporal experience, even if industry plays an increasingly large part in their computing landscapes today. Whether today's social networks and always-on, always-present mobile devices will turn out to be more like the dystopian Alpha 60, or more like a positive augmentation of human intelligence, may depend on whether people are cognizant of the ways people can relate to computers, and, through them, each other, in time.

NOTES

1. Martin Campbell-Kelly and William Aspray, *Computer: A History of the Information Machine* (Boulder, CO: Westview Press, 2004), 170.
2. Scott McCartney, *ENIAC: The Triumphs and Tragedies of the World's First Computer* (New York: Berkeley Books, 2001), 53–55.
3. Paul E. Ceruzzi, *A History of Modern Computing*, 2d ed. (Cambridge, MA: MIT Press, 2003), 77.
4. Ceruzzi, *A History*, 77.
5. Henry Ephron, Phoebe Ephron, and William Marchant, *Desk Set*, directed by Walter Lang (1957; Los Angeles, CA: 20th Century Fox, 2004), DVD.
6. Jean-Luc Godard and Paul Éluard, *Alphaville*, directed by Jean-Luc Godard (1965; The Criterion Collection, 1998), DVD.
7. E. M. Forster, *The Machine Stops*. ManyBooks.net, 2007, http://manybooks.net/titles/forstereother07machine_stops.html. Originally published in the *Oxford and Cambridge Review*, 1909, and reprinted in E. M. Forster, *The Eternal Moment and Other Stories* (London: Sidgwick & Jackson, 1928).
8. Vannevar Bush, "As We May Think," in *The New Media Reader*, ed. Noah Wardrip-Fruin and Nick Montfort (Cambridge, MA: MIT Press, 2003), 35–48, originally published in *Atlantic Monthly*, July 1945, 101–8.
9. Bush, "As We May Think," 41.
10. Forster, "The Machine Stops," n.p.
11. J. C. R. Licklider, "Man-Computer Symbiosis," in *The New Media Reader*, ed. Noah Wardrip-Fruin and Nick Montfort (Cambridge, MA: MIT Press, 2003), 73–82, originally published in *IRE Transactions on Human Factors in Electronics* HFE-1 (March 1960), 4–11.
12. Licklider, "Man-Computer Symbiosis," 77.
13. Licklider, "Man-Computer Symbiosis," 80–81.
14. J. C. R. Licklider, "The Computer as a Communication Device," *Science and Technology*, April 1968, 20–41.
15. Robert W. Tayler, "Man-Computer Input-Output Techniques," *IEEE Transactions on Human Factors in Electronics* HFE-8, no. 1 (March 1967), 1.
16. John G. Kemeny, *Man and the Computer* (New York: Scribner, 1972).
17. Thomas Kurtz, "BASIC," in *Masterminds of Programming: Conversations with the Creators of Major Programming Languages*, ed. Federico Biancuzzi and Shane Warden (Sebastopol, CA: O'Reilly Media, 2009), 79–100.
18. Douglass Engelbart and William English, "A Research Center for Augmenting Human Intellect," in *The New Media Reader*, ed. Noah Wardrip-Fruin and Nick Montfort (Cambridge, MA: MIT Press, 2003), 233–46.
19. Espen Aarseth, *Cybertext: Perspectives on Ergodic Literature* (Baltimore and London: Johns Hopkins University Press: 1997), 48.
20. Aarseth, *Cybertext*, 48.

21 The Interactive Fiction Database, http://ifdb.tads.org, is another community resource, with copious information about works of interactive fiction pieces and all past Interactive Fiction Competitions. The Boston-area People's Republic of Interactive Fiction has its site at http://pr-if.org.
22 See, for example, Steven Johnson, *Interface Culture: How New Technology Transforms the Way We Create and Communicate* (New York: Basic Books, 1997).
23 See Nick Montfort, *Twisty Little Passages: An Approach to Interactive Fiction* (Cambridge, MA: MIT Press, 2003), and "Riddle Machines: The History and Nature of Interactive Fiction," in *A Companion to Digital Literary Studies*, ed. Ray Siemens and Susan Schreibman (Hoboken, NJ: Wiley-Blackwell, 2008), 267–82.
24 Claudia Wallis, "genM: The Multitasking Generation," *Time*, March 27, 2009, http://content.time.com/time/magazine/article/0,9171,1174696,00.html.
25 Douglas Rushkoff, *Program or Be Programmed: Ten Commandments for a Digital Age* (New York: Soft Skull, 2011), 22.

19

Transmission / Influence

RACHEL HAIDU

The time of artistic influence is *past* in a manner that reveals the schisms and anxieties of academia itself. On the one hand, "influence" is a term that has been out of fashion for more than forty years in humanistic disciplines touched by theory; on the other hand, it persists as a principle of thought, abandoned but still lurking, something we might wish to unthink but instead find ourselves actively repressing. We might repudiate, disavow, or just ignore it, but influence is still part of our consciousness, often part of the way we define what we study as objects in their own right—as entities separate from ourselves that have something to say about history. Yet its persistence doesn't mitigate its distastefulness.

But what are we talking about when we complain about influence? Are we talking about what actually transpires between two (or more) authors, or what transpires through discourse? On the provisional premise that it's the latter—ultimately I would like to show how intertwined these two issues are—let me suggest a semantic experiment. Let us call what transpires between authors as they borrow, learn, and appropriate from, or react, defend against, and undermine each other, "transmission," and then explore what is left over as "influence." Time is key to organizing this distinction for two reasons. First, the work that is being shifted from "transmission" to "influence" in the model I am proposing is work that takes place in *our* time—that of viewing, producing, reading, and writing. In other words, I would like to lay claim to the ways in which influence *takes time for me* (or any other author)—time to see, diagnose, work out in writing, and convey to a reader. *That* is the time of influence. Transmission takes place in what we think of as real time. It's what happened to a young filmmaker while in film school, or while she was screening Eisenstein on her laptop; it's what happened when Manet encountered the paintings of El Greco and Velázquez, or once Picasso and Braque started really looking at each other's work. Influence is what happens on the page, when I want to describe how two authors relate; transmission is their relat*ing*, the present or past thing that we have till now thought of as "influence."

But there's more. Transmission is a helpful term because it reminds us of the reproductive, germinating (i.e., sexual), and temporal nature of what otherwise gets sublimated and sublated, cleaned up and detemporalized as "influence." Transmission evokes broadcasts, the larger, overlapping networks of audience members toward which work directs itself. It also reminds us of emissions—gaseous discharges, pollutants in the air, the "giving off or sending out (chiefly what is subtle or imponderable, light, heat, gases, odours, sounds, etc.)."[1] Then there are those involuntary nighttime secretions that evoke succubi and incubi, returning us to the specificity of the ephemeral encounter within larger systems of transitive energy. Yet: if *transmit* and *transmittĕre* are at the root of transmission, so is *mission*, from the French verb *missionner*, to send someone on a (religious) mission. That domino-type effect, in which one agent sends another on a mission, suggests a way to understand transmission's metaphorical resonances. To the degree that it calls into play the "subtle or imponderable" movements between agents, it also calls up the transitivity at work *in* those movements. Transmission refers us to both the breadth of the emission and its consequentiality, its ability, in moving from a to b, to move us in turn.

This attention to the meaning of transmission is key to reshaping what we used to think of as influence. Whereas influence, in its traditional usage, vaporizes the real-time encounter with another or her work, sublimating it into the high-end sphere of timeless analysis and judgment, transmission lands us squarely within the ways that media—among them, time-based media—have reorganized our relations to one another. Just as it is not ever the case that *only* Braque looked at Picasso and vice versa—an artwork never has an audience of one—it is not the case that the moment of reception is ever as singular as traditional "influence" would have us believe. Artworks only exist as values defined by succeeding moments of reception. Transmission both acknowledges the public or quasi-public, historico-temporal nature of the work's appeal to its multiple audiences *and* recognizes how a more specific encounter takes place within that generalization or succession. It makes that acknowledgment explicit by temporalizing (-*mission*) the travel across space (*trans*-), in which space can be the width of an ocean or merely the interval between two entities. And it points us directly toward the medium of its own movement, the vehicular and material nature of being moved in a way that moves another, in turn.

If transmission is thus fully diachronic, occurring in and across time, influence, in this new usage that I am defining, acknowledges the ways discourse aims to consume its own time. It takes time to "see" two objects acting upon each other, and indeed that is often the time of invention, fabrication, construction: the time it takes to "see" is also the time it takes to organize discursive categories, render formal analyses, locate objects in history. To declare influence *tout simple* is to ingest that time, to make it disappear down the alimentary canal of discourse, and to pretend that *its* time is not *ours*; that is, that the time in which influence happened has already taken place ("is passed") and therefore *is not*, now, "is past." Our alternative is not to give a real-time accounting of all those efforts: who wants to read that? What purpose would it serve? But rather, to systematically incorporate those efforts into a reinvention of what we mean by influence. What do we mean when we talk about what we used to mean by "influence"? Chronology, comparison, and even the autonomy of the entities engaged in one's judgment that two things are similar or different enough to warrant analysis of them together: these are critical points on the grid that has underwritten "influence." They beg to be reshuffled, introduced into discourse, and used to reinvent it.

If we do not "record" that time of our reading, it is not merely because it would bore the reader, or serve no purpose. (Or, we might also ask: why is it boring? What do we already know about it such that its unwinding over time would be unbearable?) The time of one's reading is already synthesized in the results of reading. But it is also an endless time: a time not of finite influences upon the reader but of bulky categories and endlessly recursive and shifting ideas. Let us start with the broad category of "technics," which I take to encompass not only the technologies of creation, dissemination, and reproduction but the very categories invested in *by* influence: categories of authorial identity, of types of works, of quality and kind. It is impossible, in a day of appropriation, sampling, mixing, and citation, to divorce the technologies of those activities from their meanings. If Sherrie Levine rephotographs Edward Weston it is not only her camera and his print that are in play, but—as her reception, her "influence" on photography tells us—her feminist read of what appropriation can do; her "take" on his photographic "invention," and so on. These are the technics that any discussion of any kind of influence mobilizes. Part of the point of reversing the positions of transmission and influence is to attend to the broad reach of that technics, to integrate our sense that when, for example, formal discussions mobilize categories of identity, they do so instrumentally, to achieve a discursive end (e.g., a comparison of how black artists complicate the category of blackness needs the category "black artist" in order to motivate its analysis of the forms of their work).[2] Discursive conditions, as we know, are as material as material conditions, and if the discussion of influence until recently has turned a blind eye to the former, it has been no more attentive to the latter.

Transmission

Harold Bloom, whose name apparently belongs in any contemporary discussion of influence, is notorious not only for his notion of "strong misreading" but also for having laid a groundwork for further cementing the greatness of great, white, male authors at just the moment that poststructuralism (and later, cultural studies) set about dismantling those logics. However, if it is the "space" of the canon—its unearthly isolation of authors from their fellow women and men—that offends, it is truly its "time" that should provoke militant ire. The dehistoricization enabled by the historical vacuum into which Bloom's authors slide is one thing: its confirmation that *that* dehistoricized time is the "time" of Bloom's writing is what is toxic, in the sense of actually contaminating the question of influence.[3] Space provides the first encoding in the mythification of the great author. Time is that code's second order, its assimilation into Bloom's own interpretive culture, and that which he extends to his reader. As Shakespeare, Marlowe, and Whitman enter a vacuum in which historical ties fall away, we might still be able to learn about them and the time they create through their shared language—were Bloom's analysis not implicated in the same temporalization, sharing, as it were, the value of their non-time and not his own lived temporality. As Bloom groups these authors according to formal criteria and engages with them as their equal and collaborator, he mobilizes a discursive fiction: that he is in a time that communicates with theirs across the page that is read.[4] You read Bloom as Bloom reads Shakespeare; we are each alone but all in the community of readers. There is nothing wrong with this fiction. But the specificity of those technical means of composition, dissemination, and reproduction are invested in the works themselves, are *also* invested in the time it takes to understand them and to read them across one another.

So if the authors examined in *The Anxiety of Influence* exist in the space of a historical vacuum and an absence of sociality, a "time" in which they operate upon each other that is the utterly abstracted, unreal time of synchronic formalism, then that is not merely because that's the time of canons, and Bloom can't help but affirm them. It's also because that's the time that he thought his *typewriter* was in. But let's not let the shared space of reader-writers confuse the issue, because it is not where the conceit of the timeless time of the canon originates. Let us go back (again provisionally) to the time before film and photography took hold of our imagination not only as media of reproduction but also as means by which we think of our relation to reproduction. Let us go specifically to those last glimmers of the pre-filmic. Let's think for a moment about painting around Year Zero: the painting of revolutionary France. The voice of one of its most significant Anglophone art historical interpreters, Thomas Crow, commenting on the discourse within art history and his own 1995 book, *Emulation*, writes:

> The individualist framework as a whole has been a casualty of the greatly enhanced powers of interpretation that the study of art history has lately achieved.... *Emulation*... was written to address both the gains and losses incurred by this intellectual turn. Taking the losses first, the chief of these lies in a significant sacrifice of realistic self-knowledge. The biographical model conforms to assumptions concerning the unity and integrity of the self that animate one's nearly every waking thought, however adept one may have become in theories of the author's death and the dispersal of self in a socially constructed fabric of representations. While such theories constitute a plausible world, it is simply not possible to live in it.[5]

What is remarkable here is the easy division of the world between that *in* which the art historian writes and that *of* which he writes. While Barthes, Foucault, and other progenitors of "theories of the author's death" were fairly clear that they were not writing about the empirical world of bodies, but about a world that would enable those bodies to live in and through texts more fully, Crow's recourse to biography (or in his case "dual biography") resituates the reader within a duality between the world that can be written and the world in which it is "possible" to live. Over and above the many responses to this apparently easy division—responses that have come from, for example, science fiction, Afrofuturist writing, art, and music, or indeed the entire history of modernism that strains towards the utopic, the next-ness of the world itself—there are the well-known politics of how such a binary erases the writer himself from the scene of making that world. That was of course Barthes' point (and Foucault's as well, and Derrida's, and Lacan's . . .) and one that is hardly mitigated by the effort to unveil, through "examining criss-crossing life histories . . . creative agency [manifest] as a distributed rather than individual phenomenon."[6]

It is precisely the notion of *distributed agency* that preoccupies not only Crow but also those writers who have carefully examined how poststructuralism might itself relate readers to a world of "real" actors.[7] There, almost opposing Crow and yet "presupposing corporeality," we find the work of Friedrich Kittler.[8] That is, we could take Crow's invitation to think creative agency as "distributed phenomenon" and map that onto Kittler's discourse networks, which attempt to locate *in* the materiality of form the very technics of an age, in which technics, *pace* Bernard Stiegler, are the relations of individuals not merely to their instruments but to the reproducibility of knowledge (or "mastery of nature") itself.[9] Thus, for Kittler, handwriting, phonetics, and eventually typewriting are the medial forms showing us how, around 1800, "Nature, Love, Woman . . . produced an originary discourse that Poets tore from speechlessness and translated."[10] Romanticism is produced not as style but as a function of the twinned conceptions of speech/speechlessness and writing/the unwritable, *and* the advent of a new pedagogical system, in which "phonetic reading instruction [becomes] a writing system."[11] Thus, for Kittler, does a radical new fashion in bourgeois homeschooling conjoin the fundaments of *what literature is* to its means, in which "means" are expanded to include tools and instruments, social practice, and literary form.

Their successful conjuncture—the epochal shift in both technology and form accomplished "as" Romanticism—depends, in Kittler's narrative, on the body of the Mother. For it is the actual (or one might say, biographical)[12] mother's *voice* that is required in order to execute the new pedagogical agenda, which in turn "eroticizes language ... fragmented into letters and syllables" in the name of translating Nature. She becomes both figure (or emblem) for the twinned operations of reproduction and transmission in Kittler's text, and his means of showing the absent female at the center of male-driven discourse: "Civil servants wrote (not just anything, but the determination of Mankind); the Mother did not write, she made men speak."[13] More than a mere salve to the feminist readers of Kittler's *Discourse Networks*, this move should be examined for how it allows us to understand more precisely how media of reproduction (always, media of *production*, including typewriters) are also media of transmission. Its identification of the origin of (modern) discourse with the locus of reproduction, as both figure and technology, is critical. Kittler gives us a brief history of speech around 1800: it gets broken apart into phonetic fragments that are in turn available as written fragments, which in turn becomes a pedagogical method allowing the knowledge of speech and reading to be reproduced in home-schooling sessions, as well as the modality for understanding how the signifier relates to the signified (Nature). In identifying the means of reproduction (pedagogy) with its meaning (the new modality of representation), he locates the mode of expression *within* the problem of reproduction, and shows that problem as one that is only perceptible as a story about "technology," instruments of communication that are themselves never autonomous.[14]

The "time" of transmission becomes therefore not so much past, in terms of academic fashionability, but ongoing, one mode in which we think about what technology, and the technics of being influenced, *is*. For one thing, we have redefined the seemingly "past" moment in which the film student screened—and was affected by—Eisenstein, on her laptop, as one that potentially takes place in a medium-reflexive "network." In that network, the technology of reproduction and display being now utterly aligned, the student is herself aware of the fluid barrier between herself and her instrument, and the way that signifies, in turn, the radical transmissibility of texts. To the degree that we see countless works of art representing that fluidity (from Aurélie Froment to Ryan Trecartin to James Coleman to Sherrie Levine), we are in a place little different from that in which Vertov and Eisenstein projected their viewers. On the other hand, we are in a position to say that from Vertov to Froment there is enough thought about reproduction to enhance or even authorize a new field in which what we used to call "influence" can be rethought. Those instances of what I'm calling transmission are more than self-conscious reflections on having-been-influenced. They are ways of actually rethinking the movement of influence within a world in which reproduction, having moved into utter technicity, has also seen technicity itself reshape our definition of the human, the self, the "one" under the influence.

Influence

One possibility for executing the kind of highly recursive analysis of influence that I have in mind refigures the author not as the protean subject that Shakespeare becomes in Bloom's analysis, or indeed that one might imagine resulting from such an extensive inquiry into the self-under-the-influence for which I have been arguing. On the contrary, what about imagining the figure of the Old Woman as the very culmination of all this thought about Reproduction, reproducing another "in" oneself or oneself "within" another? Here I depart from a recent, remarkable publication in *PAJ*: Yvonne Rainer's "The Aching Body in Dance." In it she details not only the "Farewell" performances of Maria Theresa ("Isadora Duncan's last surviving foster daughter") and Martha Graham, but her own return to dance after a thirty year hiatus during which she made films:

> My next dancing foray occurred in 2010 when I performed what had become my signature dance, *Trio A* (originally titled *The Mind Is a Muscle, Part 1*, created in 1966 at age thirty-two). This version of *Trio A*, subtitled *Geriatric With Talking*, encapsulated what might be called my philosophy of aging in dance, namely, "Let it all hang out." If you're going to make an appearance in front of an audience and you can't execute the material as

robustly or as accurately as you once did, then be honest; tell them what's going on moment by moment. This is exactly what I decided to do. It was language that would add the necessary consciousness to the performance, hopefully waylaying any tendency on the part of the spectators to pity or condescend. As I threaded my way through the dance, I extemporaneously told them what I was experiencing, without interrupting the flow of movement:

> This move is supposed to be a slow rise of the leg, not a battement, but why can't I get my leg up any higher than this anymore? Oh just do it and get it over with.[15]

In these lines, Rainer reminds her audience not only of the difference between her body now and the way it "was" then, or the way she remembers it behaving then, but points to the fact—as she has throughout her dance and film career—that a dancer's body has always been silent, voiceless. The whole point of the earlier dance was that you didn't hear the dancer speak: or at least, that becomes the new point of *Trio A* when you are watching and listening to *Geriatric With Talking*. Is noticing this not what we could call "influence on oneself"? Is it not exactly the line between who Rainer was and who she is "now" that exfoliates, through the spoken lines—or indeed their extemporality, their belonging only to their own time, rather than some other, discursive time? It is perhaps Rainer's "old age" that is of interest here. Not that she has in any way stopped teaching (reproducing her dances) or writing (as the *PAJ* piece demonstrates) or even making film. But *Trio A* was particularly invested in its own reproducibility (not to say its modularity: its ability to be performed at a different orientation, under different conditions, backwards, etc.), whereas *Geriatric With Talking* is invested in answering that reproducibility with the hereness of Rainer's aging body.[16]

It is not, of course, that that body becomes "less" human, or posthuman in some cyborgian (or ageist) sense. But it redirects our attention from the human to its own limits, its internal limits as it were—not the self fantasmatically or technologically extended but involved in its intensions (acts of straining, stretching). The intensive self is perfectly metaphorized in Rainer's piece by the body of a dancer suddenly unmuted, so that we can "hear" the strains, finally. This idea of a Woman Under Her Own Influence both provides a model that could be replicated and complicated in texts exploring new forms of influence, and is an emblem for the self that deserves such a new form of influence.

To the degree that I can redefine influence as that which we come to see ourselves as fabricating rather than observing, I would argue that influence demands some new ideas about the self and other. Weaving the self who sees influence into the story we tell should become an effort not merely at subjectivizing interpretation but at rendering more systematic and explicit the ways that it takes place. Influence is one way to leverage oneself into the scene of reading. The effort it takes to understand what one sets out to do when one "locates" influence can be slowed down, accounted for, and then used to rethink what influence *is*. That is, in order to rethink influence, or the relations between authors that we think we perceive (whether or not what we perceive is actually "influence"), one has to undo some of the fixed points from which a story about influence currently progresses, and rethink *them* as taking time.

Take Darby English's *How to See a Work of Art in Total Darkness*. Here the "comparison" that is tacitly underway is between a group of African American artists who are said not to have influence "on each other," but on our ideas of blackness. But of course their comparability derives from their identities, as individuals: Kara Walker is "like" Fred Wilson, who is "like" Glenn Ligon, who is "like" William Pope.L because their works share a kind of formal complexity and because they are all African American artists at the end of the twentieth century making work that wrestles with the dyad complexity/identity. English's account is compelling in part for the many ways in which his position is interspersed among the texts, dispersed by the work they do on identity and complexity. That English comes to these works in the age of Stuart Hall is everywhere throughout the text *but* in its selection of objects or the literalness with which that selection tests Hall's notions. The book brilliantly undertakes its own undoing, announces its own responsibility for the ways it tests the foundations on which its logic both enacts and complicates identity. What its sequel could do is discuss the space between the examples: the question of their comparability and the ways in which that comparability both redefines and shifts the issue of their collective and individual identities.

Another issue besides that of comparison, similarly at the basis of any judgment of influence, is chronology. At least since Borges wrote about Kafka (in 1951), we have come to understand that chronology can be read backward: that, in other words, studying the successor (e.g., of Kafka) can tell us more about the precedent than hunting down evidence that that precedent "mattered" in his successors' works. This logic has squared perfectly—thanks in part to the fashions for appropriation and citation—with developments within popular culture, which now is rife with systems and forms of reading that resituate the audience within newly fluid histories (musical sampling, web mashups, "YouTube poop," etc.). The notion that "each writer creates his precursors," however, has done little to shift our conception of history, which remains fundamentally linear, and which continues to infect even the line between "writer" and "precursor" with fundamentally stabilizing values.[17] So, just as the comparison that is necessary for diagnosing or understanding "influence" depends on a degree of likeness that can't be standardized or objectively defined, so too does the chronological impulse within influence contain even those attempts to reverse it. Any new version of influence would have to try to undo the ways that these premises rig the results of trying to think about influence.

Here the time of a redefined influence finds itself butting up against the preconceptions of historicist and materialist histories: it cannot abide either, as both relate their objects to an externality—a history that is "beyond" the object. For two authors to "influence" each other we have always had to provisionally suspend them within the space of their formal (dis)similarities, Bloom's airless space of Great Authors. The trick lies in understanding when the technologies of transmission produce a similarly airless space. Remember that technics includes pedagogies: classrooms designed, for example, to seal off their inhabitants, to create a petri dish of mutual influence. There might even be historical reasons (Cold War politics, for example, or classrooms designed to seal themselves off from intrusions by local totalitarian regimes) that something like the vacuum Bloom imagines would "actually" take place. How does the space-time of a *real* vacuum, authorized by the efforts involved in transmission, reflect on historical space-time? Here, analyses of pedagogy and pedagogical experiments seem a perfect area in which to imagine how Bloom's airless and detemporalized space-time vacuum might find a curious complication as it confronts materialist history and the kind of historicism that buttresses his own analysis. What happens *in* the airless-timeless space-time of the workshop or the academic studio, and how do those effects (of air- and time-lessness) shed light on the excessive kind of "influence" that we understand teachers to have? How can we use the extremity or excessiveness of the kind of mutual influence exerted between student and teacher to complicate not only the transitivity of transmission but the means of doing what I've been redefining as "influence"? If it is our work—discursive work—to determine not only when influence happens and what it is, but what it means to us as we recognize it, then we are in a sense completing the work of the student who is caught in the game of complicated reflexivity and projection that teacher influence exerts.

The old-fashioned notion of "influence" requires a notion of the self—and that self had better be contained, if we are going to be able to diagnose influence as "taking place" *between* two selves, or even *for* the self who is doing the writing. In approaching this premise we can look again to Crow's useful distinction regarding "theories [that] constitute a plausible world, [in which] it is simply not possible to live." In contrast to the notion of self undergirding old theories of influence, we want a theory that breaks down the "possibilities" of life within the world that is both that of our discourse on influence and that within which we find transmission to be happening. That is, we are following the earlier signs—from Kittler forward—that the self that transmits or to whom transmission happens has accepted a relation to her laptop that is different from that between the self and the instrument, that recognizes the permeability of those categories. As N. Katherine Hayles writes: "What the Turing test proves is that the overlay between the enacted and the represented bodies is no longer a natural inevitability but a contingent production, mediated by a technology that has become so entwined with the production of identity that it can no longer meaningfully be separated from the human subject."[18]

Arguably, what a new study of influence does is to point us toward those ways in which the human subject itself can no longer "meaningfully be separated from the human subject." Such a study would insist on both the discursive and the "real" coexisting within the subjects whom we compare when we discuss influence. And it examines that subject's production not as if it existed within formalism's vacuum or materialism's abundance, but within the strange kinks produced by the expanded technics of reading another (even when that other is oneself).

This is, after all, the register in which we find most of the work being produced these days. It is Rainer referring to herself dancing, then talking, in print; her own body has ceased to be clearly captured by *either* the discursive or real registers, no matter how insistently both ironic and confessional her "voice" is. Recursivity is, finally, the system by which we will find all the tools for talking about influence, from here on. It is production itself that defines the systems of reproduction that are in play, "characters" that tell us about the status of the subject, and families— even the kind that enroll and matriculate together—that tell us how the boundaries between generations are formed. If the "time" of influence is not past, it perhaps demands a tense we don't actually have in English. The next line that Rainer quotes from her performance of *Geriatric With Talking*—"And, I have to tell you that what you are just now witnessing is a state of extreme stage fright. I haven't performed for a while, so I hadn't anticipated what it would be like"—tells us as much about the time of our reading as it does the time of her body.[19] Reconciling those times is incumbent upon us because that "reconciled" time is already the time of production—it is, in other words, already past, but in a way that (*Morituri te salutamus*) salutes us.

NOTES

1 "emission, n." *OED Online.* http://www.oed.com/view/Entry/61198?redirectedFrom=emissions.
2 See Darby English, *How to See a Work of Art in Total Darkness* (Cambridge, MA: MIT Press, 2007).
3 "Authentic, high literature relies upon troping, a turning away not only from the literal but from prior tropes," writes Harold Bloom in "Preface: The Anguish of Contamination," (1997), in *The Anxiety of Influence: A Theory of Poetry*, 2d ed. (New York: Oxford University Press, 1997), xix.
4 In fact one of the striking formal components of Bloom's text is the way he complicates the reader's solitude, as one of the defining characteristics of the age of influence's anxiety. Thus one can read influence as confirming the mode of solitude that Bloom identifies with the reader, but one must also pay attention to the ways in which he qualifies that solitude with questions of sound. The auditory stream of Shakespeare's work "engulfs" the theater spectator, and its relevance to Bloom as a reader—that is, his acknowledgment of the technics undergirding his discussion of influence—is key.
5 Thomas Crow, *Emulation: David, Drouais, and Girodet in the Art of Revolutionary France*, rev. ed. (New Haven, CT: Yale University Press, 2006), 300.
6 Crow, *Emulation*, 303.

7   Any list of such writers would inevitably include the work of Bruno Latour, e.g., *Reassembling the Social: An Introduction to Actor-Network-Theory* (Oxford: Oxford University Press, 2007).

8   Wellbery writes, "The body is the site upon which the various technologies of our culture inscribe themselves, the connecting link to which and from which our medial means of processing, storage, and transmission run." David E. Wellbery, "Foreword," in Friedrich A. Kittler, *Discourse Networks 1800/1900*, trans. Michael Metteer (Stanford, CA: Stanford University Press, 1992), xiv.

9   Bernard Stiegler, *Technics and Time, 1: The Fault of Epimetheus*, trans. Richard Beardsworth and George Collins (Stanford, CA: Stanford University Press, 1998), 1–10.

10 Friedrich A. Kittler, *Discourse Networks 1800/1900*, trans. Michael Metteer (Stanford, CA: Stanford University Press, 1992), 73.

11 Kittler, *Discourse Networks*, 37. Commenting on lines of Anton Reiser's "The Sorrows of Poetry," Kittler locates "the stillbirth of poetry out of the spirit of reading. The pleasant-painful feeling that refuses to become lines of poetry results from the leveling of all signifiers; the feeling traverses the reader and, because he has attained the fluency of speech, can only leave vague generality in its wake" (*Discourse Networks*, 75).

12 In "Technology Romanticized: Friedrich Kittler's *Discourse Networks 1800/1900*," *MLN* 105, no. 3, German Issue (1990), Thomas Sebastian writes, "The success of this 'coercive act of alphabetizing' (30) was not merely initiated by a pedagogical shift to phonetics in High German orthography but rather, according to Kittler, because this measure was associated with the *body* of 'biographical' mothers" (586–87).

13 Kittler, *Discourse Networks*, 63.

14 Stiegler, *Technics and Time*, 68.

15 Yvonne Rainer, "The Aching Body in Dance," *PAJ: A Journal of Performance and Art* 36, no. 1 (2014): 4–5.

16 See the discussions of how Rainer's dance work relates to photography and film in Carrie Lambert-Beatty, *Being Watched: Yvonne Rainer and the 1960s* (Cambridge, MA, and London: October/MIT Books, 2008).

17 Jorge Luis Borges, "Kafka and His Precursors," in *Selected Non-Fictions*, ed. Eliot Weinberger (New York and London: Penguin Books, 2000), 365.

18 N. Katherine Hayles, *How We Became Posthuman: Virtual Bodies in Cybernetics, Literature, and Informatics* (Chicago: University of Chicago Press,), xiii.

19 Rainer, "The Aching Body in Dance," 4–5.

20

Silence / Beat

PAUL D. MILLER, AKA DJ SPOOKY, THAT SUBLIMINAL KID

Sample this:

Over the course of twenty-five centuries, Pythagoras has been a phantom made of fragments drifting over the ages. The fabric or the texture of his being has been made of unattributed remarks, ambiguous observations, specious fabrications, and false citations that later proved to be remarkably coherent but, amusingly enough, were never traced back to him or his original followers. Anything related to what he may or may not have said or done depends heavily on imaginative interpretation. But it's that kind of speculation that makes Pythagoras such an enticing subject for the modern writer.

Pythagoras is one of those towering figures from the dawn of Western philosophy that I had heard about when I was young. Both of my parents were in one way or another involved with academic pursuits, and various books on the history of philosophy lined our family's shelves. Human rights and feminist theory were bandied about at our dinner table, and there was always discussion about the rule of law and the way we could think about advancing the political aspects of the "social good." Philosophy underpins law. Without clear thinking and what the *ars analytica* philosophy provides, you are left with laws with no foundation in the world of ideas. You would be left, in turn, with a dialectic process where something comes from nothing. And we all know everything comes from something—a sample is nothing but a datum set in motion, no beat comes before there's a rhythm.

Dialectics is a method of thinking and interpreting the world of both nature and society; one can think of it as a way of exploring the universe with thought. In ancient Greek culture, Zeno of Elea is generally credited with starting dialectics, but it was Plato in his Dialogues that really gave it the cultural credibility it has today. Dialectics set out from an axiom that everything is in a constant state of change and flux. They are recombinant logics that point to higher truths: a thesis versus its antithesis yields their synthesis.

Let's call it the sample as data set in constant remix, an app in constant update.

Dialectics explain that change and motion involve contradiction and take place through paradoxes in scope and scale. So instead of a smooth, uninterrupted line of progress, we have a line that is interrupted by sudden and explosive periods in which slow, accumulated change (quantitative change) undergoes a rapid acceleration, and in which quantity is transformed into quality.

How does dialectical recursion work in relation to the silence of space and the beat of time? What is the beat of silence, and how, today, is music space?

Time as Silence

Quantitative vs Qualitative: dialectics is the logic of numbers applied to culture.

Dialectics is the logic of contradiction.

Pre-Socratic theory shows us that Pythagoras still occupies a unique place in the pantheon of Thales, Anaximander, Zeno of Elea, and Heraclitus. He is an echo of an echo, a philosopher who left a legacy but not a specific group of writings to hold his school of thought together—except mathematics. That is where his philosophy synchronizes with our modern world: he is, *par excellence*, the philosopher of ratio, proportion, and rationality and their unraveling. At the same time, he was a cult leader who feared the utterance of irrational numbers because, in his cosmology, they ruptured the fabric of space and time.

Word-Count. Fewer Numbers.

There are some things that we can definitively link to Pythagoras, such as his central theorem of $a^2 + b^2 = c^2$. But that's not enough: even that equation comes to us in the form of ambiguous attribution from Pythagoras's followers. This phantasm of history, Pythagoras, may or may not have lived in the sixth century B.C. on the island of Samos in the Aegean Sea; may or may not have lived in Egypt, in Babylon; is alleged to have visited India to learn at the feet of its mathematician savants; and

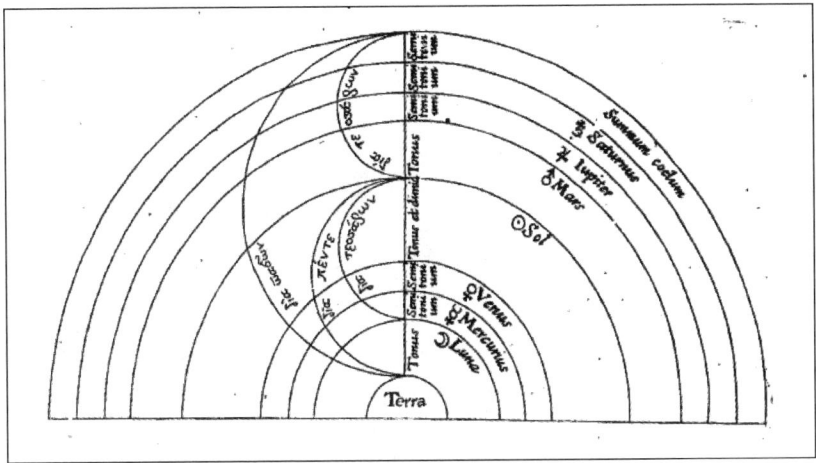

Figure 20.1. Pythagorian intervals and harmonies of the spheres. From Thomas Stanley, *The history of philosophy containing the lives, opinions, actions and discourses of the philosophers of every sect. Illustrated with the effigies of divers of them*, 2d ed. (London: printed for Thomas Bassett, at the George in Fleetstreet, Dorman Newman, at the Kings Arms, and Thomas Cockerill, at the Three Legs in the Poultery, 1687), image 219, p. 539. Accessed through EEBO: Early English Books Online database, copy from Henry E. Huntington Library and Art Gallery, reel position: Wing/1876:01.

allegedly passed away in Metapontum, southern Italy. Pythagoras may have been an indeterminate figure, but the phenomena associated with his investigations into some of the core themes of mathematics linger with us to this day. That is cool.

Many historians have in fact tried to prove the relation between Pythagoras's theorem and Pythagoras the person but have failed miserably. The only relation that historians have been able to solidly trace it to is Euclid—who lived centuries after Pythagoras. Without Euclid's contribution, it would have been impossible to create the basis of algebra and geometry that has led to modern mathematics and our algorithmic, digital culture. Historians have also presented evidence that Pythagoras may actually have traveled to Egypt and then India and learned many important mathematical theories (including Pythagoras's theorem) that were unknown in the West.

A brief search of the Google Play Store yields 300+ apps and hundreds of ebooks on Pythagoras. Ditto for the App Store in

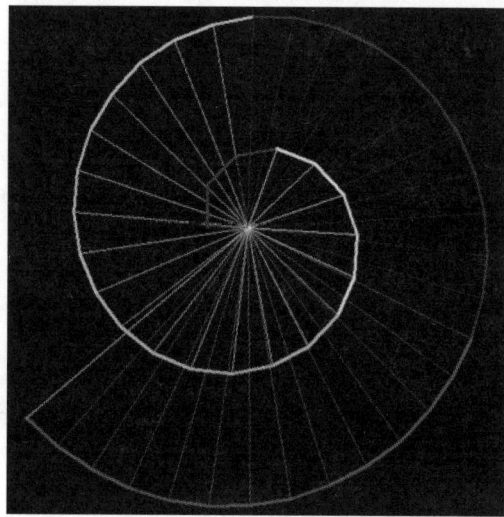

Figure 20.2. The Pythagorean spiral

Apple's iOS. Each of these books and apps are based in a hall of mirrors that leads through space and time to an uncanny situation that recalls an old phrase: the horizon recedes as we approach it. That, too, has Pythagorean implications.

The only way to describe Pythagoras is with twenty-five centuries worth of uncertain information. Yet his observations have led to a wealth of developments in mathematics as diverse as Fermat's Last Theorem and the theory of Hilbert space. With Pythagoras, numbers face an almost insurmountable paradox: they have to breathe hard truth into an empirically inclined cult—a cult that thought numerology would open the way to a true understanding of the underlying rules that govern the universe. And yet I want to figure out the way we think about informal mathematics—the way we count people waiting in line in front of us, or count how many minutes we've been on hold waiting for the electronic voice at the other end to confirm our airline reservation. The way we use Skype and Google Talk without thinking about how much computational power is used to render our voice and image and send it through fiber optic cables, wherever the other end user may be.

These kinds of daily, everyday uses of numerical realism, and the fact that so much data is being generated all the time, everywhere—this is what gives me a Pythagorean pause. In one of the defining films of the late twentieth century, *The Matrix* by Andy and Lana Wachowski, Keanu Reeves acts as a computer programmer living a double life as an underground hacker named Neo.[1] In the beginning of the film, Neo begins to receive cryptic messages on his computer monitor that lead him to a nightclub, where he gets more cryptic messages while listening to hardcore drum 'n' bass (a great track by DJ Hive called "Ultrasonic Sound" that happens to sample one of my favorite bands, Bad Brains). The key theme of Th*e Matrix* is that people live their entire existence in pods, with their brains being fed illusions to stimulate sensory input to give them the illusion of "reality." In one scene, Neo's mentor Morpheus breaks it down by saying simply to Neo that the reality he perceives is actually nothing but a "computer-generated dreamworld . . . a neural interactive simulation": the Matrix . . . by way of techniques derived from René Descartes' evil genius (*deus deceptor*, French *dieu trompeur*); Plato's "Allegory of the Cave"; Jean Baudrillard's concepts of simulacra and simulation; Lewis Carroll's *Alice's Adventures in Wonderland*; Fritz Lang's 1927 masterpiece of German expressionist cinema, *Metropolis*; Hong Kong action films; spaghetti Westerns; dystopian science fiction; M. C. Escher's paintings; and, oh, Pythagoras's ideas of reincarnation and numerology. Through the recombinant logic of a cinema of samples, we are led through one of the best films to be made in the last forty years. The reason I link *The Matrix* and films like Darren Aronofsky's *Pi* (a tale that derives its narrative interplay from the point of view of numbers making up the fabric of reality) to other "neo-noir" dystopian films like *Memento* (Christopher Nolan), *Dark City* (Alex Proyas), and *THX-1138* (George Lucas) is simple: they posit that the world as we know it exists as a series of data sets, a place where "number" and human interpretations of the data of the world are seamless fictions presented to the viewer.

Or as Pythagoras would have said: "All is Number."

Here's another example:

Hip hop mathematics. Imagine this mashup. *The Wizard of Oz*'s Scarecrow versus *Alice in Wonderland*'s Humpty Dumpty. You remember: In *The Wizard of Oz*, the 1939 film based on L. Frank Baum's classic American story (yes, you are not in Kansas anymore, and you are definitely not somewhere over the Pythagorean theorem rainbow . . .), the Scarecrow, the Tin Man, and the Cowardly Lion travel to meet the

Wizard, in search of respectively a brain, a heart, and something akin to courage.[2] In the film, the Scarecrow, after he receives his brain from the wizard, recites the Pythagorean theorem—incorrectly: "the sum of the square roots of any two sides of an isosceles triangle is equal to the square root of the remaining side." Wrong. But interestingly wrong. That's the mashup.

Time as Beat

Is the world of numbers the new intuitive hip hop? A music whose quantized beats echo through the data cloud surrounding twenty-first-century culture? This is just an update of the ancient form of dialectics that I mentioned earlier, but with a better beat.

An example:

The band Kraftwerk released the album *Computer World* (*Computerwelt*) in 1981 with a single that went on to redefine music as most of my generation knew it. With the slogan "It's more fun to compute," one of the main singles on the album was entitled "Numbers" ("Nummern").[3] The lyrics of the song were simply counting numbers in different languages. *Computer World*'s songs, along with those of Kraftwerk's 1977 album, *Trans-Europe Express*, paved the way for much of the music that sits in my record collection to this day. But the main thing that struck me when I first heard these songs, characterized by the eerie, breathtakingly artificial stylization of the vocoder and synthesizers that the band used, was the basic sense that numbers defined what some contemporary philosophers like. One could argue convincingly that Kraftwerk was the first band to really push the idea that numbers were music in a pop culture context.

The rest is history. Hip hop, techno, dub step, you name it—these are all musics that are derived from digital algorithms, and they express the way we count, name, and label the storm of data at the heart of our information economy. According to The RZA, "The truth is mathematically correct, it's never something that belongs to somebody. One plus one is two no matter where you go. Uno, uno still translates to the number two. Mathematics don't lie, so we use mathematics to express our truth more than anything."[4] The RZA is co-founder and producer of one of the most influential hip-hop groups of all time, the Wu-Tang

Figure 20.3. Bar of music

Clan, and has sounded the ideas of Zen Buddhism and the "Universal Language" of the Five-Percent Nation, or the Nation of Gods and Earths (NGE), an organization founded in 1964 in Harlem by a former member of the Nation of Islam named Clarence 13X. The NGE teaches that a "Supreme Mathematics" and "Supreme Alphabet," a set of principles created by Allah, are the keys to understanding humankind's relationship to the universe. The Supreme Mathematics relates numbers to values.[5] The RZA relates numbers and values to sound.

How does Pythagoras resonate? What are the silences of Pythagorean space within the beats of contemporary music, and vice versa?

> Let us now discuss the mental attitude
> The mental must always stay calm
> You must let nothing move you
> Be it good or bad.
>
> But when the mental and I be moved
> There is no longer good or bad
> There just "is."
> When there just "is" you have the power to form and shape. So now witness the wrath of the math.
> Tell me when you ready: I'm ready.
>
> The wrath of the math.[6]

For us denizens of the twenty-first-century info-landscape, the Pythagorean thought that "all is number" describes our sonic world. I could easily imagine that sound bite booming out over a dubstep bass-heavy sound system, time stretched and pitch shifted to match a drum machine

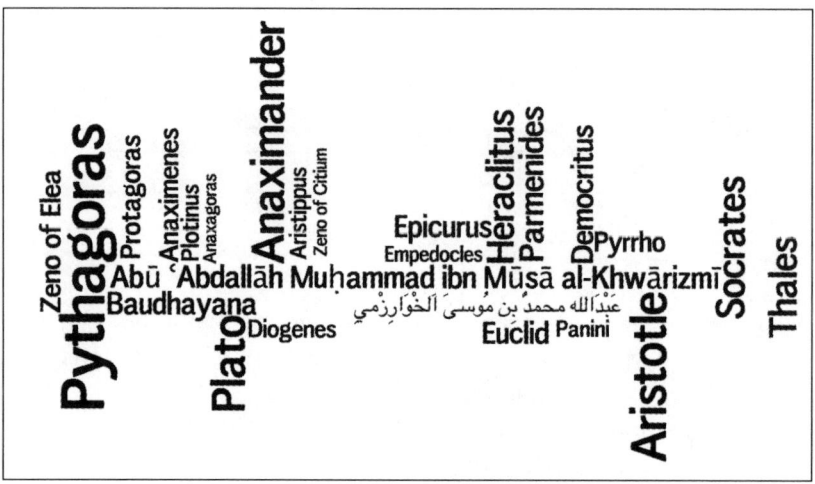

Figure 20.4. Pythagorean mashup

programmed to upload a mix to the Internet in real time so that everyone at the party can access their SoundCloud accounts to listen to the phrase looped over and over. That would be some kind of party. . . .
After all, a sample is nothing but a data set.

NOTES

1. *The Matrix*, directed by Andy Wachowski and Lana Wachowski (1999; Warner Home Video, 1999), DVD.
2. *The Wizard of Oz*, directed by Victor Fleming, with George Cukor, Mervyn LeRoy, Norman Taurog, and King Vidor (Metro-Goldwyn-Mayer [MGM], 1939), film.
3. Kraftwerk, "Numbers," written by Karl Bartos, Ralf Hütter, and Florian Schneider, track 3 of *Computer World*, released May 10, 1981, by EMI and Warner Brothers Records. Studio album, 34:21 min.
4. The RZA on the television show *The Colbert Report*, Wednesday October 14, 2009, episode 05133, "The RZA," online as of February 2014 at http://www.colbertnation.com/the-colbert-report-videos/252715/october-14-2009/the-rza. See also The RZA, with Chris Norris, *The Tao of Wu* (New York: Riverhead Trade/Penguin, 2009).
5. See the website of the Nation of Gods and Earths, *The 5% Network*, at http://www.allahsnation.net, updated by Master Allah Rule.
6. Jeru the Damaja, "Wrath of the Math," written by K. J. Davis and C. Martin, track 1 of *Wrath of the Math*, released October 15, 1996, by PayDay/FFRR/PolyGram Records. Studio album, 51:47 min.

TIME STUDIES

A Bibliographical Reading List

The following entries comprise a preliminary reading list in contemporary time studies. The list does not contain titles of journals or magazines devoted to time studies and consists primarily of monographs on the subject; it does not reproduce all of the sources quoted in the essays in this volume, so readers should also consult the essay citations as well. The following list is not exhaustive but provides an excellent starting point for anyone wishing to pursue research in this burgeoning field.

* * *

0 to 60: The Experience of Time through Contemporary Art [exhibition catalogue]. Raleigh: North Carolina Museum of Art, 2013.
Abbott, Andrew. *Time Matters: On Theory and Method*. Chicago: University of Chicago Press, 2001.
Agacinski, Sylviane. *Time Passing: Modernity and Nostalgia*. Translated by Jody Gladding. New York: Columbia University Press, 2003.
Agamben, Giorgio. "What Is the Contemporary?" In *"What Is an Apparatus?" and Other Essays*, 39–54. Translated by David Kishik and Stefan Pedatella. Stanford, CA: Stanford University Press, 2009.
Alliez, Eric. *Capital Times: Tales from the Conquest of Time*. Minneapolis: University of Minnesota Press, 1996.
Andreou, Chrisoula, and Mark D. White, eds. *The Thief of Time: Philosophical Essays on Procrastination*. New York: Oxford University Press, 2012.
Apter, Emily. "'Women's Time' in Theory." *Differences* 21, no. 1 (2010): 1–18.
Aravamudan, Srinivas. "The Return of Anachronism." *Modern Language Quarterly* 62, no. 4 (2001): 331–53.
Arntzenius, Frank. *Space, Time, and Stuff*. Oxford: Oxford University Press, 2012.
Artempo: Where Time Becomes Art [exhibition catalogue]. Venice: Palazzo Foruny, 2007.
Ast, Olga. *Infinite Instances: Studies and Images of Time*. Edited by Buzz Poole and Eli Stockwell. Brooklyn, NY: Mark Batty, 2011.

Augustine. *Confessions*. Translated by R. S. Pine-Coffin. Harmondsworth: Penguin Books, 1961.
Aveni, Anthony F. *Empires of Time: Calendars, Clocks, and Cultures*. New York: Basic Books, 1989.
Bachelard, Gaston. *Dialectic of Duration*. Translated by Mary McAllester Jones. Manchester, UK: Clinamen Press, 2000.
Bardon, Adrian. *A Brief History of the Philosophy of Time*. New York: Oxford University Press, 2013.
Bardon, Adrian, ed. *The Future of the Philosophy of Time*. New York: Routledge, 2012.
Barker, Timothy Scott. *Time and the Digital: Connecting Technology, Aesthetics, and a Process Philosophy of Time*. Hanover, NH: Dartmouth College Press, 2012.
Barrow, Mark V., Jr. *Nature's Ghosts: Confronting Extinction from the Age of Jefferson to the Age of Ecology*. Chicago: University of Chicago Press, 2009.
Baucom, Ian. *Specters of the Atlantic: Finance Capital, Slavery, and the Philosophy of History*. Durham, NC, and London: Duke University Press, 2005.
Baudrillard, Jean. *Simulacra and Simulation*. Translated by Sheila Faria Glaser. Ann Arbor: University of Michigan Press, 1994. Originally published as *Simulacres et Simulation*, Éditions Galilée, 1981.
Bender, John, and David E. Wellbery, eds. *Chronotypes: The Construction of Time*. Stanford, CA: Stanford University Press, 1991.
Benjamin, Walter. "Theses on the Philosophy of History" (1940). In *Illuminations: Essays and Reflections*, 253–64. Edited by Hannah Arendt. Translated by Harry Zohn. New York: Schocken Books, 1968.
Berardi, Franco Bifo. *After the Future*. Edited by Gary Genosko and Nicholas Thoburn. Oakland, CA: AK Press, 2011.
Bergson, Henri. *Time and Free Will: An Essay on the Immediate Data of Consciousness*. Translated by F. L. Pogson. 3d ed. London: George Allen & Company, 1913. Reprint, New York: Dover, 2001.
Berlant, Lauren. *Cruel Optimism*. Durham, NC: Duke University Press, 2011.
Blackburn, Robin. "Finance and the Fourth Dimension." *New Left Review* 39 (2006): 39–70.
Bode, Christoph, Jeffrey Kranhold, and Rainer Dietrich. *Future Narratives: Theory, Poetics, and Media-Historical Moment*. Berlin: Walter de Gruyter, 2013.
Borel, Émile. *Space and Time*. 1960. Reprint, New York: Dover, 2003.
Borst, Arno. *The Ordering of Time: From the Ancient Computus to the Modern Computer*. Translated by Andrew Winnard. Chicago: University of Chicago Press, 1994.
Bourriaud, Nicolas, ed. *Altermodern: Tate Triennial*. London: Tate Publishing, 2009.
Boym, Svetlana. *The Future of Nostalgia*. New York: Basic Books, 2002.
Brand, Stewart. *The Clock of the Long Now: Time and Responsibility*. New York: Basic Books, 2000.
Buchli, Victor, and Gavin Lucas. *Archaeologies of the Contemporary Past*. New York: Routledge, 2001.
Butterfield, Jeremy, ed. *The Arguments of Time*. Oxford: Oxford University Press, 1999.

Callender, Craig, ed. *The Oxford Handbook of Philosophy of Time.* Oxford: Oxford University Press, 2011.
Canales, Jimena. *A Tenth of a Second: A History.* Chicago: University of Chicago Press, 2009.
Canales, Jimena. *The Physicist and the Philosopher: Einstein, Bergson, and the Debate that Changed Our Understanding of Time.* Princeton, NJ: Princeton University Press, 2015.
Cardwell, Donald. *Wheels, Clocks, and Rockets: A History of Technology.* New York: W. W. Norton, 2001.
Carr, David. *Time, Narrative, and History.* Bloomington: Indiana University Press, 1986.
Chakrabarty, Dipesh. *Provincializing Europe: Postcolonial Thought and Historical Difference.* Princeton, NJ: Princeton University Press, 2000.
Cioran, E. M. *The Fall into Time.* Translated by Richard Howard. Chicago: Quadrangle Books, 1964.
Corfield, Penelope J. *Time and the Shape of History.* New Haven, CT, and London: Yale University Press, 2007.
Crary, Jonathan. *24/7: Late Capitalism and the Ends of Sleep.* London and New York: Verso, 2013.
Currie, Mark. *About Time: Narrative, Fiction and the Philosophy of Time.* Edinburgh: Edinburgh University Press, 2007.
Currie, Mark. *The Unexpected: Narrative Temporality and the Philosophy of Surprise.* Edinburgh: Edinburgh University Press, 2013.
Dastur, Françoise. *Telling Time: Sketch of a Phenomenological Chronology.* Translated by Edward Bullard. London: Athlone Press, 2000. Originally published as *Dire le temps*, Encre Marine, 1994.
Davies, Ben, and Jana Funke, eds. *Sex, Gender and Time in Fiction and Culture.* Basingstoke: Palgrave Macmillan, 2011.
Davies, Paul. *About Time: Einstein's Unfinished Revolution.* New York: Simon & Schuster, 1995.
De Landa, Manuel. *A Thousand Years of Nonlinear History.* Cambridge, MA: MIT Press, 2000.
Deleuze, Gilles. *The Time-Image.* Translated by Hugh Tomlinson and Robert Galeta. Minneapolis: Athlone Press/University of Minnesota Press, 1989.
Derrida, Jacques. *Given Time: I. Counterfeit Money.* Translated by Peggy Kamuf. Chicago: University of Chicago Press, 1992.
Derrida, Jacques. *Archive Fever: A Freudian Impression.* Translated by Eric Prenowitz. Chicago: University of Chicago Press, 1996.
Doane, Mary Ann. *The Emergence of Cinematic Time: Modernity, Contingency, the Archive.* Cambridge, MA: Harvard University Press, 2002.
Dolev, Yuval. *Time and Realism: Metaphysical and Antimetaphysical Perspectives.* Cambridge, MA: MIT Press, 2007.
Duffy, Enda. *The Speed Handbook: Velocity, Pleasure, Modernism.* Durham, NC: Duke University Press, 2009.
Dunne, J. W. *An Experiment with Time.* London: Faber and Faber, 1939.

Eco, Umberto. *Travels in Hyperreality: Essays.* Translated by William Weaver. New York: Harcourt, 1986. Originally published as *Il costume di casa* [*Faith in Fakes*], Bompiani, 1973.

Elias, Norbert. *Time: An Essay.* Translated by Edmund Jephcott. Oxford: Blackwell, 1992.

Ermath, Elizabeth Deeds. *Sequel to History: Postmodernism and the Crisis of Representational Time.* Princeton, NJ: Princeton University Press, 1992.

Eshel, Amir. *Futurity: Contemporary Literature and the Quest for the Past.* Chicago: University of Chicago Press, 2013.

Fabian, Johannes. *Time and the Other: How Anthropology Makes Its Object.* New York: Columbia University Press, 1983.

Felski, Rita. *Doing Time: Feminist Theory and Postmodern Culture.* New York: New York University Press, 2000.

Ferguson, Niall, ed. *Virtual History: Alternatives and Counterfactuals.* New York: Basic Books, 1999.

Fleming, Jennie A. *Trajectories: Marking Time in Contemporary Art.* College Park, MD: Art Gallery, University of Maryland, 2008.

Francese, Joseph. *Narrating Postmodern Time and Space.* Albany: State University of New York Press, 1997.

Fraser, J. T. *Of Time, Passion, and Knowledge, Reflections on the Strategy of Existence.* 2d ed. Princeton, NJ: Princeton University Press, 1990.

Freeman, Elizabeth. *Time Binds: Queer Temporalities, Queer Histories.* Durham, NC: Duke University Press, 2010.

Fritzsche, Peter. *Stranded in the Present: Modern Time and the Melancholy of History.* Cambridge, MA: Harvard University Press, 2010.

Frow, John. *Time and Commodity Culture: Essays in Cultural Theory and Postmodernity.* Oxford: Clarendon Press, 1997.

Galison, Peter. *Einstein's Clocks and Poincaré's Maps: Empires of Time.* New York: W. W. Norton, 2003.

Gell, Alfred. *The Anthropology of Time: Cultural Constructions of Temporal Maps and Images.* London: Bloomsbury Academic, 2001.

Gill, Carolyn Bailey, ed. *Time and the Image.* Manchester, UK, and New York: Manchester University Press, 2000.

Gillick, Liam. "Prevision. Should the Future Help the Past?" Paris: ARC Musée d'art Moderne de la Ville de Paris, 1998. Available online at http://www.liamgillick.info/home/texts/prevision.

Gleick, James. *Faster: The Acceleration of Just About Everything.* New York: Vintage, 1999.

Gould, Stephen Jay. *Time's Arrow, Time's Cycle: Myth and Metaphor in the Discovery of Geological Time.* Cambridge, MA: Harvard University Press, 1987.

Goulimari, Pelagia, ed. *Postmodernism: What Moment?* Manchester, UK: Manchester University Press, 2011.

Griffiths, Jay. *A Sideways Look at Time.* New York: Penguin Putnam, 1999.

Groom, Amelia, ed. *Time.* London: Whitechapel Gallery, in conjunction with Cambridge, MA: MIT Press, 2013.

Grosz, Elizabeth. *The Nick of Time: Politics, Evolution, and the Untimely*. Durham, NC: Duke University Press, 2004.

Grosz, Elizabeth. *Time Travels: Feminism, Nature, Power*. Durham, NC: Duke University Press, 2005.

Grosz, Elizabeth, ed. *Becomings: Explorations in Time, Memory, and Futures*. Ithaca, NY, and London: Cornell University Press, 1999.

Gumbrecht, Hans Ulrich. *After 1945: Latency as Origin of the Present*. Stanford, CA: Stanford University Press, 2013.

Gurvitch, Georges. *The Spectrum of Social Time*. Translated by Myrtle Korenbaum. Dordrecht: D. Reidel, 1964.

Halberstam, Judith. *In a Queer Time and Place: Transgender Bodies, Subcultural Lives*. New York: New York University Press, 2005.

Hartog, François. *Régimes d'historicité: Présentisme et expérience du temps*. Paris: POINTS, 2003.

Harvey, David. *The Condition of Postmodernity: An Enquiry into the Origins of Cultural Change*. Oxford: Basil Blackwell, 1989.

Heidegger, Martin. *The Concept of Time*. Translated by William McNeill. Oxford: Basil Blackwell, 1992. Originally published as *Der Begriff der Zeit: Vortrag vor'der Marburger Theologenschaft Juli 1924*, Max Niemeyer Verlag, 1989.

Heidegger, Martin. *Being and Time*. Translated by John Macquarrie and Edward Robinson. New York: Harper Perennial Classics, 2008. Originally published as *Sein und Zeit*, Max Niemeyer Verlag, 1927.

Heise, Ursula K. *Chronoschisms: Time, Narrative, and Postmodernism*. New York: Cambridge University Press, 1997.

Hellekson, Karen. *The Alternate History: Refiguring Historical Time*. Kent, OH: Kent State University Press, 2001.

Hobsbawm, Eric, and Terence Ranger, eds. *The Invention of Tradition*. Cambridge, UK: Cambridge University Press, 1983.

Hochschild, Arlie Russell. *The Time Bind: When Work Becomes Home and Home Becomes Work*. New York: Holt, 1997.

Hochschild, Arlie Russell, and Anne Machung. *The Second Shift: Working Families and the Revolution at Home*. New York: Penguin, 1989.

Holford-Strevens, Leofranc. *The History of Time: A Very Short Introduction*. Oxford: Oxford University Press, 2005.

Hoy, David Couzens. *The Time of Our Lives: A Critical History of Temporality*. Cambridge, MA: MIT Press, 2009.

Husserl, Edmund. *The Phenomenology of Internal Time-Consciousness*. Edited by Martin Heidegger. Translated by James S. Churchill. Bloomington: Indiana University Press, 1964.

Husserl, Edmund. *On the Phenomenology of the Consciousness of Internal Time (1893–1917)* [*Collected Works*, vol. 4]. Translated by John Barnett Brough. Boston: Kluwer Academic Publishers, 1991.

Huyssen, Andreas. *Twilight Memories: Marking Time in a Culture of Amnesia.* New York: Routledge, 1995.
Jameson, Fredric. *Postmodernism, or, the Cultural Logic of Late Capitalism.* Durham, NC: Duke University Press, 1991.
Jameson, Fredric. *The Seeds of Time.* New York: Columbia University Press, 1994.
Jameson, Fredric. *A Singular Modernity: Essay on the Ontology of the Present.* London: Verso, 2002.
Jameson, Fredric. "The End of Temporality." *Critical Inquiry* 29, no. 4 (2003): 695–718.
Jameson, Fredric. *Archaeologies of the Future: The Desire Called Utopia and Other Science Fictions.* London: Verso, 2005.
Jameson, Fredric. *Valences of the Dialectic.* London and New York: Verso, 2010.
Jameson, Fredric. "The Aesthetics of Singularity." *New Left Review* 92 (2015): 101–32.
Jencks, Charles. *The Language of Post-Modern Architecture.* New York: Rizzoli, 1977.
Keightley, Emily, ed. *Time, Media and Modernity.* London: Palgrave Macmillan, 2012.
Kern, Stephen. *The Culture of Time and Space, 1880–1918.* Cambridge, MA: Harvard University Press, 1983.
Koepnick, Lutz. *On Slowness: Toward an Aesthetic of the Contemporary.* New York: Columbia University Press, 2014.
Koselleck, Reinhart. *Futures Past: On the Semantics of Historical Time.* Translated by Keith Tribe. Cambridge, MA: MIT Press, 1985.
Koselleck, Reinhart. *The Practice of Conceptual History: Timing History, Spacing Concepts.* Translated by Todd Samuel Presner. Stanford, CA: Stanford University Press, 2002.
Koshar, Rudy, ed. *Histories of Leisure.* London: Berg Publishers, 2002.
Kwinter, Sanford. *Architectures of Time: Toward a Theory of the Event in Modernist Culture.* Cambridge, MA: MIT Press, 2001.
Landes, David S. *Revolution in Time: Clocks and the Making of the Modern World.* Revised edition. Cambridge, MA: Belknap/Harvard University Press, 2000.
Le Poidevin, Robin. *Travels in Four Dimensions: The Enigmas of Space and Time.* New York: Oxford University Press, 2003.
Le Poidevin, Robin, and Murray MacBeath, eds. *The Philosophy of Time.* New York: Oxford University Press, 1993.
Lee, Pamela M. *Chronophobia: On Time in the Art of the 1960s.* Cambridge, MA: MIT Press, 2004.
Levinas, Emmanuel. *Time and the Other.* Translated by Richard A. Cohen. Pittsburgh: Duquesne University Press, 1990.
Levinas, Emmanuel. *God, Death, and Time.* Translated by Bettina Bergo. Stanford, CA: Stanford University Press, 2000. Originally published as *Dieu, la mort et le temps*, Grasset & Fasquelle, 1993.
Lim, Bliss Cua. *Translating Time: Cinema, the Fantastic, and Temporal Critique.* Durham, NC: Duke University Press, 2009.
Lloyd, David. *Irish Times: Temporalities of Modernity.* Dublin: Field Day Publications, 2008.

Lorenz, Chris, and Berber Bevernage, eds. *Breaking Up Time: Negotiating the Borders between Present, Past and Future*. Göttingen and Bristol, CT: Vanderhoeck & Ruprecht, 2013.
Lucas, Gavin. *The Archaeology of Time*. New York: Routledge, 2005.
Lucas, J. R. *A Treatise on Time and Space*. London: Methuen, 1973.
Luciano, Dana. *Arranging Grief: Sacred Time and the Body in Nineteenth-Century America*. New York: New York University Press, 2007.
Luckhurst, Roger, and Peter Marks, eds. *Literature and the Contemporary: Fictions and Theories of the Present*. New York: Longman, 1999.
Lukacher, Ned. *Time-Fetishes: The Secret History of Eternal Recurrence*. Durham, NC: Duke University Press, 1998.
Lütticken, Sven. *History in Motion: Time in the Age of the Moving Image*. Berlin: Sternberg Press, 2013.
M/M (Paris). *Live Recorded Delay: An Archive of Il Tempo Del Postino*. Berlin: Sternberg Press, 2008. Part of M/M's installation at "theanyspacewhatever" exhibition at the Guggenheim Museum, New York, October 24, 2008–January 7, 2009.
Ma, Jean. *Melancholy Drift: Marking Time in Chinese Cinema*. Hong Kong: Hong Kong University Press, 2010.
Manning, Erin. *Relationscapes: Movement, Art, Philosophy*. Cambridge, MA: MIT Press, 2009.
Massumi, Brian. *Semblance and Event: Activist Philosophy and the Occurrent Arts*. London and Cambridge, MA: MIT Press, 2011.
Maudlin, Tim. *Philosophy of Physics: Space and Time*. Princeton, NJ: Princeton University Press, 2012.
May, Jon, and Nigel Thrift, eds. *Timespace: Geographies of Temporality*. London: Routledge, 2001.
McCallum, E. L., and Mikko Tuhkanen, eds. *Queer Times, Queer Becomings*. Albany: State University of New York Press, 2011.
McCrossen, Alexis. *Marking Modern Times: A History of Clocks, Watches, and Other Timekeepers in American Life*. Chicago: University of Chicago Press, 2013.
McNeill, William H. *Keeping Together in Time: Dance and Drill in Human History*. Cambridge, MA: Harvard University Press, 1995.
Meifert-Menhard, Felicitas. *Playing the Text, Performing the Future: Future Narratives in Print and Digiture*. Berlin: Walter de Gruyter, 2013.
Meillassoux, Quentin. *After Finitude: An Essay on the Necessity of Contingency*. Translated by Ray Brassier. New York and London: Continuum Books, 2008.
Mendilow, A. A. *Time and the Novel*. New York: Humanities Press, 1972.
Meyerhoff, Hans. *Time in Literature*. Berkeley: University of California Press, 1955.
Miller, Tyrus. *Time Images: Alternative Temporalities in Twentieth-Century Theory, Literature, and Art*. Newcastle upon Tyne: Cambridge Scholars Publishing, 2009.
Miller, Tyrus, ed. *Given World and Time: Temporalities in Context*. Budapest: Central European University Press, 2008.

Moxey, Keith. *Visual Time: The Image in History.* Durham, NC: Duke University Press, 2013.
Muecke, Stephen. *Ancient and Modern: Time, Culture and Indigenous Philosophy.* Sidney: University of New South Wales Press, 2004.
Muñoz, José Esteban. *Cruising Utopia: The Then and There of Queer Futurity.* New York: New York University Press, 2009.
Nealon, Jeffrey T. *Post-Postmodernism, or, the Cultural Logic of Just-in-Time Capitalism.* Stanford, CA: Stanford University Press, 2012.
Newman, Karen, Jay Clayton, and Marianne Hirsch, eds. *Time and the Literary.* New York: Routledge, 2002.
Noys, Benjamin. *Malign Velocities: Accelerationism and Capitalism.* Washington, DC: Zero Books, 2014.
Osborne, Peter. *The Politics of Time: Modernity and Avant-Garde.* London: Verso, 1995.
Osborne, Peter. *Anywhere or Not at All: Philosophy of Contemporary Art.* New York: Verso, 2013.
Postone, Moishe. *Time, Labor, and Social Domination: A Reinterpretation of Marx's Critical Theory.* Cambridge, UK: Cambridge University Press, 1993.
Price, Huw. *Time's Arrow and Archimedes' Point: New Directions for the Physics of Time.* Oxford: Oxford University Press, 1996.
Quinones, Ricardo J. *The Renaissance Discovery of Time.* Cambridge, MA: Harvard University Press, 1972.
Radstone, Susannah. *The Sexual Politics of Time: Confession, Nostalgia, Memory.* Abingdon, Oxon: Routledge, 2007.
Reynolds, Simon. *Retromania: Pop Culture's Addiction to Its Own Past.* New York: Faber and Faber, 2011.
Richards, E. G. *Mapping Time: The Calendar and Its History.* Revised edition. New York: Oxford University Press, 2000.
Richardson, Brian, ed. *Narrative Dynamics: Essays on Time, Plot, Closure, and Frames.* Columbus: Ohio State University Press, 2002.
Ricoeur, Paul. *Time and Narrative.* Translated by Kathleen McLaughlin, Kathleen Blamey, and David Pellauer. 3 vols. Chicago: University of Chicago Press, 1984–1988.
Ricoeur, Paul. *Memory, History, Forgetting.* Translated by Kathleen Blamey and David Pellauer. Chicago: University of Chicago Press, 2004.
Ridderbos, Katinka, ed. *Time: The Darwin College Lectures.* Cambridge, UK: Cambridge University Press, 2002.
Roediger, David R., and Philip S. Foner. *Our Own Time: A History of American Labor and the Working Day.* New York: Verso, 1989.
Rosa, Hartmut. *Social Acceleration: A New Theory of Modernity.* New York: Columbia University Press, 2013.
Rosen, Philip. *Change Mummified: Cinema, Historicity, Theory.* Minneapolis: University of Minnesota Press, 2001.
Rosenberg, Daniel, and Anthony Grafton. *Cartographies of Time: A History of the Timeline.* Princeton, NJ: Princeton Architectural Press, 2010.

Rosenberg, Daniel, and Susan Harding, eds. *Histories of the Future*. Durham, NC: Duke University Press, 2005.
Ross, Christine. *The Past Is the Present; It's the Future Too: The Temporal Turn in Contemporary Art*. New York: Continuum, 2012.
Rossum, Gerhard Dohrn-van. *History of the Hour: Clocks and Modern Temporal Orders*. Translated by Thomas Dunlap. Chicago: University of Chicago Press, 1996.
Rüsen, Jörn, ed. *Time and History: The Variety of Cultures*. New York: Berghahn Books, 2007.
Said, Edward W. *Beginnings: Intention and Method*. New York: Columbia University Press, 1985.
Sandison, Alan, and Robert Dingley, eds. *Histories of the Future: Studies in Fact, Fantasy and Science Fiction*. London: Palgrave, 2000.
Schall, Jane, ed. *Tempus Fugit: Time Flies*. Seattle: University of Washington Press, 2001.
Schenk, Sabine. *Running and Clicking: Future Narratives in Film*. Berlin: Walter de Gruyter, 2013.
Schuppli, Madeleine, Claudia Jolles, Felicity Lunn, and Philippe Pirotte, eds. *Yesterday Will Be Better: Mit der Erinnerung in die Zukunft/Taking Memory into the Future*. Bielefeld: Kerber/Aargauer Kunsthaus Aarau, 2010.
Scott, David. *Omens of Adversity: Tragedy, Time, Memory, Justice*. Durham, NC: Duke University Press, 2014.
Sharma, Sarah. *In the Meantime: Temporality and Cultural Politics*. Durham, NC: Duke University Press, 2014.
Shaviro, Steven. *No Speed Limit: Three Essays on Accelerationism*. Minneapolis: University of Minnesota Press, 2015.
Shelton, Beth Anne. *Women, Men, and Time: Gender Differences in Paid Work, Housework and Leisure*. Westport, CT: Greenwood Press, 1992.
Singles, Kathleen. *Alternate History: Playing with Contingency and Necessity*. Berlin: Walter de Gruyter, 2013.
Smith, Quentin, and L. Nathan Oaklander. *Time, Change and Freedom: An Introduction to Metaphysics*. New York: Routledge, 1995.
Smith, Terry. *What is Contemporary Art?* Chicago: University of Chicago Press, 2009.
Smith, Terry, Okwui Enwezor, and Nancy Condee, eds. *Antinomies of Art and Culture: Modernity, Postmodernity, Contemporaneity*. Durham, NC, and London: Duke University Press, 2008.
Stewart, Garrett. *Framed Time: Toward a Postfilmic Cinema*. Chicago: University of Chicago Press, 2007.
Stiegler, Bernard. *Technics and Time, 1: The Fault of Epimetheus*. Translated by Richard Beardsworth and George Collins. Stanford, CA: Stanford University Press, 1998.
Stiegler, Bernard. *Technics and Time, 2: Disorientation*. Translated by Stephen Barker. Stanford, CA: Stanford University Press, 2008.
Stiegler, Bernard. *Technics and Time, 3: Cinematic Time and the Question of Malaise*. Translated by Stephen Barker. Stanford, CA: Stanford University Press, 2010.

Strogatz, Steven H. *Sync: How Order Emerges from Chaos in the Universe, Nature, and Daily Life*. New York: Hachette Books, 2004.
Terdiman, Richard. *Present Past: Modernity and the Memory Crisis*. Ithaca, NY: Cornell University Press, 1993.
Thompson, E. P. *The Making of the English Working Class*. Harmondsworth: Penguin, 1968.
Toulmin, Stephen, and June Goodfield. *The Discovery of Time*. Chicago: University of Chicago Press, 1965.
Virilio, Paul. *Speed and Politics*. Translated by Mark Polizzotti. New edition. Los Angeles: Semiotext(e), 2006.
Virilio, Paul. *The Futurism of the Instant: Stop-Eject*. Translated by Julie Rose. Cambridge, UK: Polity Press, 2010.
Virilio, Paul. *Lost Dimension*. Translated by Daniel Moshenberg. New edition. Los Angeles: Semiotext(e), 2012.
Weheliye, Alexander G. *Phonographies: Grooves in Sonic Afro-Modernity*. Durham, NC: Duke University Press, 2005.
West-Pavlov, Russell. *Temporalities*. New York: Routledge, 2013.
White, Hayden. *Tropics of Discourse: Essays in Cultural Criticism*. Baltimore: Johns Hopkins University Press, 1978.
White, Hayden. *The Content of the Form: Narrative Discourse and Historical Representation*. Baltimore: Johns Hopkins University Press, 1990.
White, Hayden. *Figural Realism: Studies in the Mimesis Effect*. Baltimore: Johns Hopkins University Press, 2000.
Wilcox, Donald J. *The Measure of Times Past: Pre-Newtonian Chronologies and the Rhetoric of Relative Time*. Chicago: University of Chicago Press, 1987.
Williams, Evan Calder. *Combined and Uneven Apocalypse: Luciferian Marxism*. London: Zero Books, 2011.
Wittenberg, David. *Time Travel: The Popular Philosophy of Narrative*. New York: Fordham University Press, 2013.
Wittmann, Marc, and Virginie van Wassenhove, eds. "The Experience of Time: Neural Mechanisms and the Interplay of Emotion, Cognition and Embodiment." Special issue, *Philosophical Transactions of the Royal Society of London, Series B* 364, no. 1525 (2009).
Wood, David. *The Deconstruction of Time*. Evanston, IL: Northwestern University Press, 2001.
Wood, David. *Time after Time*. Bloomington and Indianapolis: Indiana University Press, 2007.
Zerubavel, Eviatar. *Hidden Rhythms: Schedules and Calendars in Social Life*. Chicago: University of Chicago Press, 1981.
Zielinski, Siegfried. *Deep Time of the Media: Toward an Archeology of Hearing and Seeing by Technical Means*. Translated by Gloria Custance. Cambridge, MA: MIT Press, 2006.

ABOUT THE CONTRIBUTORS

Aubrey Anable is Assistant Professor of Film Studies in the School for Studies in Art and Culture at Carleton University. Her articles have appeared in the journals *Television & New Media*, *Mediascape*, and *Ada: A Journal of Gender, New Media, and Technology*, and the *Social Text Blog*. She is currently completing a book manuscript, *Playing with Feelings: Video Games and Affect*, that examines video games as nodes in everyday affective systems and as crucial sites through which to read the contemporary structure of feeling emerging from and alongside the last sixty-plus years of computerized living.

Ben Anderson is Reader in Human Geography at Durham University (Department of Geography). His monograph on theories of affect—*Encountering Affect: Capacities, Apparatuses, Conditions*—was published in 2014. Supported by a 2013 Phillip Leverhulme Prize, he is currently conducting a genealogy of the government of emergencies in the UK that focuses on the birth of the emergency state and the invention and formalization of ordinary techniques for governing emergencies.

Joel Burges is Assistant Professor of English at the University of Rochester, where he is also affiliated with Film and Media Studies, Digital Media Studies, the Graduate Program in the Digital Humanities, and the Graduate Program in Visual and Cultural Studies. He was a 2014–2015 External Faculty Fellow at the Susan and Donald Newhouse Center for the Humanities at Wellesley College. Burges is the author of essays and reviews that have appeared in *New German Critique*, *Post45*, *Twentieth-Century Literature*, and *Cinema Journal Teaching Dossier*. He is completing a book entitled *Out of Sync & Out of Work: Mediating Time in the Culture of Obsolescence, 1973–Present*. He is also collaborating on a digital annotation tool for moving-image media, which will be published in 2016. His next book is tentatively entitled *Literature after TV*.

Jimena Canales is the Thomas M. Siebel Chair in the History of Science at the University of Illinois at Urbana-Champaign. She is the author of *A Tenth of a Second: A History* (2011), *The Physicist and the Philosopher: Einstein, Bergson, and the Debate That Changed Our Understanding of Time* (2015), and numerous scholarly and journalistic texts on the history of modernity, focusing primarily on science and technology. Her publications have appeared in specialized journals (*Isis, Science in Context, History of Science, British Journal for the History of Science*, and *MLN*, among others) and in visual, film, and media studies periodicals such as *Architectural History, Journal of Visual Culture, Thresholds, Aperture*, and *Artforum*.

Mark Currie is Professor of Contemporary Literature at Queen Mary University of London. His research is focused on theories of narrative and culture, particularly in relation to time. He is the author of *Postmodern Narrative Theory* (1998; 2d ed. 2011), *Difference* (2004), *About Time: Narrative Fiction and the Philosophy of Time* (2007), *The Unexpected: Narrative Temporality and the Philosophy of Surprise* (2013), and *The Invention of Deconstruction* (2013). His recent work is focused on the relation between fictional narrative and philosophical writings about time, and more generally, on questions of futurity in intellectual history. He is currently writing a book called *On Uncertainty*, which aims to explore concepts of uncertainty in the physical and social sciences in relation to questions about novelty and value in literature.

Amy J. Elias is Professor of English at the University of Tennessee, Knoxville. She has published numerous articles, book chapters, and journal special issues concerning contemporary literature, media, cultural theory, narrative theory, and history/time studies, and her books include *Sublime Desire: History and Post-1960s Fiction* (2001), which won the Perkins Award from the International Society for the Study of Narrative, and *The Planetary Turn: Relationality and Geoaesthetics in the 21st Century* (2015), co-edited with Christian Moraru. She is the founding president of A.S.A.P.: The Association for the Study of the Arts of the Present and founding co-editor of its scholarly journal, *ASAP/Journal*. She is completing a monograph titled *Dialogue at the End of the World*.

Elizabeth Freeman is Professor of English at the University of California, Davis, and co-editor of *GLQ: A Journal of Lesbian and Gay Studies*. She is the author of *Time Binds: Queer Temporalities, Queer Histories* (2010) and *The Wedding Complex: Forms of Belonging in Modern American Culture* (2002), as well as guest editor of a special issue of *GLQ*, "Queer Temporalities" (2007). She is currently working on a manuscript tentatively titled "It Goes without Saying: Sense-Methods in the United States's Very Long Nineteenth Century."

Jared Gardner is Professor of English and Film Studies at the Ohio State University, where he also directs the Popular Culture Studies program. He is the author of *Master Plots: Race and the Founding of an American Literature* (1998), *Projections: Comics and the History of 21st-Century Storytelling* (2012), and *The Rise and Fall of Early American Magazine Culture* (2012).

Rachel Haidu is Associate Professor in the Department of Art and Art History and Director of the Graduate Program in Visual and Cultural Studies at the University of Rochester. She is the author of *The Absence of Work: Marcel Broodthaers 1964–1976* (2010) and numerous essays on artists such as Chantal Akerman, James Coleman, Sharon Hayes, Thomas Hirschhorn, Sol LeWitt, Yvonne Rainer, and Gerhard Richter. Her book-in-progress, entitled *The Knot of Influence*, proposes new models of artistic influence with particular attention to historiographic concerns and the influx of performance and technologies of reproduction in contemporary art.

Stanley Hauerwas is Gilbert T. Rowe Emeritus Professor of Christian Ethics at Duke Divinity School. He has written more than forty theological books on topics ranging from virtue ethics, mental illness, and disability, to narrative and ordinary language philosophy. His memoir, *Hannah's Child: A Theologian's Memoir* (2010) received widespread critical praise. His more recent publications include *Approaching the End: Eschatological Reflections on Church, Politics, and Life* (2013) and *Working with Words: On Learning to Speak Christian* (2011).

Ursula K. Heise is the Marcia H. Howard Professor in the Department of English and at the Institute of the Environment and Sustainability at

UCLA. Her books include *Chronoschisms: Time, Narrative, and Postmodernism* (1997), *Sense of Place and Sense of Planet: The Environmental Imagination of the Global* (2008), and *Nach der Natur: Das Artensterben und die moderne Kultur* (*After Nature: Species Extinction and Modern Culture*, 2010). Her latest book, *Imagining Extinction: The Cultural Meanings of Endangered Species*, will be published in 2016.

Heather Houser is Associate Professor of English at the University of Texas at Austin. Her publications include *Ecosickness in Contemporary U.S. Fiction: Environment and Affect* (2014) and essays in *American Literary History*, *Public Culture*, *American Literature*, and *Contemporary Literature*, among other venues. Her book-in-progress, tentatively titled "Environmental Art and the Infowhelm," concerns the aesthetics of information management across environmental media.

David James is Reader in Modern and Contemporary Literature at Queen Mary University of London. He is author of *Contemporary British Fiction and the Artistry of Space* (2008) and, most recently, *Modernist Futures* (2012). He has edited several collections, including *The Legacies of Modernism* (2011) and *Andrea Levy: Contemporary Critical Perspectives* (2014), along with special issues of the journals *Contemporary Literature* and *Modernist Cultures*. He is currently editing *The Cambridge Companion to British Fiction since 1945*.

Jesse Matz is Professor of English at Kenyon College. He is author of *Literary Impressionism and Modernist Aesthetics* (2001) and a number of articles on the problem of time in modern and contemporary literature and culture. His current book-project is "Modernist Time Ecology," a study of the impulse to cultivate the temporal environment in texts and other projects from *A Christmas Carol* to *It Gets Better*.

Mark McGurl is Professor of English at Stanford University and Director of the Center for the Study of the Novel. His most recent book is *The Program Era: Postwar Fiction and the Rise of Creative Writing* (2009).

Paul D. Miller, aka DJ Spooky, That Subliminal Kid, is Professor of Music Mediated Art at the European Graduate School and an internationally known Washington, D.C.–born electronic and experimental hip hop musician. He has released numerous recordings solo or in collaboration with other artists, including *Songs of a Dead Dreamer* (1996), *Riddim Warfare* (1998), *Drums of Death* (2005), and *The Secret Song* (2009), and his work on film includes *Rebirth of a Nation* (2008) and the score for the film *Slam* (1998). He is the author or co-author of *Rhythm Science* (2004), *Sound Unbound: Sampling Digital Music and Culture* (2008), and *Book of Ice* (2011).

Nick Montfort, Associate Professor of Digital Media at MIT, wrote or co-wrote *Twisty Little Passages: An Approach to Interactive Fiction* (2003), *Racing the Beam: The Atari Video Computer System* (2009), and *10 PRINT CHR$(205.5+RND(1));: GOTO 10* (2013). He is co-editor of *The New Media Reader* (2003).

James Phelan is Distinguished University Professor of English at the Ohio State University. He is the author and editor of many books on the novel and on narrative theory, including *Reading the American Novel, 1920–2010* (2013), *Narrative Theory: Core Concepts and Critical Debates* (with David Herman, Peter J. Rabinowitz, Robyn Warhol, and Brian Richardson, 2012), *Experiencing Fiction* (2007), and *Living to Tell about It* (2005).

Anthony Reed is Associate Professor of English and African American Studies at Yale University. He is the author of the book *Freedom Time: The Poetics and Politics of Black Experimental Writing* (2014), winner of the 2015 William Sanders Scarborough Prize from the Modern Language Association. He is currently working on a book-length study of the recorded collaborations between poets and musicians. His articles on black literature and popular culture have appeared in such journals as *African American Review*; *Callaloo*; *Black Camera*; *Souls: A Critical Journal of Black Politics, Culture, and Society*; *Camera Obscura*; and *Feminism, Culture, and Media Studies*.

Michelle Stephens is Professor of English and Latino and Hispanic Caribbean Studies at Rutgers University, New Brunswick. Her first book, *Black Empire: The Masculine Global Imaginary of Caribbean Intellectu-*

als in the United States, 1914 to 1962, was published in 2005. Her second book, *Skin Acts: Race, Psychoanalysis and the Black Male Performer*, a study of the film work of African American stars Paul Robeson and Harry Belafonte, and the musical performances of blackface minstrel Bert Williams and reggae performer Bob Marley, was published in 2014. She is currently working with Brian Russell Roberts on a co-edited collection entitled *Archipelagic American Studies: Decontinentalizing the Study of American Culture*.

Sandra Stephens is Associate Professor in Time and Motion Arts and Gallery Director at PrattMWP in Utica, New York. Stephens creates video and video installation works to explore issues related to cultural and individual identity and has exhibited nationally, throughout museums and galleries in the U.S and internationally in various locations including Circulo de Bellas Artes in Madrid, Triennale di Milano in Milan, Centre de Cultura Contemporania in Barcelona, and Atelier-Haus/Galerie ZeitZone in Berlin. She has also curated various shows throughout the States and is currently working on *conversationXchange*, a series of exhibitions that create a platform for cultural exchange between artists based in the Caribbean and those within the diaspora where artists present their own work, appropriate each other's work, and collaborate together.

INDEX

Aarseth, Espen, 318
Abbott, H. Porter, 240
Abschaffung der Arten, Die (Dath), 62–63
"Aching Body in Dance, The" (Rainer), 330–31, 335
ACT UP, 139
adaptation, 20, 55–59
Adorno, Theodor W., 296; *Aesthetic Theory*, 88–89
Aesthetic Theory (Adorno), 88–89
aesthetic time, 25, 76, 226–28. *See also* art and aesthetics
affect theory, 16–17, 140
African American culture, 294, 296–97, 299–300, 300–301, 307–8n14. *See also* authenticity
African American spirituals, 295
African diaspora, 26, 300–306
Afro-Cuban rhythms, 301
À la recherche du temps perdu (Proust), 240
aleatory variability, 106–7
Alphaville (Godard film), 312
altermodernism, 73–79
Amazon.com, 213
American Gigolo (film), 5
amnesia, 138
anachrony, 133–37; politics of, 137–40; race and sex in, 135; use of term, 129–30, 133–34. *See also* synchrony
analepsis, 240–46, 248, 251–52
Andrejevic, Mark, 196
Animal's People (Sinha), 57
Annales historians, 110–20

Anthropocene, use of term, 146–47
anticipation, 21, 44, 100–105; and desire, 83, 89–90; as keyword, 97; and unexpected, 99–100, 109
antislavery activism, 295
Anxiety of Influence, The (Bloom), 327
applications (apps), 195–96, 198, 202–5, 203. *See also* gaming
Aravamudan, Srinivas, 125
Archer, David, 150
Aristotle, 117; *Physics*, 1
Armstrong, Tim, 228
Aronofsky, Darren: *Pi* (film), 341
art and aesthetics: aesthetic time, 25, 76, 226–28; and altermodernism, 73–79; conceptual, 264–65; of historical time, 76, 92; influence of, 26–27; literary (*see* literary works); and modernism, 27, 66, 68–69, 88–89, 227, 228; periodization in, 20, 66–68, 74–75, 77; postcolonial, 25; and postmodernism, 66, 232; prosthetic-aesthetic time, 231–33; as representation, 19–20; self-reflexivity in, 70. *See also specific works*
artificiality, 298–300; artificial time, 220; vs. authenticity, 26, 294–96
A-series time, 162. *See also* tensed time
Aspray, William, 310
Atlantic Monthly (journal), 313
Atonement (McEwan), 241, 246–48, 250, 251–52
Attali, Jacques, 297
Auerbach, Erich, 42–43

Augustine, Saint, 100–102, 117; *Confessions*, 1, 286–91
authenticity, 296–97; vs. artificiality, 26, 294–96
automobile industry and obsolescence, 86–87

Back to the Future (Zemeckis film), 247–48
Bagayoko, Amadou, 303
Baker, Nicholson: *The Mezzanine*, 87
Banet-Weiser, Sarah, 196
Barker, Jennifer, 272
Barth, Karl, 282–83, 285, 328
BASIC programming language, 309, 316–17
batch time, 310–12; batch processing, defined, 309–10. *See also* interactive time
Baucom, Ian, 1, 10
Baudelaire, Charles, 74, 86
Baudrillard, Jean, 6–7
Before Midnight (film), 219, 220–21
Before Sunrise (film), 213–14, 219
Being-in-the-world, concept of, 13
Beloved (Morrison), 241, 245–46, 250–51
Benjamin, Walter, 9, 24, 45, 131, 133, 136, 294
Benthien, Claudia, 258
Berger, John, 76
Bergson, Henri, 4, 16, 118
Berlant, Lauren, 17, 140
Betrayal (Pinter), 232
biodiversity, decline of, 52–55, 57, 58, 61
black culture. *See* African American culture
black diaspora. *See* African diaspora
blackness, 300, 301, 306, 326, 332
black oppression, 296
Blanchot, Maurice, 182, 184–85
Bloch, Marc, 18
Bloom, Harold, 326–27, 334; *The Anxiety of Influence*, 327
Blue Marble (NASA photograph), 151, 153

Blumenberg, Hans, 118
Body Worlds (von Hagens exhibits), 261, 267
Bourdieu, Pierre, 130–31, 193
Bourriaud, Nicolas, 73–74
Braudel, Fernand, 119, 145–46
Braverman, Harry, 194
Brenner, Robert, 84–85
Brill, Sara, 229
Built-In Obsolescence (Chedda portrait), 261–62, 263
Bukatman, Scott, 165, 229
Burger, Neil: *Limitless* (film), 25, 234–36
Bush, Vannevar, 313, 315

Calder Act (1918), 21
Campbell-Kelly, Martin, 310
capitalism: capitalist realism, use of term, 220; global, 11, 77; and immaterial labor, 195; as perpetual present, 90; synchronic time in, 131–32; temporality of, 195; values of, 6
carbon calculators, 148–50
Care in the Family (Clarke-Stewart), 214–15
Castells, Manuel, 210–11
Castenholz, William B., 90
Castorp, Hans, 231
Century (Everson film), 91–92
Ceruzzi, Paul, 311–12
Chakrabarty, Dipesh, 46, 137, 147
Chang, Jeff, 299
Chedda, Camille, 274; *Built-In Obsolescence* (portrait), 261–62, 263
child development research, 214–15
Christ and Time (Cullman), 281–84
Christianity: apocalypticism, 43; as Biblical history, 281–82; Constantinian, 285; eschatological time, 42; figure-fulfillment model, 42; as Gnosticism, 285; and hope, 292; need for church, 285; roots in time, 25–26, 281, 282, 284, 285, 289–90. *See also* theological time

chronobiopolitics, 22, 130
chronology. *See* clocks and timekeeping
cinema: aesthetic/prosthetic time in, 225–26, 227–28, 233–36; filmic reasoning, 233; filmic time, 124, 225–26; human time in, 227; infinite zoom technique, 235–36; parallel editing in, 10–11; seriality in, 173. *See also specific films*
Clarke-Stewart, Allison: *Care in the Family*, 214–15; *Interactions between Mothers and Their Young Children*, 214
climate change, 52, 54–55, 57–58, 144–45, 148; model visualizations, 152–57, *154*, *156*
Clock, The (Marclay art installation), 4–6, 7–9, 13, 18
clocks and timekeeping: atomic, 22, 113, 164; and civilization, 121; clock time, 114–17; clock time vs. lived time, 117–18, 120, 123–26; clock time vs. natural time, 121; creation of, 8–9; decimal system, 113–14; in film, 5; importance of, 115, 119, 123; leap-second, 113; wristwatches, 116
coevalness, use of term, 294
Cognition of the Literary Work of Art, The (Ingarden), 104
cognitive prosthesis, use of term, 233
Cohn, Dorrit, 242
Cold War strategy, 211
comics, time in, 165–67, 171–72
computers: digital computing, 82, 86, 193, 199–200; early, 310–12; and gender, 314–15; interfaces, 317; and typing ability, 314, 316–17. *See also* interactive time
Computer World (Kraftwerk album), 342
Conceptual art, 264–65
Condon, William, 139
Confessions (Augustine), 1, 286–91
conjunctural time, 9
Connor, Steven, 13
consumerism, 6–7, 83–84, 86–90, 296

contemporaneity: meaning of term, 3–4, 68, 69; vs. modernity, 66, 75–77; multiplicity/simultaneity in, 3–13, 18; present as virtue in, 10
contempt-embarrassment, 83
contingency, necessity of, 107
Coordinated Universal Time, 210
correlationism, 105–6
Crary, Jonathan, 15, 164, 218; *24/7*, 232
critical theory, 137
Crow, Thomas, 334; *Emulation*, 327
Crutzen, Paul, 146
Cruz, Jon, 295
Crying of Lot 49, The (Pynchon), 241, 242–46, 250, 252
cubism, 76, 164, 226
Cullman, Oscar: *Christ and Time*, 281–84
cultural appropriation, 297
cultural studies, 301. *See also* African American culture; popular culture

Danto, Arthur C., 121
Darwin, Charles, 51, 55, 58
Dath, Dietmar: *Die Abschaffung der Arten*, 62–63
Dawkins, Richard, 56
de-extinction, 59–63
Deleuze, Gilles, 12, 16
Derrida, Jacques, 108–9; *Of Grammatology*, 101–2, 104–5; *Speech and Phenomena*, 108
Desk Set (film), 312, 316
dialectics, 337–38
Diffenbaugh, Noah, 155
digital computing. *See under* computers
Discourse Networks (Kittler), 329
disembodiment, 25, 257, 262–67, 269, 270, 272, 274. *See also* embodiment; reembodiment
distributed agency, concept of, 328
Dixon, Keith, 156
Doane, Mary Ann, 10–11, 227
Doumbia, Miriam, 303

durational time, 16, 22, 35–36, 44, 45. *See also* techno-duration

Eco, Umberto, 6–7
ecological crisis, 52–58. *See also* climate change
economic crisis of 2007–2008, 84
Edwards, Paul, 152
Ehrlich, Anne, 55
Ehrlich, Paul, 55
Einstein, Albert, 23, 118, 161, 163–64, 167. *See also* theory of relativity
Eisenstein, Sergei, 233
Elias, Norbert, 18–19
embodiment, 25, 257–62, 264, 266–67, 269, 270–74, 277. *See also* disembodiment; reembodiment
emergency time, 23, 178–81; vs. everyday time, 177, 184–89; as exception, 179; and precarity, 185–86; premediation for, 188; preparations for, 187–88; states of emergency, 185; and urgency, 180–81
emergent terms, 21, 97–99
Emulation (Crow), 327
Engelbart, Doug, 317
English, Darby: *How to See a Work of Art in Total Darkness*, 332
ENIAC computer, 311, 315
Enlightenment, 258
environmental crisis, 52–58. *See also* climate change
Enwezor, Okwui, 74
ethnosymphony, use of term, 295
Euclid, 339
eugenic movement, 135, 138
event-machine, 108–9
Everson, Kevin Jerome: *Century* (film), 91–92
everyday time, 23, 181–84; as banality, 182; vs. emergency time, 177, 184–89; and ordinariness, 183–84
evolutionary change. *See* adaption; extinction

Exorcist, The (film), 5
extinction, 20, 51–55; de-extinction/re-animation, 59–63; linguistic, 53–54; mass, 51–52
"Extinction Optical Depth by Aerosol in 2012" (Ginoux model visualization), 153–54, *154*

Fabian, Johannes, 1, 134–35, 294
Fascism, 132–33
Faulkner, William: *The Sound and the Fury*, 231
Field, Christopher, 155
figure-fulfillment model, 42–43
film. *See* cinema; *specific films*
Final Hour, The (Griffiths film), 124
Fisher, Mark, 220
flashback. *See* analepsis
flashforward. *See* prolepsis
flashmobs, 139
folk music revival, 295
folk-spirit, 294, 295, 296, 300
Forster, E. M., 315; "The Machine Stops," 312–13
François, Jacob: *Body in Space*, 265, *266*; *Spatialization of Time*, 273–74, *274*
Frankfurt School, 132–33, 136, 196
Frenz, Lothar, 59
Freud, Sigmund, 135, 138
Friedman, Susan Stanford, 68–69, 76
Fritzsche, Peter, 10, 19
Frow, John, 7
future: and emergency time, 179–80, 189; memory of, 105, 109; of present, 9–10; relationship to past, 21, 36–37; retrofuturism, 20, 38–39, 42; as slipstream, 20, 41–44; use of term, 1

Gaillard, Cyprien, 40
game theory, 106
gaming, 172–73, 198–205, *201–3*
Gaonkar, Dilip Parameshwar, 73, 76–77

Genette, Gérard, 240, 242, 244, 245, 247, 249
geologic time, 45, 144–45
Geologic Time Scale (GTS), 146
Geophysical Fluid Dynamics Laboratory (GFDL), 153, 156
GFDL. *See* Geophysical Fluid Dynamics Laboratory
GHGs. *See* greenhouse gases
Gibbs, Anna, 130–31
Gibson, William: *Pattern Recognition*, 83
Gillman, Susan, 77
Ginoux, Paul, 153
Glass Palace, 45–46
Gleick, James, 219
Global Footprint Network (GFN), 148–50, *149*
globalization, 11, 14, 17, 45, 77
GMT. *See* Greenwich Mean Time
Godard, Jean-Luc: *Alphaville* (film), 312
Googie architecture, 38–39
greenhouse gases (GHGs), 57, 146, 148, 149–50, 151, 157
Greenwich Mean Time (GMT), 131, 210
Gregg, Melissa, 195
Gregorian calendar, 113, 131
Gregory, Paul, 85
Griffiths, D. W., 11; *The Final Hour* (film), 124
Groom, Amelia, 19–20
Grosz, Elizabeth, 230–31
Groundhog Day (Ramis film), 225
Grusin, Richard, 43, 188
GTS. *See* Geologic Time Scale
Guattari, Félix, 12, 16
Guffey, Elizabeth, 39

Habermas, Jürgen, 11, 68, 76
Hagens, Gunther von: *Body Worlds* (exhibits), 261, 267
Halocene, use of term, 146
Hansen, James, 145
Haraway, Donna, 229

Hartog, François, 19, 38
Harvey, David, 7
haunting, trope of, 137
Hayles, N. Katherine, 334
Hayot, Eric, 75, 76
Heidegger, Martin, 4, 13, 116, 123–24
Heraclitus, 117
"Here" (McGuire comic), *171*, 171–72
Herzog, Werner: *Wo die grünen Amiesen träumen* (film), 56
High Noon (film), 5
hip-hop music: African-born artists, 301; and African diaspora, 300–306; Afro-Cuban rhythms in, 301; artificiality in, 299–300; authenticity in, 298–99; interrelations in, 27; mathematics and, 341–43; mixtapes, 302; resemblance in, 300–304; sampling in, 303–4, 337–38, 344; time as beat in, 342–44; transition from live to recordings, 299
historical time, 18–21, 25, 26, 45; and aesthetics, 76, 92; and Christianity, 284; and chronology, 119–20; and modernism, 75; vs. perpetual present, 36, 90–91; politics of, 83, 134; and premediation, 44; rhythm of, 83, 87, 89–90; and temporal sensation, 83, 88, 89
Hochschild, Arlie Russell: *The Second Shift*, 216; *The Time Bind*, 216; on TQM, 217
Hollerith cards, 311
Hölscher, Lucian, 19
Homer: *Iliad*, 241
home-schooling, 328–29
Homme à la manivelle, L' (film), 124
How to See a Work of Art in Total Darkness (English), 332
HSBC advertisement campaigns, 98–99
human disability, 228, 229–30
humanities: keywords in, 97; temporality in, 2

humanness, 145–47; and adaptation, 58–59; and extinction, 52, 54; and identity formation, 145; and otherness, 157; and re-animation, 63

human time, 22–23, 26, 125, 144, 150; in cinema, 227; and computer programming, 309, 317, 320; in literature, 226; and mobile time, 193; and narrative, 240; and planetary time, 152; as prosthetic, 228, 230, 234. *See also* inhuman time

Husserl, Edmund, 101–2, 116

Hutcheon, Linda, 70

Huyssen, Andreas, 7

IEEE Transactions on Human Factors in Electronics (journal), 314

Iliad (Homer), 241

imagined communities, 131, 210

immanence, 11–12

immaterial labor, 192, 193–95, 196, 200, 205

imperialism, Euro-American, 295

influence, 330–35; use of term, 323–24

Ingarden, Roman: *The Cognition of the Literary Work of Art*, 104

inhuman time, 22–23, 144–45, 147–48, 152, 157. *See also* human time

innovation, 14, 21, 88–93; in art, 76; technological, 37, 60, 82–84; value of, 71. *See also* obsolescence

inoperative community, concept of, 12

inscription model, 102–4

instantaneity, 22

Interactions between Mothers and Their Young Children (Clarke-Stewart), 214

interactive time, 312–17; in fiction, 318–19; time-sharing, defined, 309–10; ubiquitous, 319–20; use of term, 315. *See also* batch time

International Commission on Stratigraphy, 146

International Time Bureau (ITB), 115

International Time Commission (ITC), 115

IRE Transactions on Human Factors in Electronics (journal), 314

irruption *ex nihilo*, 107–8

ITB. *See* International Time Bureau

ITC. *See* International Time Commission

Iton, Richard, 302

Jain, Sara, 229–30

James, William, 16

Jameson, Fredric, 6, 7, 8, 17, 39, 40, 45, 70–71, 78, 137

Jencks, Charles, 7

Jenkins, Henry, 39

Jones, Duncan: *Source Code* (film), 25, 225, 227–28, 229, 232, 233–34

Jurassic Park (Spielberg film), 60

Kaiwar, Vasant, 77

Kemeny, John, 316–17

Kern, Stephen, 210, 228

Keynes, John Maynard, 23, 192

keywords: defined, 1–2; in humanities, 97; modernism as, 66, 72, 74, 75, 77, 79; postmodernism as, 70, 72; on specificity of time, 3; of time, 145, 209; use of, 67–68

Kirkman, Robert: *The Walking Dead* (comic), 172

Kittler, Friedrich, 328–29; *Discourse Networks*, 329

Klein, Naomi, 45–46

K'Naan: "Um Ricka" (collaboration with Wale), 302–5

Koepnick, Lutz, 15

Koselleck, Reinhart, 35, 36–37, 38, 44, 45, 120–21, 134

Kracauer, Siegfried, 197

Kraftwerk: *Computer World* (album), 342

Kugelberg, Johan, 298–99

Kurtz, Thomas, 316–17

Kuti, Fela, 303

labor time: and clocks, 9, 123, 131–32; and Fascism, 132–33; flextime, 194; vs. leisure time, 23, 192–93, 205; limits on, 194; and synchronic time, 131–32; and zaniness, 204–5. *See also* immaterial labor
Lacan, Jacques, 266
Lahsen, Myanna, 156
Landes, David S., 115
Landsberg, Alison, 233
Langevin, Paul, 118
Latour, Bruno, 122
Lazarus, Neil, 76, 78
Lazzarato, Maurizio, 194–95
Leakey, Richard, 54
Lefebvre, Henri, 182–83
Le Goff, Jacques, 18, 119, 163
leisure time, 196–205; and Fascism, 132–33; Keynes' prediction on, 23, 192; vs. labor time, 23, 192–93, 205; and mobile technology, 197–205; and mobile time, 193, 197–201, 204, 205
Lethem, Jonathan, 68
Lewin, Roger, 54
Licklider, J. C. R., 316; "Man-Computer Symbiosis," 314
Limitless (Burger film), 25, 234–36
Lippard, Lucy, 264–65
literary works, 25, 104; analepsis in, 240–46, 248, 251–52; innovations in, 249–52; literary time, 124, 125–26, 241–42; prolepsis, 240–41, 246–49, 250, 251–52. *See also specific works*
liturgical time, 290, 292. *See also* theological time
lived time, 26, 116, 117–23; vs. clock time, 117–18, 120–21, 123–26; vs. world time, 118
Lockhurst, Roger, 4
Lomax, Alan, 295
Lomax, John, 295
Long Now Foundation, 232
Luciano, Dana, 130
Luhmann, Niklas, 163

"Machine Stops, The" (Forster), 312–13
Magic Mountain, The (Mann), 231
Magli, Patrizia, 259–60
Malabou, Catherine, 105
"Man-Computer Symbiosis" (Licklider), 314
Mann, Thomas: *The Magic Mountain*, 231
Manovich, Lev, 233
mapmaking, 115
Marclay, Christian: *The Clock*, 4–6, 7–9, 13, 18
Marks, Laura U., 272
Marks, Peter, 4
Marx, Karl, 92, 94n9, 115, 119, 193
mass culture. *See* popular culture
mass ornament, 132–33
Massumi, Brian, 16, 258, 260, 262, 263–64, 266, 269
Matrix, The (Wachowski and Wachowski film), 341
Mazumdar, Sucheta, 77
McEwan, Ian: *Atonement*, 241, 246–48, 250, 252
McGuire, Richard: "Here" (comic), *171*, 171–72
McHale, Brian, 71, 72, 250
McLuhan, Marshall, 211, 233
McMahon, Maria Driscoll, 280n49
McTaggart, John, 162
meanwhile time, 9
Meillassoux, Quentin, 105–8, 109, 220
Memex, 313, 315
memory, 38, 103–5, 109, 286–87; amnesia, 138; forgetfulness, 287; muscular, 262–63, 269–70; prosthetic, 233; rememory concept, 250–51; and trauma, 291–92
Memory, History, Forgetting (Ricoeur), 104
Merleau-Ponty, Maurice, 269
Merton, Robert K., 117
Metropolis (film), 40
Mezzanine, The (Baker), 87
Michigan Technic (magazine), 211–12
Miéville, China, 40

milk delivery, obsolescence of, 87–88
Miller, Arthur J., 164
Miller, Karl Hagstrom, 294
Miller, Kathleen, 186
mobile time, 193, 197–201, 204, 205
modernism: and aesthetics, 27, 66, 68–69, 88–89, 227, 228; altermodernism, 73–79; alternative, 73–79; artistic, 66, 68–69; and black music, 296; cosmopolitan, 284–85; defined, 96; global polycentrism in, 76; high, 11; and historical time, 75; as keyword, 66, 72, 74, 75, 77, 79; late, 71; and presentism, 10; and prosthetic-aesthetic time, 231; synchronic time in, 131 (*see also* synchrony); taking for granted, 77–78; vs. contemporaneity, 66, 75–77; as worldly time, 26
Moorcock, Michael, 39
"More Today than Yesterday" (song), 240, 241
Morra, Joanne, 230
Morrison, Toni: *Beloved*, 241, 245–46, 250–51
multiplicity, 3–13, 14, 24, 27
multitasking, 26, 198–99, 319–20
Mumford, Lewis, 115
Murbridge, Eadweard: sequential photographs of running horse, 164–65, *165*
Murdoch, Iris: *The Sea, the Sea*, 103

Nancy, Jean-Luc, 12
NASA, 151, 153
Nation of Gods and Earths (NGE), 343
Negro spirituals. *See* African American spirituals
Nelson, George, 86
neoliberalism, 17, 30n30, 192–93, 199
New Formations (journal), 230
New Science (Vico), 134
Ngai, Sianne, 204–5
NGE. *See* Nation of Gods and Earths
Niedecker, Lorine, 144

Nietzsche, Friedrich, 138
Ning, Wang, 71–72
Nixon, Rob, 55, 157
NLS (oN Line System), 317
NOAA. *See* U.S. National Oceanic and Atmospheric Association

obsolescence, 14, 21, 84–88; planned, 85–90; and rhythm of consumption, 83–84, 87–88, 92; role in advanced capitalism, 84–85; and technological change, 82–84. *See also* innovation
Occupy Wall Street movement, 140
Of Grammatology (Derrida), 101–2, 104–5
Omens of Adversity (Scott), 291–92
Osborne, Peter, 12
Otherness, 157, 272, 273, 283, 294–95
other-worldly time, 286–90

Packard, Vance: *The Waste Makers*, 86
Paley, William, 114
Palumbi, Stephen, 59
parent-child bond, 214–15
Parmenides, 117
past: of present, 8, 9–10; relationship to future, 21, 36–37; as retrofuture, 20, 36–40, 42; use of term, 1
Pattern Recognition (Gibson), 83
People Revisited (Stephens video installation), 255–56, 255–57, 260, 262, 264, 267–70, *268*
personal equation, 117, 118
phenomenology, 21, 102, 116
Physics (Aristotle), 1
Pi (Aronofsky film), 341
Picasso, Pablo, 164
Pinter, Harold: *Betrayal*, 232
planetary time, 22–23, 36, 45, 144–45, 151, 152, 156
planned obsolescence. *See under* obsolescence
Plato, 337
pollution, 52, 54–55

popular culture, 71–72, 132, 174, 183, 196–97, 301–2, 333
portention and retention, 104
postmodernism: artistic, 66, 70–73; as global, 12; as keyword, 70, 72; as outmoded term, 69; and prosthetic-aesthetic time, 232; and space/time, 6; thirdspace, 7; time theory, 232; as Western time, 71
Postone, Moishe, 90, 195
post-World War II period, 2, 22, 35, 82, 118–19
power-chronography, use of term, 15
prediction, 21, 98, 99–100
premediation, 43–44
present: and duration, 35–36; future of, 9–10; long, 82–83; as mediated affect, 17; and modernity, 10; multiplicity of, 4; now, defined, 4; past of, 8, 9–10; perpetual, 21, 36, 90–91; as presence, 24; presentism, use of term, 11, 20, 35–36, 38; semantics of, 3; as synthesis, 100–101, 104; temporal, defined, 100; use of term, 1, 4, 12–13; vanishing, 101
production: flexible vs. rigid, 212; immaterial labor, 192, 193–95, 196, 200, 205; overproduction, 72; productivity apps, 195–96, 198, 202–5, *203*; product life cycle, 83, 90; software applications, 195–96; total quality management, 217
Program or Be Programmed (Ruhskoff), 320
progressive time, 134, 136, 137
prolepsis, 240–41, 242, 246–49, 250, 251–52
prophecy vs. prognosis, 37–38, 44
prosthetic time, 25, 228–31; prosthetic-aesthetic time, 231–33, 236
Proust, Marcel, 231; *À la recherche du temps perdu*, 240
Pudovkin, V. I., 124
punch cards, 311, 317

Pynchon, Thomas, 39–40; *The Crying of Lot 49*, 241, 242–46, 250, 252
Pythagoras, 337, 338–41, *339*, *340*, 343–44

quality time, 213–19, 221
quantity time, 220–21

Rabaté, Jean-Michel, 67
race, 26, 135; racialized spatiotemporal schema, 294, 295, 296, 297, 298, 299
Rainer, Yvonne: "The Aching Body in Dance," 330–31, 335
Ramis, Harold: *Groundhog Day* (film), 225
rap music. *See* hip-hop music
"Rapper's Delight" (Sugarhill Gang), 299
Raup, David, 51–52
reaction time, 117
realness. *See* artificiality; authenticity
real time, 209–13, 218–19, 221
re-animation, 59–63
reconciled time, 335
recursivity, 335
reembodiment, 25, 257, 267–77. *See also* disembodiment; embodiment
Religious Peace of Augsburg, 37
rememory, concept of, 250–51
retention. *See* memory
retrofuturism, 20, 36–40, 42
Reynolds, Simon, 38
Ricci, Matteo, 121
Ricoeur, Paul, 226, 240; *Memory, History, Forgetting*, 104
Robbins, Bruce, 253n6
Roberts, John, 184
Robespierre, 37
Robinson, Kim Stanley: *2312*, 60–61
Rodin, Auguste, 165
role-playing games, 319
Roman Catholic Church, 37
Romanticism, 328–29
Ronson, Mark, 303
Rose, Tracey: *Span II* (performance art), 267

Rushkoff, Douglas: *Program or Be Programmed*, 320
Rüsen, Jörn, 19
Ryle, Gilbert, 118

sacred time, 25–26. *See also* theological time
Saussure, Ferdinand de, 129
Scarry, Elaine, 180
Schmitt, Carl, 179
Scott, David, 2; *Omens of Adversity*, 291–92
Sea, the Sea, The (Murdoch), 103
Second Shift, The (Hochschild), 216
Segal, Naomi, 270
Seltzer, Mark, 229, 230
Sennett, Richard, 186
Sepkoski, Jack, 51
seriality, 23, 161, 162–63, 170–72, 175n7
Serres, Michel, 122
Sharma, Sarah, 15
Sharp, Sharon, 39
Silverman, Kaja, 266
simultaneity, 3–13, 18, 23, 77, 118, 133, 161, 163
singularity, 11–12
Sinha, Indra: *Animal's People*, 57
Slaughterhouse-Five (Vonnegut), 232
slipstream, 20, 41–44
Sloterdijk, Peter, 45
slow movement, 15, 232
Smith, Marquard, 230
Smith, Terry, 4, 69, 74
social media: algorithmic analysis, 195–96; games on, 200–201; self-branding in, 196; as sociosynchronic, 133; temporality of, 218–19
Soja, Edward, 7
Somerville, Siobhan, 125
Sound and the Fury, The (Faulkner), 231
Source Code (Jones film), 25, 225, 227–28, 229, 232, 233–34
sovereignty, defined, 179

space: and cosmopolitanism, 284; movement through, 164; and postmodernism, 7; supplanted by time, 117
Span II (Rose performance art), 267
special theory of relativity. *See* theory of relativity
Speech and Phenomena (Derrida), 108
speech recognition, 314
speed, culture of, 14
Spielberg, Steven: *Jurassic Park* (film), 60
Spinoza, Baruch, 16
Spiral Starecase (band), 240, 241
Stafford, Barbara, 155
standard/daylight savings time, 21
steampunk, 40, 135
Steffen, Will, 147
Stephens, Sandra: *People Revisited* (video installation), 255–56, 255–57, 260, 262, 264, 267–70, *268*; *Skin* (video installation), 270–72, 270–73
Sterling, Bruce, 41
Stevens, Brooks, 90
Stiegler, Bernard, 328
Stoermer, Eugene, 146
Storer, Thomas, 125
Sugarhill Gang, 299
"Surface Air Temperature Anomalies" (Dixon model visualization), 154, *156*
synchrony, 130–33; and engroupment, 133; in industrial capitalism, 131–32; politics of, 137–40; use of term, 129–30. *See also* anachrony

Tabbi, Joseph, 236
Taylor, Charles, 285
Taylor, Jason deCaires: *Vicissitudes* (sculpture), 275–76, 275–77
Taylor, R. W., 314–15
Taylorism, 123, 132, 194, 195, 216
technics, 325–26, 329–30, 333
techno-duration: presentism in, 20, 35–36, 38; vs. real, 45–46
television, seriality in, 173

temporality: and adaption, 55–59; emergent terms, 21, 97–99; epochal, 97, 109–10; experience of, 1; and extinction, 51–55; fuzzy, 249; in humanities, 2; multiplicity of, 3–13, 14, 24, 27; and rhythm changes, 83–84, 87–88; seriality of, 23, 161; simultaneity of, 3–13, 18, 23. See also future; past; present; time
tensed time, 162, 165, 170–71, 173, 174
tenseless time, 166, 168, 169, 170–71, 174
theological time, 26, 281–83, 286–90
theory of relativity, 118, 161, 164, *167*
thinking-feeling, 16–17
Thomas Paine's Death Mask, 263
Thompson, E. P., 120
Through the Arc of the Rainforest (Yamashita), 58–59
Tiller Girls dance troupe, 132, 197
time: categories of, 145; as culture, 24–27; as elusive, 1; as history, 18–21 (*see also* historical time); inscription model of, 102–4; measuring, 21–24; as silence, 338–42; standards of, 116. See also future; past; present; temporality; *specific types, e.g.* leisure time
Time Bind, The (Hochschild), 216
timekeeping. See clocks and timekeeping
timelines, use of, 119, 137
time studies: emergence of, 14–18; overview, 1–13
time travel, 225–26
Titanic, sinking of, 210
total quality management (TQM), 216–17
TQM. See total quality management
trace, concept of, 105, 108
Tran, Jonathan: *The Vietnam War and Theologies of Memory*, 290–91, 292
transmission, 326–30; use of term, 323–25
Treitschke, Heinrich von, 119
Trumpener, Katie, 75
Turner, Fred, 199–200
Twelve Monkeys (film), 40
24/7 (Crary), 232

2312 (Robinson), 60–61
Twine, Richard, 259

ubiquitous, use of term, 319–20
"Um Ricka" (K'Naan and Wale), 302–5
uncertainty, 106
unconscious, 136, 166
unexpected, 105–10; and anticipation, 99–100, 109; as keyword, 97, 98–99
universal time, 115
U.S. National Oceanic and Atmospheric Association (NOAA), 153, *156*

Vicissitudes (Taylor sculpture), 275–76, 275–77
Vico, Giambattista: *New Science*, 134
video games. See gaming
Vietnam War and Theologies of Memory, The (Tran), 290–91, 292
Virilio, Paul, 14–15, 209, 212
Vonnegut, Kurt: *Slaughterhouse-Five*, 232

Wachowski, Andy and Lana: *The Matrix* (film), 341
Wale: "Um Ricka," 302–5; "Um Ricka" (collaboration with K'Naan), 302–5
Walking Dead, The (Kirkman comic), 172
war model, concept of, 14
Washington Post strike, 93
Waste Makers, The (Packard), 86
Weheliye, Alexander, 300
White, Hayden, 19, 42–43
Whitehead, Alfred North, 16
Witte, Karsten, 132–33
Wittram, Reinhard, 36
Wo die grünen Amiesen träumen (Herzog film), 56
Wolin, Sheldon, 290
Woodward, Kathleen, 187
work-life balance, 217
work time. See labor time
worldly time, 26, 283–86

Wreck-It Ralph (animated film), 93
writing and time, 102–4
Wu-Tang Clan, 342–43
Wynter, Sylvia, 258–59

X-Men (Marvel Comics), 138

Yamashita, Karen Tei: *Through the Arc of the Rainforest*, 58–59

Yoder, John Howard, 282, 283, 285, 290

zaniness and labor, 204–5
Zemeckis, Robert: *Back to the Future* (film), 247–48
Zeno of Elea, 337
Ziser, Michael, 151
Žižek, Slavoj, 157, 178